S0-BWW-823

WITHDRAWN

Stability and Origin of Biological Information

The First Aharon Katzir-Katchalsky Conference

a workshop on

Stability and Origin of Biological Information

at the
Weizmann Institute of Science,
Rehovot, Israel, June 23–28, 1973

Held under the auspices of
The Israel Academy of Sciences and Humanities,
The European Molecular Biology Organization and
The International Union of Pure and Applied Biophysics

and Sponsored by
The Israel National Council for Research and Development and
The Robert Bosch Foundation, Germany

Stability and Origin of Biological Information

PROCEEDINGS OF
THE FIRST AHARON KATZIR-KATCHALSKY CONFERENCE

Edited by I. R. MILLER
Laboratory of Membranes and Bioregulations
Weizmann Institute of Science, Rehovot

A HALSTED PRESS BOOK

JOHN WILEY & SONS, New York · Toronto
ISRAEL UNIVERSITIES PRESS, Jerusalem

© 1975 by Keter Publishing House Jerusalem Ltd.

All rights reserved

ISRAEL UNIVERSITIES PRESS
is a publishing division of
KETER PUBLISHING HOUSE JERUSALEM LTD.
P.O.Box 7145, Jerusalem, Israel

Published in the Western Hemisphere by
HALSTED PRESS, a division of
JOHN WILEY & SONS, INC., NEW YORK

Library of Congress Cataloging in Publication Data

Aharon Katzir-Katchalsky Conference, 1st, Weizmann
Institute of Science, 1973
Stability and origin of biological information.

"A Halsted Press book."
1. Information theory in biology—Congresses.
2. Biological control systems—Congresses. I. Miller,
I II. Title. [DNLM: 1. Neurophysiology—
Congresses. 2. Molecular biology—Congresses.
3. Genetics—Congresses. 4. Immunology—Congresses.
W3 AH285 1973s/WL102 A285 1973s]
QH507.A37 1973 574.1′8 75–20105
ISBN 0–470–60391–7

Distributors for the U.K., Europe, Africa and the Middle East
JOHN WILEY & SONS, LTD., CHICHESTER

Distributors for Japan, Southeast Asia and India
TOPPAN COMPANY, LTD., TOKYO AND SINGAPORE

Distributed in the rest of the world by
KETER PUBLISHING HOUSE JERUSALEM LTD.
IUP cat. no. 26003 0
ISBN 0 7065 1457 2

Set, printed and bound by Keterpress Enterprises, Jerusalem
PRINTED IN ISRAEL

BNA

Contents

Foreword vii

Genes, Viruses and Human Cancer: A Molecular Approach
S. Spiegelman 1

Stability and Thermodynamic Properties of Dissipative Structures in Biological Systems
I. Prigogine and R. Lefever 26

Hamilton–Jacobi Approach to Fluctuation Phenomena
Kazuo Kitahara 58

The Structure of Reaction Networks
George F. Oster 70

Source and Transmission of Information in Biological Networks
H. Atlan 95

Evolutionary Games
M. Eigen 119

The Extracellular Evolution of Structure in Replicating RNA Molecules
S. Spiegelman, D. R. Mills and F. R. Kramer 123

Antigen Design and Immunological Recognition
Michael Sela 173

Receptor Interactions and Mitogenesis in Lymphoid Cells
Gerald M. Edelman 190

T-Cell Recognition of Cell Surface Antigens
Michael Feldman, Hartmut Wekerle and Irun R. Cohen 193

Structural Analogies between the Immune System and the Nervous System
Niels Kaj Jerne 201

Dissipative Structures in Neural Nets
J. D. Cowan 205

The Transduction of Chemical Signals into Electrical Information at Synapses
ROBERT WERMAN 226

Chemotaxis in Bacteria as a Simple Sensory System
D. E. KOSHLAND, JR. 245

Workshop on Stability and Origin of Biological Information: General Discussion 258

Foreword

One of the activities of the Katzir-Katchalsky Center is an annual Conference commemorating the tragic death of Aharon Katzir-Katchalsky. These Conferences are to be held in alternate years at the Weizmann Institute of Science in Israel and in other centers of research and higher learning throughout the world. We present here the Proceedings of the First Conference held at the Weizmann Institute in Rehovot on July 23–28, 1973. The outbreak of the Yom Kippur War three months later on October 6 delayed publication of the Proceedings to some extent.

The selection of a topic for the first Conference was made only after long deliberations. We first considered the feasibility of presenting a representative cross-section of the wide interests of Aharon Katzir-Katchalsky. However, we immediately became convinced that this would be an impossible task. The scope of Aharon's interests and knowledge was much too great. The fields to which he actively contributed during the last three decades ranged from music, literature, and arts in general, philosophy and education, to various aspects of biology, biophysics, polymer and physical chemistry. In his youth he gave lectures on music which today are still remembered and appreciated by people who became well-known musicologists; he wrote poetry that showed great promise, and a book on the butterflies of Israel which continues to delight nature lovers. His knowledge of diverse fields was very profound and he was able to discuss the different topics with the experts on their own level. Since we could not hope to bring together experts in this wide variety of fields and have a meaningful discussion, we settled for a more limited scope based on his active interests during his last years.

It was Professor Manfred Eigen who suggested the topic "Stability and Origin of Biological Information." This topic perfectly fitted Aharon's active involvement in the general endeavor to elucidate the origin of life, and also his contributions to the application of irreversible and network thermodynamics to the understanding of biological processes and information transmission in biological systems. At the same time, it was clear that even this is a very broad topic and thus we attempted to select with care speakers and participants who would bridge the gap between the different disciplines. I think that we were successful, or perhaps we were simply fortunate in this respect. Biologists and biochemists discussing the different possible sources of molecular development and diversification and information transfer in the immunological system were able to find a common ground with theoreticians trying to describe analogous processes by means of the language of irreversible and network thermodynamics. The understanding between the experimentalists and theoreticians was equally good when discussing events on the molecular level (e.g., principles of enzyme regulation) as when discussing the expression of information in the nervous system. A major premise underlying our discussion is that the amount of information that can be created and stored in a system increases with its complexity. Out of this arose one of the paramount questions discussed at this Conference, namely, how did the complex systems with their enormous information content develop from the simple, primitive systems, or in other words, under what conditions can Darwinian development take place on the molecular level.

The different sectional discussions do not follow their respective papers, but are incorporated into the General Discussion which took place on the last day of the Conference. In so doing we may have lost some of the opinions expressed, however, we hope that we have gained by rendering the discussions into a more concise form. The high standard of the discussions was assued by the excellent team of speakers and participants at the Conference. However, as was pointed out by one of the participants (Professor Ora Kedem), one person was unfortunately missing and he would have been capable of still further raising the scientific level of the discussions. This was the late Aharon Katzir-Katchalsky.

Rehovot, 1974 I. R. Miller

Scientific and Organizing Committee

Chairman: M. Eigen, Max-Planck-Institut für biophysikalische Chemie, Göttingen

Vice-Chairman: E. Katzir-Katchalsky, Weizmann Institute of Science

Scientific Secretary: I. R. Miller, Weizmann Institute of Science

Members: H. Atlan, Weizmann Institute of Science
Y. Bar-Hillel, Israel Academy of Sciences and Humanities
S. Bendor, Israel Academy of Sciences and Humanities
H. Eisenberg, Weizmann Institute of Science
M. Sela, Weizmann Institute of Science
A. Silberberg, Weizmann Institute of Science
R. Werman, The Hebrew University of Jerusalem

Acknowledgement

The editing of this book could not have been accomplished without the efficient help of Mrs. Ruth Goldstein in organizing the meeting and arranging the material. The help of Dr. Diana Bach and Dr. Melanie Spodheim in deciphering the recordings and their transcripts was instrumental for editing the discussions.

Genes, viruses and human cancer: A molecular approach

S. SPIEGELMAN

Institute of Cancer Research
Department of Human Genetics and Development
College of Physicians & Surgeons, Columbia University, New York, 10032, U.S.A.

Prologue

In what follows I will attempt to illustrate how the concepts of genetics can be combined with the methodologies of macromolecular biochemistry to illuminate the probable etiology of cancer, man's most distressing illness.

The central purpose will be to describe recent efforts aimed at determining how much of the knowledge gained from experimental viral oncology in animals is transferable to an understanding of the etiology and pathogenesis of human cancer. Emphasis will be placed on the use of the techniques and methodologies of molecular biology to search in human tumors for the presence of "malignant" nucleic acid homologous to that found in tumor viruses.

Evidence will be presented for the existence in human neoplasias of RNA molecules related in sequence to those found in RNA tumor viruses known to cause corresponding cancers in experimental animals. The diseases studied include human breast cancer, leukemias, lymphomas, and sarcomas. Further experiments demonstrated that the RNA molecules identified in these human neoplasias are viral in size and are associated in a particle with "reverse transcriptase", an enzyme found in all of the RNA tumor viruses examined.

Finally, we will describe experiments with human neoplastic cells that test the validity of the "virogene-oncogene" theory, which proposes that every cell contains in its DNA a segment coding for malignancy. The data obtained suggest that this concept is not applicable to the human cancers we have examined. Under the circumstances, one can perhaps entertain more hopeful pathways leading to the control and cure of these malignancies.

Introduction

To a geneticist the most striking feature characterizing the conversion of a normal to a cancer cell is that the malignant phenotype is permanently transmissible to progeny cells. A plausible explanation would invoke a change in genotype and we are faced then with a problem in cell population genetics. Indeed it was on this basis that we proposed (Shapiro, Spiegelman and Koster, 1937) some thirty-six years ago, when *in vitro* proliferating tissue cells were not available, that the neoplastic problem could best be approached indirectly by studying genotypic changes in populations of bacteria growing as colonies on agar surfaces. The intervening years witnessed the development of microbial genetics which in turn yielded the information and material that catalyzed the emergence of molecular genetics. The use of bacteria and their viruses produced concepts and methodologies that provided insights and experimental techniques leading to novel approaches to cell genetics and to gene function. The application of these to the cancer problem is the principal focus of the present paper.

A priori, the genotypic change presumed to underly the cellular transformation to malignancy could be ascribed to any one, or a combination, of the following: 1) somatic mutation in a structural or regulatory gene; 2) chromosomal translocation, or similar mechanical rearrangements; 3) imbalance due to loss or extra duplications of one or more of the chromosomal complement; 4) the insertion of RNA viral information into the genome in analogy with the transducing phages, a mechanism which would however require making a DNA copy of the viral RNA and which has been called the provirus hypothesis (Temin, 1964). To these we can add a non-genotypic but heritable phenotypic modification; and 5) a self-reinforcing derepression by extrinsic or intrinsic factors of a preexistent silent malignant segment of genetic information. If this segment also codes for an RNA tumor virus the concept becomes equivalent to the virogene hypothesis (Todaro and Huebner, 1972).

In what follows we describe experiments designed to decide amongst these alternatives. Only the fourth, the transduction mechanism, leads to the prediction that malignant cells will contain in their genome viral-related information that will not be detectable in normal cells. If, in a given instance, data supporting this mechanism can be provided, all the other four possibilities are in essence eliminated for the particular neoplasia examined.

Animal viral oncology

We had perhaps best begin by explicitly noting a bias stemming from our previous experiences with RNA viruses and our assessment of the information available on the relation between viruses and cancer. Both DNA and RNA tumor viruses have been identified in laboratory experiments with animals. Nevertheless, we decided to concentrate our experimental efforts on the RNA oncornaviruses, a choice made for two reasons. One is that the RNA oncogenic agents predominate quantitatively in the animal systems studied. The second is that it is the RNA tumor viruses that have been repeatedly demonstrated to cause tumors in their indigenous hosts. Thus, from the viewpoint of the natural history of the disease, the RNA oncogenic viruses appeared to us as the more probable "natural" candidates for etiologic agents of cancer in the animal kingdom, including man.

Over the past fifty years, animal viral oncology has amassed a considerable fund of information concerning the so-called B-type and C-type RNA tumor viruses (Gross, 1970). These agents have been implicated in a variety of cancers in a wide spectrum of animals. In many instances, convincing evidence has been obtained by inoculation of purified viruses into susceptible animals in experiments that satisfy Koch's criteria for the identification of the causative agent of a particular disease.

Our working assumption was based on the improbability that cancer in man would have so different an etiological basis that the information provided by animal experiments would be irrelevant to the human disease. If this be the case, one might wonder why it has been so difficult to provide acceptable evidence for a viral agent in human cancer. A plausible explanation emerges from an examination of the historical development of animal viral oncology. Forty to fifty years ago, it was as difficult to identify viruses in animal tumors as it is today in the human neoplasias. The experimental picture changed dramatically as the result of a decision by geneticists that had a profound serendipic effect on the subsequent development of cancer research. A group of investigators set out to study the genetics of cancer and, in the course of doing so, developed highly inbred strains of mice characterized by a high frequency of various neoplasias, including mammary tumors, leukemias, etc. The intent was of course to mate these with low frequency strains and from the distribution of the characteristic cancer phenotype deduce the genetic factors underlying susceptibility and resistance to the disease. What occurred unexpectedly, however, was that in the course of developing the inbred stocks of high cancer frequency, a genetic background was inadvertently created permitting the tumor viruses to multiply to levels that made their identification inevitable.

Those who are concerned with human neoplasias are for the most part faced with the same difficulties encountered by the early animal oncologists when only outbred animals were available. It follows that the detection of putative human viral agents requires more sensitive devices than those which sufficed to establish their presence in the inbred animal systems. In searching for such tools, we quite naturally turned to molecular hybridization and the other methodologies developed by molecular biologists in the past several decades.

Our investigations evolved through a number of stages that are conveniently identified by the questions we posed in the following sequence for experimental resolution.

1) Do human neoplasias contain RNA molecules possessing detectable homology to the RNA of tumor viruses known to cause similar cancers in animals?

2) If a positive outcome is obtained, do the RNA molecules identified in tumors possess the size and physical association with reverse transcriptase that characterize the RNA of the animal oncornaviruses?

3) If such RNA exists in human tumors, is it encapsulated in a particle possessing the density and size of the RNA tumor viruses?

4) Does the RNA of human tumor particles possess homology to the RNA of the viruses causing the corresponding disease in animals?

5) The virogene concept proposes that all animals prone to cancer carry in their germ line a complete copy of the information required to convert a cell from normal to malignant and for the production of tumor virus particles. Is this concept valid for randomly bred populations and in particular for the human disease?

The animal models as a point of departure

To better understand the biological rationale underlying the experiments to be described, it is useful to summarize briefly some of the salient features of the animal model systems used as a guide. Table 1 lists a representative group of animal viruses, including those actually employed to generate the molecular probes used. There are two avian viruses, myeloblastosis virus (AMV) and Rous sarcoma virus (RSV), that cause leukemias and sarcomas in chickens. In addition, we have the murine leukemia virus (MuLV) and murine sarcoma virus (MuSV) that induce similar diseases in mice, and finally the murine mammary tumor virus (MMTV), which is the unique etiologic agent for mammary tumors.

When these are examined for sequence homologies amongst their nucleic acids, a rather informative pattern emerges. It will be noted that the two chicken agents (AMV and RSV) have sequences in common but do not show detectable homology with any of the murine agents. Turning to the murine viruses, we find that the nucleic acids of the leukemia, lymphoma, and sarcoma agents show homology to one another but not to either of the two avian agents or to the mammary tumor virus. Finally, the mouse mammary tumor virus has a singular sequence homologous only to itself.

It is important to understand that a plus sign does not indicate identity but simply sufficient similarity to be detectable by the relaxed hybridization conditions used in these initial exploratory experiments. Similarly, a negative sign does not imply a complete absence of sequence homology but rather that none was detectable by the procedures employed.

If analogous virus particles are associated with the corresponding human diseases, certain predictions might be hazarded on the basis of the specificity patterns exhibited in Table 1, and these may be listed as follows:

1) In view of the lack of homology between the avian and murine agents, it is

Table 1
COMPARISON OF SOME REPRESENTATIVE ONCORNAVIRUSES

Virus	Indigenous host	Homology*					Disease
		AMV	RSV	MuLV	MSV	MMTV	
AMV	Chicken	+	+	−	−	−	Leukemia
RSV(RAV)	Chicken	+	+	−	−	−	Sarcoma
MuLV	Mouse	−	−	+	+	−	Leukemia, lymphoma
MSV(MuLV)	Mouse	−	−	+	+	−	Sarcoma
MMTV	Mouse	−	−	−	−	+	Breast cancer

*The results of molecular hybridizations between [^{3}H]DNA complementary to the various RNAs and the indicated RNAs. The plus sign indicates that hybridizations were positive and the negative sign indicates that none could be detected.

unlikely that human agents, should they exist, would show homology to the avian group.

2) It follows that the murine tumor viruses would represent a more hopeful source of the molecular probes to search for similar information in the analogous human cancers.

3) If particles are found to be associated with human leukemias, sarcomas, and lymphomas, their RNAs might show homology to one another and possibly to that of the murine leukemia virus.

4) If RNA particles are identified in human breast cancer, these should not exhibit homology to the RNA of the viruses causing the human lymphoid neoplasias or to RLV RNA but might exhibit some homology to the RNA of the murine mammary tumor virus.

On the basis of availability and the considerations outlined above, it is clear why the murine agents were initially chosen for producing the necessary molecular probes to look for corresponding information in the human disease. Furthermore, the desire to monitor the biological consistency of our findings dictated that we examine a number of human neoplasias simultaneously. This would permit us to quickly determine whether our findings in man mirrored biologically what was known from the animal experimental models. For this purpose, from the onset, we focused our attention in parallel on the lymphoid neoplasias and on breast cancer.

Molecular hybridization with radioactive DNA probes

To answer the first question posed on whether human tumors contain viral-related RNAs, we naturally turned to the use of DNA-RNA hybridization (Hall and Spiegelman, 1961), which we designed some dozen years ago to answer questions of almost precisely this nature in the case of virus-infected bacteria. This method depends on the ability of any piece of single-stranded DNA to find its complementary RNA and form, under the proper conditions, a double-stranded hybrid structure. The reaction is highly specific and has proved to be of considerable value in molecular biology over the past decade.

The radioactive DNA needed was synthesized by supplying detergent-disrupted virus preparations with magnesium and the four deoxyriboside-triphosphates required, one of them being labeled with tritium. When the synthesis is terminated, the protein and the RNA present are eliminated and the residual radioactive DNA purified to completion. Each [^{3}H]DNA preparation must be rigorously monitored for the specificity of its hybridizability to its appropriate template. Thus, each probe is challenged with its homologous RNA as well as with the other unrelated viral RNAs. Further, normal and cancer tissue RNAs from animal models were used to check the reliability of the hybridizations. Only if a probe demonstrated the required specificity in these test hybridizations was it used for the subsequent experiments with human material.

There are certain features built into these exploratory experiments that are worthy of explicit mention. We of course did not know at the outset whether there would be *any* homology between the murine viruses and any putative human agents. At the time RLV and MMTV were among the few mammalian oncornaviruses available in adequate supply and we could only *hope* that they would have sufficient homology to be used for our purpose. However, even if this hope were realized, it would hardly be likely that the homology would be extensive and therefore hybridization experiments had to be carried out under conditions sufficiently relaxed to yield a reaction with a limited level of homology. Further, the method of detection of hybrid structures would have to possess a low background noise level so that weak positive reactions would be readily identified and distinguished from negative outcomes. Finally, the $C_{r_0}t$ (concentration of nucleotides in moles per liter × time in seconds) values of the annealing reactions were deliberately adjusted to be adequate for multiple (10 or more) rather than individual copies per cell. This was done in order to magnify any existent difference between normal and malignant tissue. We were not initially concerned with determining whether normal cells could produce the same or related malignant messages at very low levels.

All of these considerations entered into the design of the detailed methodology used and is diagrammed in Fig. 1. After monitoring for specificity, the purified viral-specific [^{3}H]DNA is mixed with cytoplasmic RNA prepared from a variety of human tumors and annealed under the conditions indicated. The hybridizations were always carried out with a vast excess of tumor RNA. Since the viral-specific [^{3}H]DNA is small compared with the RNA, any complexes formed between them will behave physically more like RNA than like DNA. Such complexes are readily

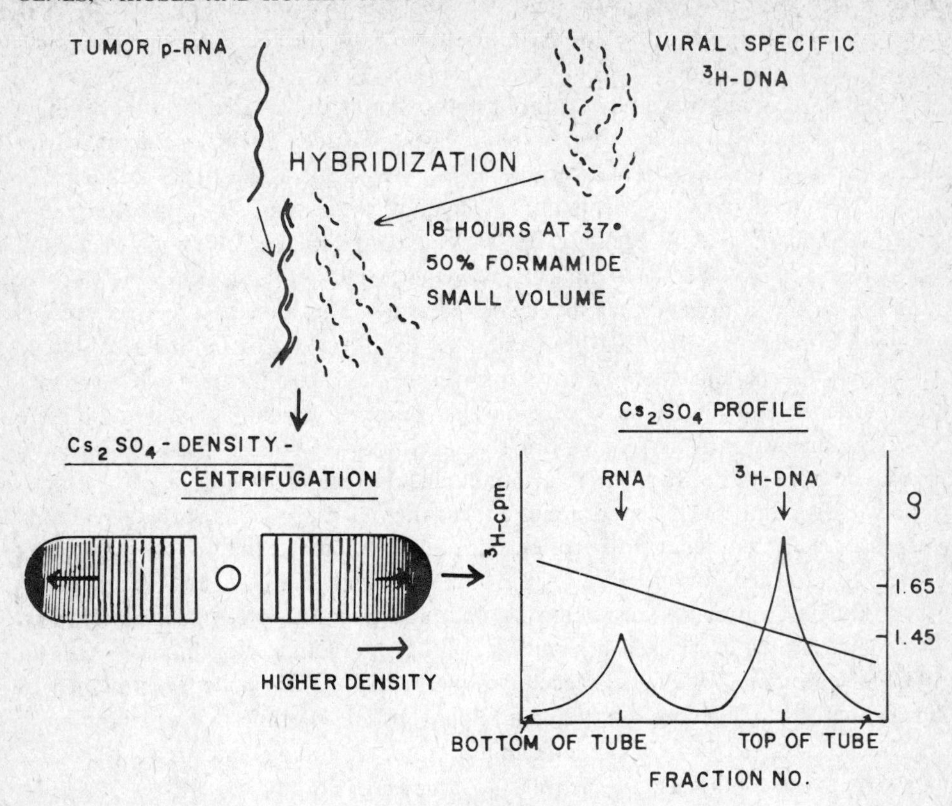

Figure 1
Molecular hybridization and detection with viral-specific [^{3}H]DNA and tumor RNA (see text for further details).

detected by isopycnic separation in equilibrium density gradients of Cs_2SO_4. At the end of the centrifugation, the distribution of [^{3}H]DNA is examined across the gradient. Any uncomplexed DNA will remain at a density corresponding to about 1.45. DNA molecules that have annealed partially or completely to large RNA molecules will band at or near the density of RNA ($\rho = 1.65$).

The movement of the [^{3}H]DNA to the RNA region is then the signal that the probe used has found complementary sequences in the tumor RNA with which it is being challenged. It is important to emphasize that this test is essentially a qualitative one and does not measure how good the fit is in the duplex. A positive outcome says only that sufficient homology exists between the tritiated DNA probe and some RNA molecules in the tumor RNA sample to form a complex stable enough to be detected by the test used. Once such structures are identified, the fidelity of the pairing can be examined by a temperature-melt analysis. These more sophisticated examinations demand more hybrids than we could initially produce. It was therefore necessary to postpone these more informative experiments until more efficient probes became available.

We understandably began by exploring those human malignancies for which there were suitable animal models and viruses. These included adenocarcinoma of the breast (Schlom, Spiegelman and Moore, 1971; Axel, Schlom and Spiegelman, 1972a and 1972b), the leukemias (Hehlmann, Kufe and Spiegelman, 1972a), the sarcomas (Kufe, Hehlmann and Spiegelman, 1972), and lymphomas (Hehlmann, Kufe and Spiegelman, 1972b). The leukemias examined encompassed both acute and chronic varieties of the lymphatic and myelogenous types. The human sarcomas studied included fibro, osteogenic, and liposarcomas. The series on lymphomas contained Hodgkin's disease, Burkitt's tumors, lymphosarcomas, and reticulum cell sarcomas. Control adult and fetal tissues were always examined in parallel and these were invariably negative. In the case of breast cancer, the two benign diseases, fibroadenoma and fibrocystic disease, were also included and found to be negative.

Table 2 summarizes in diagrammatic form the outcome of this survey of human neoplasias with the animal virus probes. The plus signs indicate that the corresponding [^{3}H]DNA formed complexes with the indicated tumor RNAs and the negative signs that no such complexes were detected. The proportion of positives amongst those labeled plus in these earlier studies (Axel, Schlom and Spiegelman, 1972a and 1972b; Hehlmann, Kufe and Spiegelman, 1972a and 1972b; Kufe, Hehlmann and Spiegelman, 1972; Schlom, Spiegelman and Moore, 1971) ranged from about 67% for breast cancer to 92% for the leukemias. As our technology improved so has our per cent positive amongst the neoplastic samples.

What is most noteworthy of the pattern exhibited in Table 2 is its concordance with the predictions deducible from the murine system. Thus, human breast cancer

Table 2

HOMOLOGIES AMONG HUMAN NEOPLASTIC RNAs AND ANIMAL TUMOR VIRAL RNAs*

	Human neoplastic RNAs*			
Viral RNAs	Breast cancer	Leukemia	Sarcoma	Lymphoma
MMTV	+	−	−	−
RLV	−	+	+	+
AMV	−	−	−	−

*The results of molecular hybridization between [^{3}H]DNA complementary to the various viral RNAs and pRNA preparations from the indicated neoplastic tissues. The plus sign indicates that hybridizations were positive and the negative sign that none could be detected (Hehlmann, Kufe and Spiegelman, 1972b).

contains RNA homologous only to that of the murine mammary tumor virus (MMTV). Human leukemias, sarcomas, and lymphomas all contain RNA showing sufficient homology to that of the murine leukemia virus (RLV) to make a stable duplex. These lymphoid neoplasias contain no RNA homologous to the MMTV RNA. Finally, none of the human tumors contain RNA detectably related to that of the avian myeloblastosis virus (AMV). Recently Gallo and his associates (1973) have confirmed the homology of leukemic RNA to that of RLV and in the process have shown even more relatedness to the RNA of a simian sarcoma virus. In summary, the specificity pattern of the unique RNA found in the human neoplasias is in complete agreement with what has been described for the corresponding viral-induced malignancies in the mouse.

Needless to say, the existence of this remarkable concordance does not establish a viral etiology for these diseases in man. The next step required performance of experiments designed to answer the second and third questions raised in the introduction, i.e., those relating to the size of the RNA being detected and whether it is associated with a reverse transcriptase in a particle possessing other features of complete or incomplete virus particles. We now turn our attention to the description of these techniques and their application to the analysis of human malignant tissue.

Simultaneous detection test for reverse transcriptase and high molecular weight RNA

What we sought was a method of detecting the presence of particles similar to the RNA tumor viruses in human material that would be simple, sensitive, and sufficiently discriminating so that a positive outcome could be taken as an acceptable signal for the presence of a virus-like agent. To achieve this goal, we devised a test that depended on the simultaneous detection of two diagnostic features of the animal RNA tumor viruses.

The oncornaviruses exhibit two identifying characteristics. They contain a large (1×10^7 daltons in molecular weight) single-stranded RNA molecule having a sedimentation coefficient of 60 to 70*S*, often referred to as high molecular weight (HMW) RNA. They also have reverse transcriptase (Baltimore, 1970; Temin and Mizutani, 1970), an enzyme that can use the viral RNA as a template to make a complementary DNA copy (Spiegelman et al., 1970).

The possibility of a concomitant test for both the enzyme and its template was suggested by studies of the early reaction intermediates (Spiegelman et al., 1970; Rokutanda et al., 1970). The initial DNA product was found complexed to the large 70*S* RNA template. These structures can be detected by the unusual position of the newly synthesized small tritiated DNA products in Cs_2SO_4 density gradients, glycerol velocity gradients, and acrylamide gel electrophoresis (Bishop et al., 1971). The most informative assay is to subject the isotopically labeled product to sedimentation analysis prior to the removal of the RNA. If the labeled DNA product behaves as if it is a 70*S* molecule, and if it can be shown that it does so because it is complexed to a 70*S* RNA molecule, then evidence is provided for the presence of a reverse transcriptase using a 70*S* RNA template. One may then tentatively conclude that the material examined contains particles similar to the RNA tumor viruses.

It was on this basis that Schlom and Spiegelman (1971) developed the simultaneous detection test, which was used to demonstrate (Schlom, Spiegelman and Moore, 1972) the presence in human milk of particles containing 70*S* RNA and a reverse transcriptase. The test was modified (Gulati, Axel and Spiegelman, 1972) to be applicable to tumor tissue using the mouse mammary tumor as the experimental model.

Figure 2 diagrams the procedure used. Tumor cells are first broken by the use of a Dounce homogenizer and the nuclei and large cell debris removed by low speed centrifugation. The supernate is subjected to trypsin digestion to inactivate any nucleolytic enzymes and the trypsin neutralized by trypsin inhibitor. The supernate is then centrifuged at 150,000*g* to yield a cytoplasmic pellet containing virus particles, if present. The pellet is then treated with a nonionic detergent (0.1% of NP-40) to disrupt possible viral particles and used in an endogenous reverse transcriptase reaction. Actinomycin D is always included (100–200 μ/ml) to inhibit DNA-instructed DNA synthesis. The product of the reaction with its RNA template is freed of protein and then analyzed in a glycerol velocity gradient to determine the sedimentation

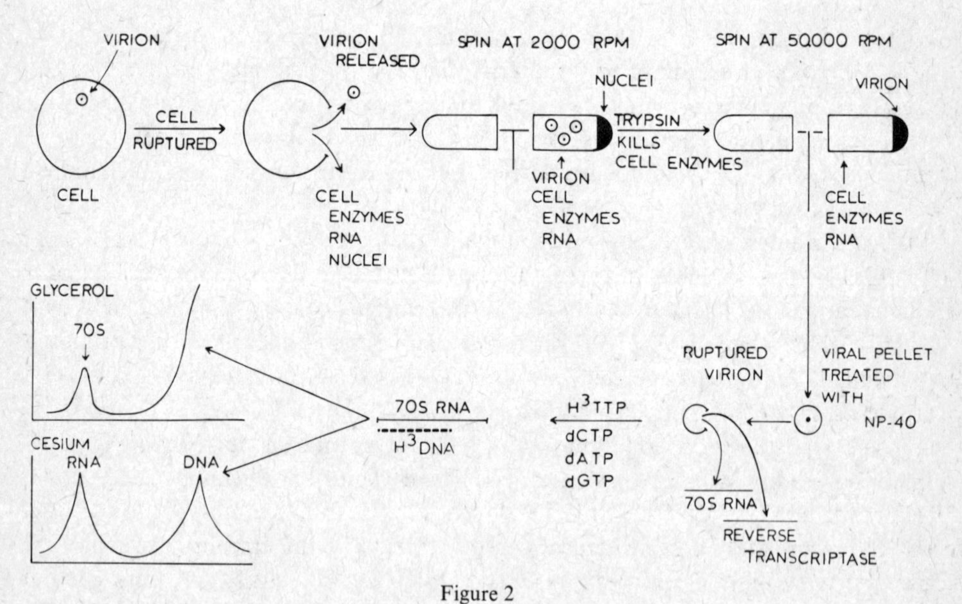

Figure 2
Simultaneous detection test for 70*S* RNA and reverse transcriptase in neoplastic tissue (see text for further details).

coefficient of the [^{3}H]DNA as well as in a Cs_2SO_4 equilibrium gradient to determine its density.

The presence of particles encapsulating 70*S* RNA and reverse transcriptase will be indicated by the appearance of a peak of newly synthesized DNA traveling at a speed corresponding to a 70*S* RNA molecule. That its apparent large size is due to its being complexed to a 70*S* RNA molecule can be readily verified by subjecting the purified nucleic acid to ribonuclease prior to velocity examination and this should result in the disappearance of the "70*S*" [^{3}H]DNA. Similarly, if the reaction is positive, newly synthesized DNA should appear in the RNA region of a Cs_2SO_4 density gradient, and this again should be eliminated by prior treatment with ribonuclease.

The simultaneous detection test was first applied to human breast cancer (Axel, Gulati and Spiegelman, 1972) and Fig. 3 exemplifies a positive reaction with material obtained from a malignant adenocarcinoma and treated as outlined in Fig. 2. Here we see the 70*S* DNA complex observed with the P-100 fraction and it is evident that

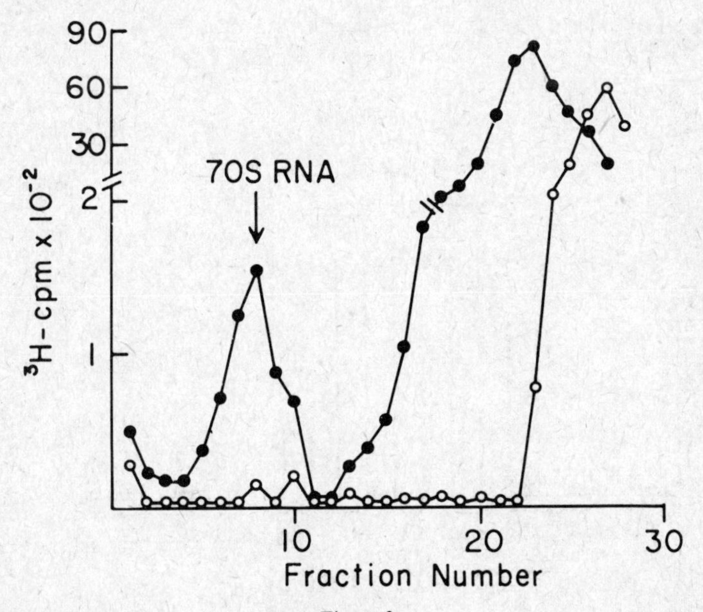

Figure 3

Effect of ribonuclease on the detection of the high-molecular-weight RNA–[^{3}H]DNA complex. Breast tumor tissue (adenocarcinoma 756-5-72) was processed as described in Fig. 2. The viral pellet (P-100) was suspended in 0.01 M Tris–HCl (pH 8.3) and divided into two equal parts. A standard RNA-instructed DNA polymerase reaction was performed on one part of the P-100 fraction; after incubation for 15 min at 37°, the nucleic acid complex was extracted with phenol-cresol and was sized on a 30–10% linear glycerol gradient (black circles). After disruption with detergent, the other half of the P-100 fraction was incubated in the presence of RNase A (50 μg/ml) and RNase T_1 (50 μg/ml) for 15 min at 25°. A standard RNA-instructed DNA polymerase reaction was then performed (open circles).

this complex is an RNA-dependent one since prior treatment with ribonuclease eliminates the [^{3}H]DNA from the 70*S* region of the velocity gradient. In this series, 38 adenocarcinomas and ten nonmalignant controls were examined in this manner. It was found that 79% of the malignant samples were positive for the simultaneous detection reaction and all of the control samples from normal and benign tissue were negative.

The data obtained therefore indicated that one can, with a high probability, find in human breast cancers particulate elements that encapsulate RNA-instructed DNA polymerase and a 70*S* RNA. It was of obvious interest to see whether these particles possessed the density characteristic of an RNA tumor virus. To this end, aliquots of pellet fractions, which gave positive simultaneous detection tests, were first subjected to sucrose equilibrium centrifugation in a linear gradient of 25 to

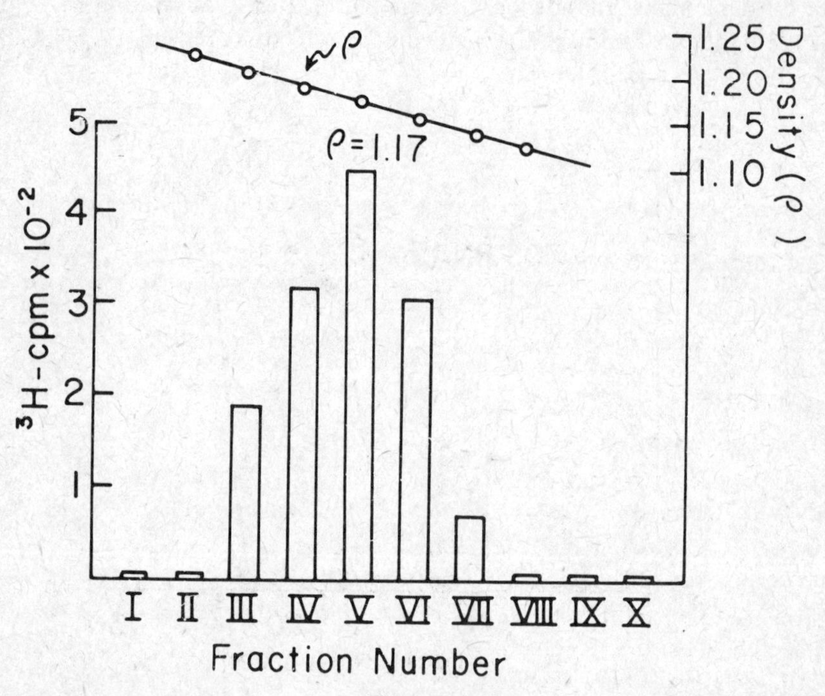

Figure 4

Density of 70*S* RNA and RNA-instructed DNA polymerase activity in extracts of human mammary tumors by sucrose gradient centrifugation. A P-100 pellet was prepared from human mammary tumor. The pellet was suspended in Tris–NaCl–EDTA buffer and layered on a linear gradient of 50–25% sucrose in the same buffer and spun at 25,000 rpm in a SW27 (Spinco) rotor at 4° for 195 min. The gradient was dripped from below and ten equal fractions were collected; fractions were then diluted with Tris–NaCl–EDTA buffer to a sucrose concentration of less than 15%. Each fraction was then spun at 100,000*g*. The amount of 70*S* [^{3}H]DNA synthesized by an endogenous RNA-instructed DNA polymerase was determined by glycerol velocity centrifugation.

50% sucrose. The gradient was then divided into 10 equal fractions, which were diluted to 15% sucrose and spun at 100,000*g* for 1 hr to pellet any particles present. Each of these pellets was then tested to determine the distribution in the density gradient of the reaction that yields a 70*S* RNA-instructed DNA synthesizing activity. It is clear from the results shown in Fig. 4 that particles possessing the reverse transcriptase activity and its 70*S* RNA template localize at a density between 1.16 and 1.19 g/ml, the density characteristic of the oncogenic viruses.

Application of the simultaneous detection test to the leukemias

In the study of the leukemias (Baxt, Hehlmann and Spiegelman, 1972), peripheral leukocytes were prepared from the buffy coats of both leukemic and nonleukemic control patients. Cells were disrupted and fractionated as described in Fig. 2. Representative experiments examining the effects of ribonuclease treatment of the product and omission of one of the deoxytriphosphates during the reaction are shown in Fig. 5. We see the telltale 70*S* peaks of DNA synthesized by the pellet fractions from the leukocytes of patients with acute lymphoblastic and acute myelogenous leukemias. The elimination of the complex by prior treatment with ribonuclease (Fig. 5A) shows that the tritiated DNA is indeed complexed to a 70*S* RNA molecule. Further, the omission of dATP (Fig. 5B) leads to a failure to form the 70*S* complex, a result expected if the reaction is in fact leading to the synthesis of a proper heteropolymer. In similar experiments, it was shown that omission of either dCTP or dGTP also resulted in the absence of the 70*S* RNA-[^{3}H]DNA complex, all of which argues against nontemplated end addition reactions.

In some cases leukemic cells were obtained in amounts adequate to permit a more complete characterization of the product. Hybridization of the human product to the appropriate viral RNAs provides the most revealing information since it tests sequence relatedness to known oncogenic agents. We summarize in Table 3 the results of examining the peripheral leukocytes of 23 patients, all in the active phases of their disease, including both acute and chronic leukemias. Of the 23 leukemic patients examined, 22 showed clear evidence that their peripheral leukocytes contained particles mediating a reaction leading to the appearance of endogenously synthesized DNA in the 70*S* region of a glycerol gradient. Nine of these were tested for ribonuclease sensitivity and in all cases the complexes were destroyed. In nine others, the DNA was recovered from the complex and annealed to RLV RNA and to either MMTV RNA or AMV RNA. In all nine, hybridizations occurred with RLV RNA and not to either of the unrelated MMTV RNA or AMV RNA. In four patients, enough DNA complex was formed to permit a complete characterization of the product. In all four, the DNA complexes were destroyed by ribonuclease and the purified DNA hybridized uniquely to RLV RNA.

It is noteworthy that positive outcomes were observed in more than 95% of the

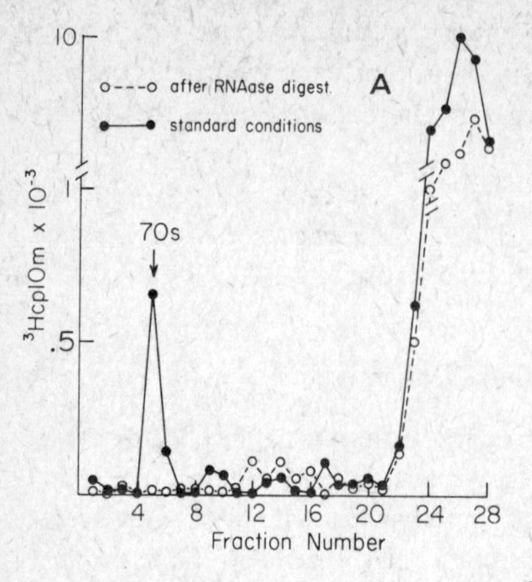

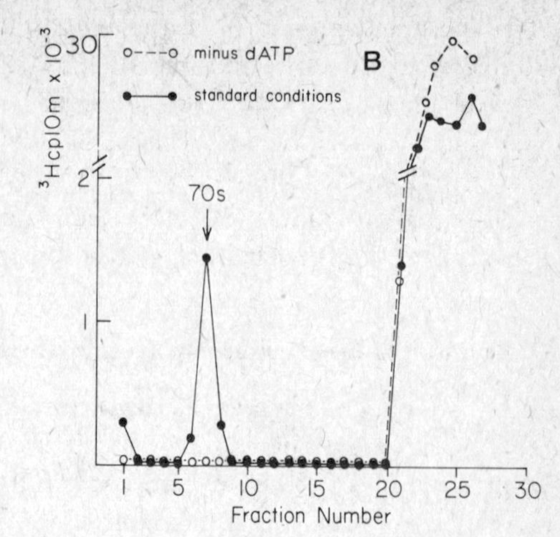

Figure 5

Detection of 70*S* RNA–[^{3}H]DNA complex in human leukemic cells. 1 g of leukemic WBC was washed in 5 ml of 0.01 M NaCl, 0.01 M Tris–HCl, pH 7.4, resuspended in 4 ml of 5% sucrose, 0.005 M EDTA, 0.01 M Tris–HCl, pH 8.3, and ruptured with three strokes of a Dounce homogenizer. The nuclei were removed by low speed centrifugation (2000*g*, 5 min, 2°C). The supernatant was brought to a final concentration of 1 mg/ml trypsin (Worthington) and incubated at 37°C for 30 min. A ten-fold excess of lima bean trypsin inhibitor (Worthington) was added (final concentration 3 mg/ml) and the solution again centrifuged at 2000*g* for 5 min at 2°C. The supernatant was then centrifuged at 45,000 rpm for 60 min at 2°C. The resulting cytoplasmic pellet was resuspended in 0.5 ml of 0.01 M Tris–HCl, pH 8.3, brought to 0.1% Nonidet P-40 (Shell Chemical Co.) and incubated at 0°C for 15 min. DNA was synthesized in a typical reverse transcriptase reaction mixture (final volume 1 ml) containing: 50 μmol of Tris–HCl, pH 8.3, 20 μmol NaCl, 6 μmol $MgCl_2$, 100 μmol each of dATP, dGTP, dCTP, and 50 μmol [^{3}H]dTTP (Schwarz Biochemical, 800 cpm per pmol). 50 μg/ml actinomycin D was added to inhibit DNA-instructed DNA synthesis. After incubation at 37°C for 15 min, the reaction was adjusted to 0.2 M NaCl and 1% SDS, and deproteinized by phenol–cresol extraction. The aqueous phase was layered on a 10 to 30% gradient of glycerol in TNE buffer (0.01 M Tris–HCl, pH 8.3, 0.1 M NaCl, 0.003 M EDTA) and centrifuged in a SW-41 rotor Spinco at 40,000 rpm for 180 min at 2°C. Fractions were collected from below and assayed for TCA-precipitable radioactivity. In this, as in all sedimentation analyses, 70*S* RNA of the avian myeloblastosis virus was used as a marker.

A. One aliquot of product was run on the gradient as a control and the other was pretreated with 20 μg of RNase 1 (Worthington) for 15 min at 37°C prior to sedimentation analysis.

B. Reactions with and without dATP (Baxt, Hehlmann and Spiegelman, 1972).

leukemic patients whether they were acute or chronic, lymphoblastic or myelogenous. Thus, despite their disparate clinical pictures and differing cellular pathologies, these various types of leukemias have similar, though probably not identical, virus-related information.

In contrast with these results are those obtained with a control series of eighteen white blood samples from nonleukemic patients. These included nine with elevated white blood cell counts (in the range of 25,000/mm^3) due to a variety of disorders,

Table 3
SIMULTANEOUS DETECTION OF 70*S* RNA AND REVERSE TRANSCRIPTASE IN LEUKEMIC CELLS (BAXT, HEHLMANN AND SPIEGELMAN, 1972)

Leukemias	Simultaneous detection, cpm	RNase sensitivity	Hybridization to RLV RNA	Hybridization to AMV RNA or MMTV RNA
Acute lymphatic				
Od	400	+	NT	NT
Sm	95	+	NT	NT
Acute lymphatic/lymphosarcoma				
Hl	805	+	+	–
Hz	200	NT	+	–
Bx	185	+	NT	NT
St	105	+	NT	NT
Acute myelogenous				
Ge	170	+	NT	NT
De	985	+	+	–
Be	305	+	NT	NT
Vi	1295	+	+	–
Sh	1010	NT	+	–
S	115	NT	+	–
Du	415	NT	+	–
He	400	NT	+	–
Bl	605	+	NT	NT
Ga	215	+	+	–
C	285	+	NT	NT
Pe	0	NT	NT	NT
Ne	1400	NT	+	–
Chronic lymphatic				
Ri	200	+	NT	NT
Chronic myelogenous				
Ec	405	NT	+	–
Har	390	NT	+	–
An	600	NT	+	–

NT = Not tested

including polycythemia vera. None of these exhibited evidence of a RNA–DNA complex that was ribonuclease-sensitive and contained DNA that hybridized specifically with RLV RNA. Thus, none of the eighteen nonleukemic individuals exhibited evidence of the specific 70*S* RNA-directed DNA synthesis found in virtually all of the leukemic patients examined.

Implications of simultaneous detection tests on human breast cancer and the leukemias

The experiments we have just summarized on human breast cancer and the leukemias were designed to probe further the etiological significance of our exploratory investigations (Axel, Schlom and Spiegelman, 1972a; Hehlmann, Kufe and Spiegelman, 1972a), which identified in these neoplasias RNA homologous to those of the corresponding murine oncornaviruses. The data obtained with the simultaneous detection test established that at least a portion of the tumor-specific virus-related RNA we were detecting was a 70*S* RNA template physically associated with a reverse transcriptase in a particle possessing a density between 1.16 and 1.19 g/ml, three of the diagnostic features of the animal RNA tumor viruses. Further, the DNA synthesized in the particles from both classes of neoplasias hybridized uniquely to the RNA of the corresponding oncornavirus. Note that this last result is complementary to and completes the logic of our experimental approach. We started out by using animal tumor viruses to generate [^{3}H]DNA probes that were used to find related RNA in human neoplastic tissue. We concluded by using analogous human particles to generate [^{3}H]DNA probes, which were then used to determine sequence relatedness to the RNA of the relevant oncornaviruses. None of the human probes hybridized to the avian viral RNA. The probe generated by the particles from human breast cancer was homologous only to the RNA of mouse mammary tumor virus, whereas the human leukemic probe was related in sequence only to RLV RNA, the murine leukemic agent. The biologically logical consistency of these results adds further weight to their relevance to the human disease.

On the problem of germ-line transmission of viral information

We now come to grips with the popular virogene (Todaro and Huebner, 1972) concept, which derives from animal experiments and argues that all animals prone to cancer contain in their germ line at least one complete copy of the information necessary and sufficient to convert a cell from normal to malignant and produce the corresponding tumor virus. This hypothesis presumes that the malignant segment usually remains silent and that its activation by intrinsic or extrinsic factors leads to the appearance of virus and the onset of cancer.

There are various ways of testing the validity of the virogene hypothesis, but the pathways differ in the technical complexities entailed. One approach commonly used attempts to answer the question: "Does every normal cell contain at least one complete copy of the required viral-related malignant information?" The methodologies employed included the techniques of genetics, chemical viral induction, and molecular hybridization. However, for a variety of reasons, none of these gave, or could give, conclusive answers. Genetic experiments do not readily distinguish between susceptibility genes and actual viral information. Further, even if genetic data succeeded in identifying some structural viral genes, it would still be necessary to establish that *all* the viral genes are represented in the genome. Attempts to settle the question by demonstrating that *every* cell of an animal can be chemically induced to produce viruses have thus far, for obvious reasons, not been tried. The best that has been achieved along these lines is to show that *cloned* cells do respond positively. However, the proportion of clonable cells is small and *clonability may well be a signal for prior infection with a tumor virus.*

Finally, the quantitative limitations of molecular hybridization make it almost impossible to provide definitive proof that each cell contains one complete viral copy in its DNA. Although it is easy to show that 90% of the information is present, it is the last 10% that constitutes the insurmountable barrier and 10% of 10^7 daltons amounts to a far from trivial 10^6 daltons, the equivalent of about three genes.

A useful way to obviate these technical difficulties is to invert the problem. Instead of asking whether one complete copy exists in normal cells, the question can be phrased in the following terms: "Does the DNA of a malignant cell contain viral-related sequences that are *not* found in the DNA of its normal counterpart?" Phrasing the issue in this manner leads to the design of experiments that avoid the uncertainties generated by the demonstrated fact that many indigenous RNA tumor viruses share, completely or partially, *some* sequences with the normal DNA of their natural hosts (Ruprecht, Goodman and Spiegelman, 1973). The crucial point is of course whether *all* of the viral sequences are to be found in normal DNA. The approach we adopted requires removal of those viral sequences that are contained in non-neoplastic DNA by exhaustive hybridization of the viral probe to normal DNA in vast excess. Any unhybridized residue can then be used to determine whether malignant DNA contains viral-related sequences not detectable in normal tissue.

We first investigated this question in the case of the human leukemias (Baxt and Spiegelman, 1972) and the strategy, as diagrammed in Fig. 6 (A and B), may be outlined in the following steps:

a) Isolate from leukemic cells the fraction enriched for the particles encapsulating the 70*S* RNA and RNA-directed DNA polymerase;

b) Use this fraction to generate [^{3}H]DNA endogenously synthesized in the presence of high concentrations of actinomycin D to inhibit host and viral DNA-directed DNA synthesis;

c) Purify the [^{3}H]DNA by hydroxylapatite and Sephadex chromatography with care being exercised to remove by self-annealing and column chromatography all self-complementary material in the tritiated probe;

d) Use the resultant [^{3}H]DNA to detect complementary sequences in normal and leukemic leukocyte DNA;

e) If viral-related sequences are detected in *both*, remove those found in normal leukocytes by exhaustive hybridization to normal DNA; and

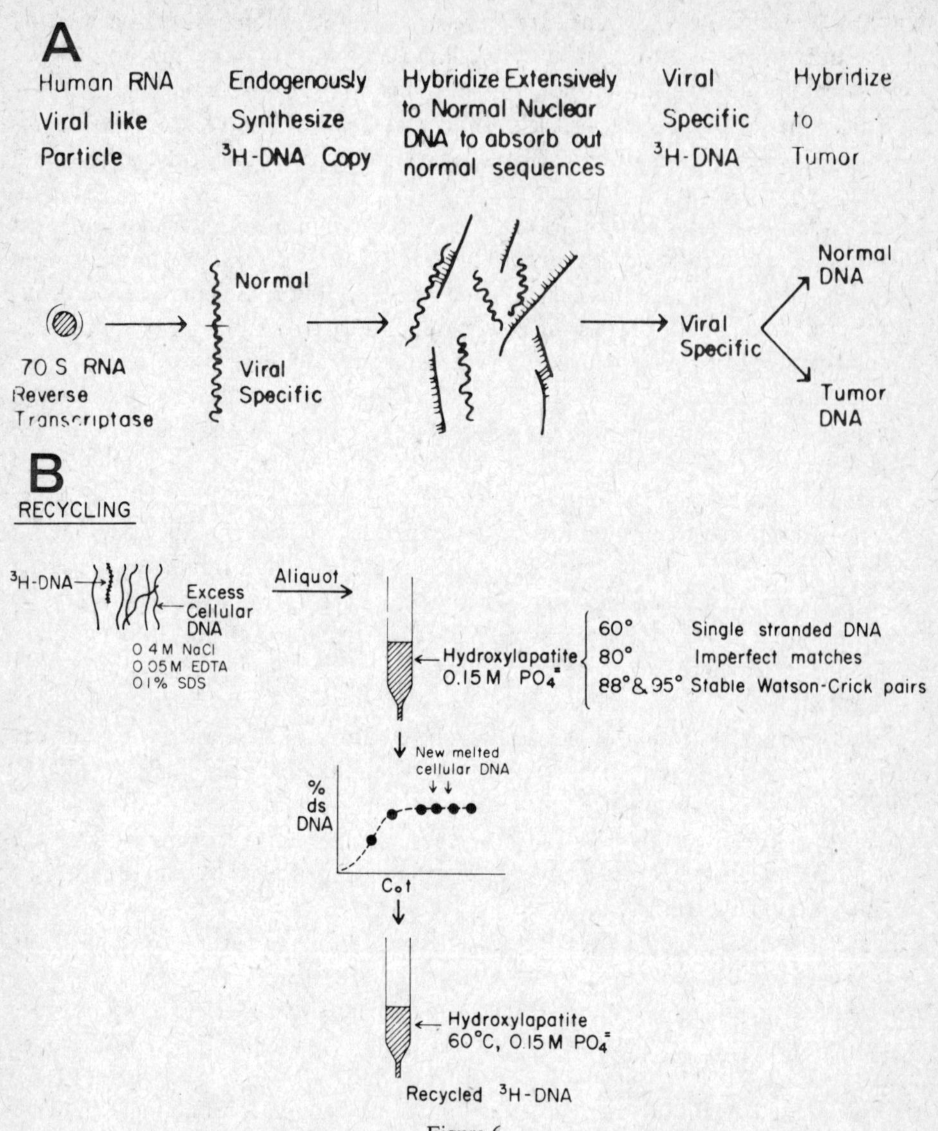

Figure 6

A. Generation of [^{3}H]DNA by human leukemic particles and hybridization of sequences shared with normal DNA.

B. Separation of leukemia-specific sequences by hydroxylapatite chromatography.

See text for further details.

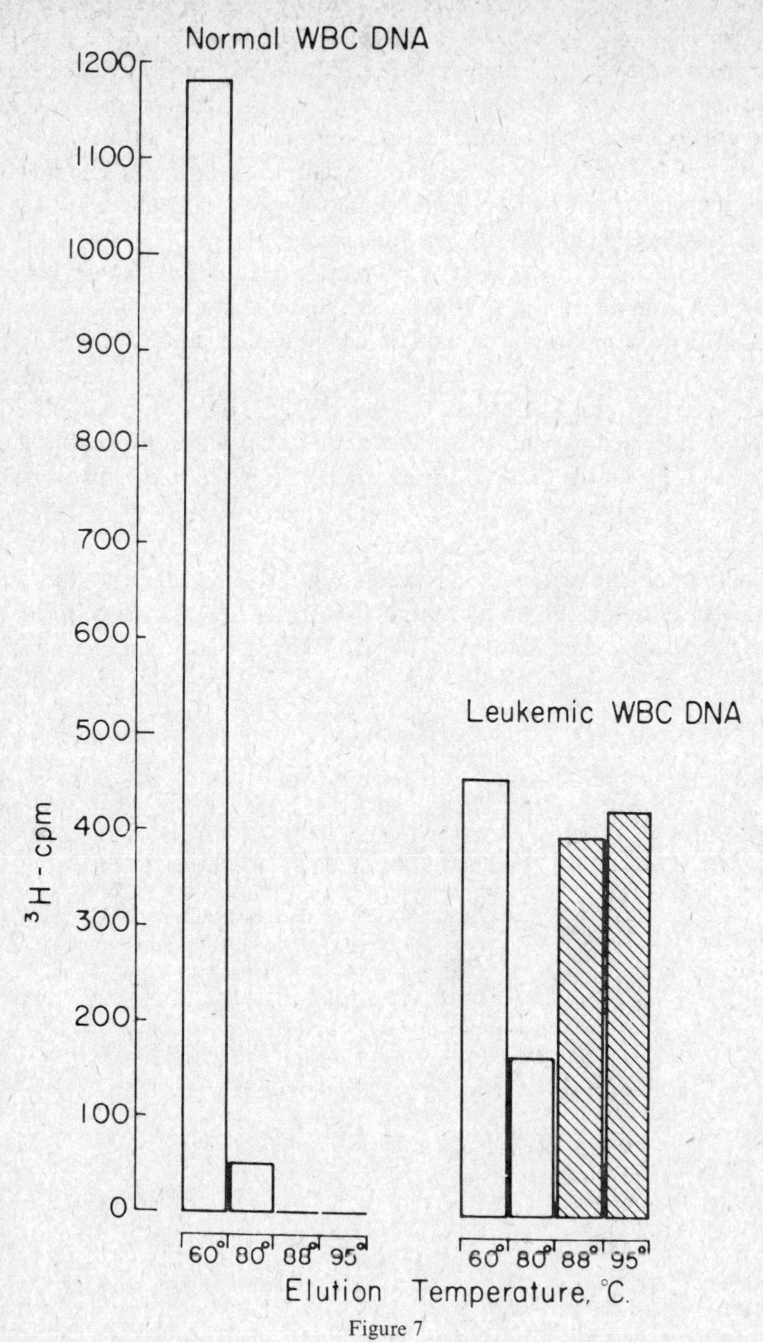

Figure 7

Hydroxylapatite elution profile of a hybridization reaction of recycled leukemic [^{3}H]DNA to nuclear DNA from normal leukocytes and from leukemic leukocytes of the patient from which the [^{3}H]DNA was derived.

f) Test the residue for specific hybridizability to leukemic DNA.

In carrying out the recycling and test hybridizations, it is imperative that conditions be chosen to account for the possibility that the leukemia-specific sequences are present in only one copy per genome, a possibility which is in fact realized (Baxt and Spiegelman, 1972). To this purpose, the concentration in moles per liter (C_0) of DNA and the time (t in seconds) of annealing is adjusted to C_0t values of 10,000, which are adequate to locate single sequences.

A typical outcome of hybridizing such recycled tritiated DNA to normal and leukemic DNA is shown in Fig. 7. It is evident that no complexes stable at temperatures above 88° are formed with normal DNA. On the other hand, 57% of the recycled [^{3}H]DNA probe forms well-paired duplexes with leukemic DNA. A series of such experiments was performed with particle-generated [^{3}H]DNA and nuclear DNA obtained from 8 untreated patients with either acute or chronic myelogenous leukemia. In every case (Table 4), the [^{3}H]DNA, after being subjected to exhaustive annealing to normal DNA, yielded a residue that forms stable duplexes only with leukemic DNA, in agreement with the experiment in Fig. 7.

In estimating the implication of this result, it must be recalled that the leukemia-specific sequences found (Baxt and Spiegelman, 1972) in leukemic cells are present as nonreiterated copies per genome. The sensitivity used to examine normal cells

Table 4

EXHAUSTIVE HYBRIDIZATION OF [^{3}H]DNA PROBE SYNTHESIZED BY LEUKEMIC PARTICLES WITH NORMAL-LEUKOCYTE NUCLEAR DNA, FOLLOWED BY HYBRIDIZATION OF THE NONHYBRIDIZING RECYCLED LEUKEMIC [^{3}H]DNA PROBE TO NORMAL DNA AND TO LEUKOCYTE NUCLEAR DNA FROM THE SAME LEUKEMIC PATIENT

		Leukemic [^{3}H]DNA hybridized to normal-leukocyte DNA		Recycled leukemic [^{3}H]DNA hybridized to leukocyte DNA			
				leukemic		normal	
		cpm	% hybridization	cpm	% hybridization	cpm	% hybridization
Ne	(AML)	3020	61	523	56	0	0
Ha	(CML)	1350	40	1020	51	0	0
Co	(CML)	2580	51	431	35	3	0
Mi	(CML)	510	45	101	36	0	0
Go	(AML)	1100	43	303	48	4	0
Si	(AML)	4520	49	1130	46	1	0
El	(AML)	390	42	45	69	0	0
Mac	(AML)	1450	49	510	52	0	0

Background was 30 cpm and all counts recorded represent cpm above background. CML = chronic myelogenous leukemia. AML = acute myelogenous leukemia. First column indicates initials of patients.

for the leukemia-specific sequences was such that 1/50th of an equivalent of that found in leukemic cells would have been readily detected. Consequently, one may conclude that the vast majority of normal cells do not contain this particular stretch of malignant information and it cannot therefore be represented in the germ line of nonleukemic individuals.

Evidence from studies of identical twins

Although the comparison of leukemic patients with normals suggests that healthy individuals do not contain the leukemia-specific sequences, the data do not rule out the possibility that those who do come down with the disease do so because they in fact inherit the required information in their germ line. One way to resolve this issue is to study the situation in identical twins. Since identical twins are monozygous, i.e., derive their genomes from the same fertilized egg, any chromosomally transmitted information must be present in both. If the leukemic member of the pair contains the particle-related DNA sequences, and does so because he inherited them through his germ line, then these same sequences must be found in the leukocyte DNA of his healthy sibling. To perform the experiment, it was necessary to locate identical twins with completely convincing evidence for monozygosity and where only one one of them was leukemic. Further, the twins had to be of adult age since at least a unit of whole blood is required to provide enough leukocyte DNA to carry out the required hybridization.

Two sets of identical twins satisfying all these requirements were found and an experiment similar to the one outlined above was performed with each pair (Baxt et al., 1973). In each instance, particles containing the reverse transcriptase and 70*S* RNA were again isolated from the leukocytes of the leukemic members and used to generate the [^{3}H]DNA endogenously. The [^{3}H]DNA was purified and sequences shared with normal DNA removed by exhaustive hybridization in the presence of a vast excess of normal DNA from random healthy blood donors. This was then followed by hydroxylapatite chromatography to separate paired from unpaired [^{3}H]DNA. It is important to emphasize that in the recycling step, the normal DNA used came from the leukocytes of healthy, random blood donors and not from the normal twin. To have used the latter would have obviously confused the issue. The residue of the tritiated DNA that did not pair with the normal DNA was then used to test for the presence of a sequence in the leukocyte DNA of the patient and that of his healthy sibling.

The results obtained with the two sets of twins are described in Fig. 8 and it is evident that the same situation holds between the members of the twin pairs as was observed in the comparison of unrelated leukemic patients and random normals (Fig. 7 and Table 4). The leukemic twin contains particle-related sequences that cannot be detected in the leukocytes of his healthy sibling.

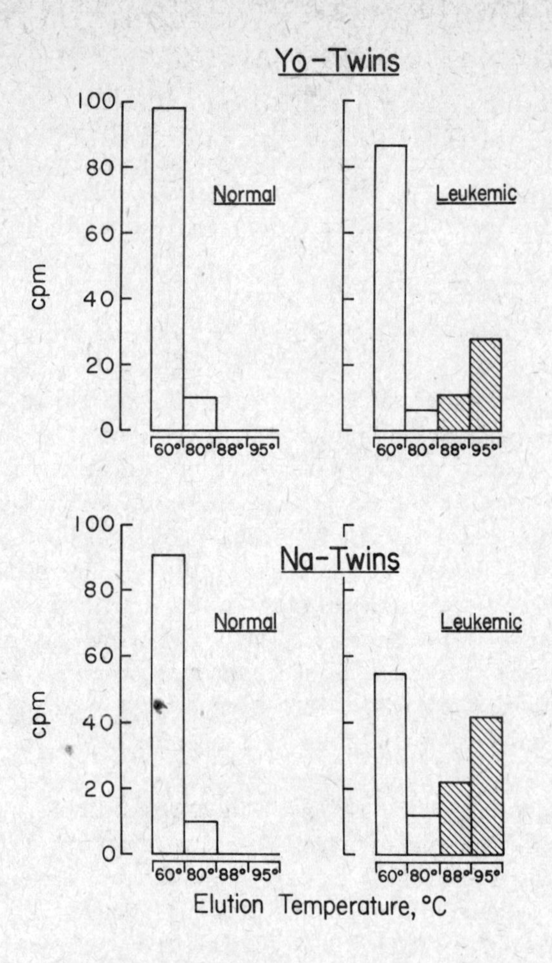

Figure 8
Hydroxylapatite elution profile of a hybridization reaction of the recycled leukemic twin [^{3}H]DNA probe to nuclear DNA from normal leukocytes, normal twin leukocytes, and leukocytes from the same leukemic twin.

The annealing reaction mixtures contained 20 A_{260} units of cellular DNA, 0.004 pmol of [^{3}H]DNA, and 15 μmol NaH PO_4 (pH 7.2) in a final volume of 0.1 ml. The reaction was brought to 98° for 60 sec and 0.04 mmol of NaCl was added. The reaction mixture was then incubated at 60° × 50 hr. The reaction was stopped by the addition of 1 ml of 0.05 M NaH PO_4 (pH 6.8). The sample was then passed over a column of hydroxylapatite of 20-ml-bed volume at 60°. The column was washed with 40 ml of 0.15 M NaH PO_4 (pH 6.8) at 60°, 80°, and 95°. Fractions of 4 ml were collected, the A_{260} of each fraction was read, and the DNA was precipitated with 2 μg/ml of carrier yeast RNA and 10% trichloroacetic acid. The precipitate was collected on Millipore filters, which were dried and counted. In all cases, more than 80% of the nuclear DNA reannealed. A background count of 8 cpm was subtracted in all instances (Baxt et al., 1973).

The fact that we could establish a sequence difference between identical twins implies that the additional information found in the DNA of the leukemic members was inserted after zygote formation. This finding argues against the applicability of the virogene hypothesis to this disease since it would demand that the leukemia-specific sequences found in the DNA of the individual with the disease must surely also exist in the genome of his identical twin. These results are also inconsistent with the possibility that individuals who succumb to leukemia do so because they inherit the complete viral genome.

We have made kindred studies of human lymphomas, including Hodgkin's disease and Burkitt's lymphoma (Kufe, Peters and Spiegelman, 1973), and have established also in these neoplasias that a DNA sequence can be identified in the neoplastic tissue that cannot be detected in normal DNA.

Concluding comments

In the introductory paragraphs we listed a number of genetic mechanisms which could be invoked to explain the change in genotype leading to malignancy. These included somatic mutations and chromosomal rearrangements or imbalances. It is difficult to see how any of these, either singly or in combination, could lead to the appearance of new genetic information related to that contained in a virus-like particle found in the tumor cells.

It should be noted further that the amount of unique genetic information contained in malignant DNA is far from trivial, corresponding to between 3 and 30 gene equivalents. The presence of unique sequences in malignant DNA is clearly also not consistent with the de-repression of preexistent malignant sequences transmitted through the germ line. In particular our data do not support the virogene version, which requires germ-line transmission of complete copies of tumor virus information. The only etiologic mechanism supported by the findings reported here is the one that requires post-zygotic insertion into the genome of viral information, a process known to occur with RNA tumor viruses in animal cells and one that is readily reproduced with tissue cultures (Goodman et al., 1973) and whole animals (Sweet et al., in press) under laboratory conditions.

It is important to emphasize that our conclusions are valid only for the human leukemias and lymphomas, the neoplasias and associated particulate elements actually subjected to experimental examination. What is evident is that these findings suggest more optimistic pathways for the control of these diseases. The data imply that in at least some instances we may not be forced to master the control of our own genes in order to cope with these neoplasias. Finally, the existence of the leukemia and lymphoma-specific sequences provides the clinician with a hitherto unsuspected parameter that could potentially be a useful adjunct in diagnosis and in monitoring therapy.

Acknowledgments

This study was supported by Grant CA-02332 and Contract No. NO1-CP3-3258 within the Virus Cancer Program of the National Cancer Institute, National Institutes of Health.

References

AXEL, R. GULATI, S.C. and SPIEGELMAN, S., 1972. "Particles containing RNA-instructed DNA polymerase and virus-related RNA in human breast cancers." *Proc. Nat. Acad. Sci. USA* **69:** 3133–3137.

AXEL, R., SCHLOM, J. and SPIEGELMAN, S., 1972a. "Presence in human breast cancer of RNA homologous to mouse mammary tumour virus RNA." *Nature* **235:** 32–36.

AXEL, R., SCHLOM, J. and SPIEGELMAN, S., 1972b. "Evidence for translation of viral-specific RNA in cells of a mouse mammary carcinoma." *Proc. Nat. Acad. Sci. USA* **69:** 535–538.

BALTIMORE, D., 1970. "RNA-dependent DNA polymerase in virions of RNA tumour viruses." *Nature* **226:** 1209–1211.

BAXT, W., HEHLMANN, R. and SPIEGELMAN, S., 1972. "Human leukaemic cells contain reverse transcriptase associated with a high molecular weight virus-related RNA." *Nature New Biol.* **240:** 72–75.

BAXT, W.G. and SPIEGELMAN, S., 1972. "Nuclear DNA sequences present in human leukemic cells and absent in normal leukocytes." *Proc. Nat. Acad. Sci. USA* **69:** 3737–3741.

BAXT, W., YATES, J.W., WALLACE, H.J., JR., HOLLAND, J.F. and SPIEGELMAN, S., 1973. "Leukemia-specific DNA sequences in leukocytes of the leukemic member of identical twins." *Proc. Nat. Acad. Sci. USA* **70:** 2629–2632.

BISHOP, D.H.L., RUPRECHT, R., SIMPSON, R.W. and SPIEGELMAN, S., 1971. "Deoxyribonucleic acid polymerase of Rous sarcoma virus: Reaction conditions and analysis of the reaction product nucleic acids." *J. Virol.* **8:** 730–741.

GALLO, R.C., MILLER, N.R., SAXINGER, W.C. and GILLESPIE, D., 1973. "Primate RNA tumor virus-like DNA synthesized endogenously by RNA-dependent DNA polymerase in virus-like particles from fresh human acute leukemic blood cells." *Proc. Nat. Acad. Sci. USA* **70:** 3219–3224.

GOODMAN, N.C., RUPRECHT, R.M., SWEET, R.W., MASSEY, R., DEINHARDT, F. and SPIEGELMAN, S., 1973. "Viral-related DNA sequences before and after transformation by RNA tumor viruses." *Int. J. Cancer* **12:** 752–760.

GROSS, L., 1970. *Oncogenic Viruses*. Second edition. Oxford: Pergamon Press, pp. 1–991.

GULATI, S. C., AXEL, R. and SPIEGELMAN, S., 1972. "Detection of RNA-instructed DNA polymerase and high molecular weight RNA in malignant tissue." *Proc. Nat. Acad. Sci. USA* **69:** 2020–2024.

HALL, B.D. and SPIEGELMAN, S., 1961. "Sequence complementarity of T2-DNA and T2-specific RNA." *Proc. Nat. Acad. Sci. USA* **47:** 137–146.

HEHLMANN, R., KUFE, D. and SPIEGELMAN, S., 1972a. "RNA in human leukemic cells related to the RNA of a mouse leukemia virus." *Proc. Nat. Acad. Sci. USA* **69:** 435–439.

HELLMANN, R., KUFE, D. and SPIEGELMAN, S., 1972b. "Viral-related RNA in Hodgkin's disease and other human lymphomas." *Proc. Nat. Acad. Sci. USA* **69:** 1727–1731.

KUFE, D., HEHLMANN, R. and SPIEGELMAN, S., 1972. "Human sarcomas contain RNA related to the RNA of a mouse leukemia virus." *Science* **175:** 182–185.

KUFE, D.W., PETERS, W.P. and SPIEGELMAN, S., 1973. "Unique nuclear DNA sequences in the involved tissues of Hodgkin's and Burkitt's lymphomas." *Proc. Nat. Acad. Sci. USA* **70:** 3810–3814.

ROKUTANDA, M., ROKUTANDA, H., GREEN, M., FUJINAGA, K., RAY, R.K. and GURGO, C., 1970. "Formation of viral RNA-DNA hybrid molecules by the DNA polymerase of sarcoma-leukemia viruses." *Nature* **227:** 1026–1029.

RUPRECHT, RUTH M., GOODMAN, N. C. and SPIEGELMAN, S., 1973. "Determination of natural host taxonomy of RNA tumor viruses by molecular hybridization: Application to RD-114, a candidate human virus." *Proc. Nat. Acad. Sci. USA* **70:** 1437–1441.

SCHLOM, J. and SPIEGELMAN, S., 1971. "Simultaneous detection of reverse transcriptase and high molecular weight RNA unique to oncogenic RNA viruses." *Science* **174:** 840–843.

SCHLOM, J., SPIEGELMAN, S. and MOORE, D.H., 1971. "RNA-dependent DNA polymerase activity in virus-like particles isolated from human milk." *Nature* **231:** 97–100.

SCHLOM, J., SPIEGELMAN, S. and MOORE, D.H., 1972. "Detection of high molecular weight RNA in particles from human milk." *Science* **175:** 542–544.

SHAPIRO, A., SPIEGELMAN, S. and KOSTER, H., 1937. "A rapid method for the study of genetics in large populations." *J. Genetics* **34:** 237–245.

SPIEGELMAN, S., BURNY, A., DAS, M.R., KEYDAR, J., SCHLOM, J., TRAVNICEK, M. and WATSON, K., 1970. "Characterization of the products of RNA-directed DNA polymerases in oncogenic RNA viruses." *Nature* **227:** 563–567.

SWEET, R.W., GOODMAN, N.C., CHO, J.-R., RUPRECHT, R.M., REDFIELD, R.R. and SPIEGELMAN, S., 1974. "The presence of unique DNA sequences following viral induction of leukemia in mice." *Proc. Nat. Acad. Sci. USA,* **71:**1705–1709.

TEMIN, H. M., 1964. "Nature of the provirus of Rous sarcoma." *Nat. Cancer Inst. Monogr.* **17:**557–570.

TEMIN, H. M. and MIZUTANI, S., 1970. "RNA-dependent DNA polymerase in virions of Rous sarcoma virus." *Nature* **226:**1211–1213.

TODARO, G. J. and HUEBNER, R. J., 1972. "The viral oncogene hypothesis: New evidence." *Proc. Nat. Acad. Sci. USA* **69:**1009–1015.

Stability and thermodynamic properties of dissipative structures in biological systems*

I. Prigogine** and R. Lefever

Faculté des Sciences, Université Libre de Bruxelles, 1050 Bruxelles, Belgium

1. Introduction

Our objective in this paper is to deal with some stability and thermodynamic properties characteristic of the dynamic processes taking place in living systems.

Since the earliest publications on this subject, the idea has been emphasized that non-equilibrium is a possible source of order and self-organization (Schrödinger, 1945; Prigogine and Wiame, 1946; Prigogine, 1947). However, the existence of a non-equilibrium environment, although necessary, is far from being a sufficient condition of self-organization in biological or open systems in general.

In fact, the state of matter near equilibrium can be obtained through extrapolation from equilibrium. It is found that constraints which prevent the system from going to equilibrium increase the entropy in some cases, while in other they decrease the entropy. Examples may be found in previous publications (Prigogine, 1947). In such situations we can hardly speak about self-organization.

Far from equilibrium, on the contrary, certain open systems become unstable and undergo a complete change of their macroscopic properties with respect to the equilibrium situation (Prigogine and Nicolis, 1967; Glansdorff and Prigogine, 1971; Prigogine, 1969). Particularly, it has been shown that the symmetry-breaking instabilities earlier described by Turing (1952), whose conditions of occurrence have also been investigated in a general formal way (Gmitro and Scriven, 1966; Othmer

* The results presented here will also partly be reported in a special volume of the *Advances in Chemical Physics* devoted to the 1972 EMBO meeting in Brussels on "Membranes, Dissipative Structures and Evolution."

** Also Center for Statistical Mechanics and Thermodynamics, The University of Texas at Austin, Austin, Texas 78712, U.S.A.

and Scriven, 1969), enter within this class of phenomena. The kinetic properties of such systems and their behavior beyond instability have been studied on models (Prigogine and Lefever, 1968; Lefever, 1968; Lefever and Nicolis, 1971; Lavenda, Nicolis and Herschkowitz-Kaufman, 1971; Prigogine and Nicolis, 1971). It has also been shown that the properties of several well-known biochemical processes are indicative of systems functioning beyond such a point of instability (Prigogine, 1971; Prigogine, et al., 1969). In fact, the number of biological and chemical problems in which the existence of instabilities has been investigated is continuously increasing. Strikingly also, it is at the most macroscopic levels of life, with phenomena like biological evolution (Eigen, 1971; Prigogine, Nicolis and Babloyantz, 1972), the functioning of the brain (Wilson and Cowan, 1972), or developmental biology (Martinez, 1972), that the characteristic properties of non-equilibrium instabilities have from the beginning seemed of special interest.

The main point to remember is that, in all these cases, there appears to exist a thermodynamic threshold which corresponds to a clear distinction between the class of *equilibrium structures*[1] and those structures which have been called *dissipative structures* because they spontaneously appear only in response to *large* deviations from thermodynamic equilibrium. This response corresponds to the amplification and stabilization of fluctuations by the flows of energy and matter; the organizations which then appear in this way have a negligible probability of occurrence at equilibrium. Thus in order to achieve a satisfactory understanding of systems which function in the non-linear domain of thermodynamics and exhibit a self-organizing character, we require a detailed knowledge of two sets of parameters:

1) molecular parameters which determine the dynamic processes inside the system and also the nature of the fluctuations which might occur;

2) the boundary conditions which are imposed on the system and control its interactions with the outside world; those are parameters which refer to the intensity of the fluxes through the system and also to its geometry.

The influence of this latter set of parameters should never be regarded a priori as spurious or simply contingent. Far from equilibrium, even with a *given* set of molecular kinetic properties, a large variety of dissipative structures may be generated simply by changing the properties of the boundaries. For example, while at equilibrium a chemical reaction is insensitive to the dimensions of the reaction vessel, out of equilibrium on the contrary, by a simple change in the dimension of the reaction volume, one may switch from a time-independent steady state to a regime periodic in time. An example will be considered in Sec. 5.

All these remarks clearly suggest the essential role played by stability considerations. This is the reason why in Sec. 2 we present some general definitions of stability,

[1] This class comprises the structures which are obtained by a continuous modification of the equilibrium regime when the boundary conditions steadily deviate from their equilibrium value. These structures form what has also been called "the thermodynamic branch".

introduce the concepts of Lyapunov function and enumerate some basic problems of structural stability.

Before we switch to the thermodynamic stability theory we first discuss in Sec. 3 the dynamic meaning of entropy. Recent developments in non-equilibrium statistical mechanics have made clear the microscopic meaning of entropy and entropy production. As a consequence we can make more precise the non-equilibrium region to which a thermodynamic description may be consistently applied.

In Sec. 4 we then discuss thermodynamic stability theory and we show that in many cases we can construct a Lyapunov function which has a simple physical significance. We then briefly discuss two examples of structural stability: (1) the behavior of an enzymatic reaction with positive feedback in an array of cells interacting by diffusion, and (2) the coupling between pattern formation and cellular division as it has been described recently by Martinez (1972).

The last sections have a largely exploratory character. We present some preliminary considerations on the relation between the structure (or topology) of chemical networks and specific entropy production. We show that modifications of chemical networks due to catalytic effects may have a simple physical meaning in terms of "adaptation" of the open system to external conditions in the sense that they permit the propagation of the value of chemical potentials given by external conditions inside the reacting system. For reasons indicated briefly (see also Nicolis, 1971), this effect should be of importance in the thermodynamic interpretation of prebiological evolution.

2. Definition of stability, Lyapunov functions, structural instabilities

We shall limit ourselves here to a few definitions and results in stability theory, which are relevant to the next sections. A very complete presentation of stability problems in ordinary differential equations and methods of investigation can be found in the classical textbooks by Andronov et al. (1966), Minorsky (1962) or Cesari (1962). General reviews on stability problems in biology have also appeared by Nicolis et al. (1971, 1973).

2.1. *Stability in the sense of Lyapunov and orbital stability*

Let us consider a system described by the set of differential equations:

$$\dot{x}_i = F_i(x_1, x_2, \ldots, x_n) \qquad (i = 1, \ldots, n) \tag{2.1}$$

The resting state x_i^0, such that all $F_i = 0$, will be stable in the sense of Lyapunov if any solution $U(t)$ of (2.1) (i.e., any series of functions $x_i = U_i(t)$ which satisfies identically (2.1), taking into account the initial and boundary conditions) which is close to it at some time instant t_0 remains in its neighborhood for all times $t \geqq t_0$. More precisely, if given $\varepsilon > 0$ and t_0, there exists a neighborhood $\eta(\varepsilon, t_0)$ such that any solution $U_i(t)$ for which $|U_i(t_0) - x^0| < 0$ satisfies $|U_i(t) - x^0| < \varepsilon$ for $t \geqq t_0$, then the resting state x^0 is said to be stable. It is *asymptotically stable* if moreover $\lim_{t \to \infty} |U_i(t) - x^0| = 0$. These definitions can be extended to motions in a straightforward way.

When the system (2.1) is autonomous, i.e., when time does not appear as an explicit variable in the right-hand side of the equation, those motions which trace closed trajectories or orbits in the phase space are of special interest; indeed they correspond to a periodic behavior of the variables x_i in time. It should be noted that, because of the invariance of translation of autonomous systems, an infinity of motions or solutions differing from each other by their phase corresponds to a given closed trajectory C. One then has the following definitions of stability: a given orbit C of (2.1) is *orbitally stable* if given $\varepsilon > 0$, a representative P of a trajectory which at time t_0 is within a distance $\eta(\varepsilon)$ from C remains at a distance ε from C for all $t \geqq t_0$. Otherwise C is said to be *orbitally unstable.* If moreover the distance between P and C tends to zero when t tends to infinity, C is said to be *asymptotically orbitally stable.*

It is important to distinguish stability in the sense of Lyapunov from orbital stability. The condition for orbital stability only requires that two orbits C and C' remain in the neighborhood of each other; stability in the sense of Lyapunov of a solution requires moreover that the representative points P and P' associated with different solutions remain close to each other if they initially were.

A supplementary distinction must also be made concerning asymptotic stability. Clearly a periodic solution cannot be asymptotically stable if there exist other periodic solutions in its neighborhood; indeed a periodic function cannot be asymptotic to another periodic function. However if $U(t)$ is a periodic solution of (2.1), then $U(t + \tau)$ is another solution of the same system for all values of τ. There thus exist periodic solutions infinitesimally close to $U(t)$ and as a result a periodic solution of an autonomous system cannot be asymptotically stable. An orbit which is asymptotically orbitally stable is called a *limit cycle* according to Poincaré.

From these definitions it should be clear why stability theory is so important in macroscopic physics. We deal always with systems which are formed by many interacting entities (such as electrons, atoms, or molecules). Therefore fluctuations are unavoidable. The description of physical situations in terms of differential equations such as (2.1) neglects these fluctuations. Therefore the condition of validity of this description is precisely that these equations satisfy stability conditions.

2.2. *Lyapunov functions*

From these stability definitions, a criterion of stability can be deduced in a straightforward manner. If we consider the positive definite function x^2 (square of the distance in the space of states between two neighboring solutions) and are able to show that its time derivative satisfies the inequality

$$(\dot{x^2}) \leqq 0 \tag{2.2}$$

for all values of t, then the evolution of (2.1) is stable (≤ 0) or asymptotically stable (<0). It must be noted that (2.2) constitutes only a *sufficient* condition of stability; indeed, for example, an oscillatory behaviour of x^2 would enter within the general definition of stability without obeying (2.2) for all times.

A quadratic function like x^2 which leads to a stability condition of the form (2.2) is called a *Lyapunov function.* While the existence of a Lyapunov function leads only to a sufficient condition of equilibrium, it is easy for autonomous differential equations to give the necessary and sufficient conditions of stability.

2.3. *Linear stability*

Suppose that we have a system described by a set of autonomous ordinary differential equations (i.e., when time does not appear as an explicit variable in the left-hand side of a system like (2.1)) which admits a steady-state solution $\{x_i{}^0\}$. Linear stability of this solution will be ensured if any small deviation δx_i of a variable i ($\delta x_i/x_i^0 \ll 1$) from its steady-state value regresses when $t \to \infty$. In the neighborhood of $\{x_i^0\}$ the response to arbitrary small fluctuations $\{\delta x_i^0\}$ is given by the linear set of equations

$$\frac{d}{dt}\delta x_i = \sum_{j=1}^{n} \alpha_{ij}\delta x_j \tag{2.3}$$

where α_{ij} defined as $\alpha_{ij} = (\partial F/\partial x_j)_0$ is evaluated at the steady state. Equations (2.3) admit solutions of the form:

$$\delta x_j = \delta x_j^0 e^{\omega t} \tag{2.4}$$

which are called normal-mode solutions. Stability implies that the eigenvalue $\omega = \omega_r + i\omega_i$ corresponding to each normal mode has a negative real part ω_r. This condition is fulfilled whenever all the roots of the secular equation

$$|\omega\,\delta_{ij} - \alpha_{ij}| = 0, \qquad \delta_{ij} = \begin{cases} 0 & (i \neq j) \\ 1 & (i = j) \end{cases} \tag{2.5}$$

obtained by substituting (2.4) in (2.3) satisfy the condition:

$$\omega_r < 0 \tag{2.6}$$

An explicit form of this stability condition (2.6) corresponds to the well-known Hurwitz–Routh condition (see, for example, Cesari (1962)). It ensures a *necessary* and *sufficient* stability condition in the sense of Lyapunov of the steady-state solution x_i^0.

It is however important to notice that this type of linear analysis furnishes no indication concerning the *amplitudes* of the fluctuations δx_i. We must not expect therefore that it always satisfactorily predicts the behavior of open systems, particularly if fluctuations of *finite* amplitude have a non-zero probability to occur. This has been illustrated on the following chemical example (Turner, 1973, 1974):

$$\begin{aligned} A + X &\rightleftarrows 2X \\ X + E &\rightleftarrows C \\ C &\rightleftarrows E + B \end{aligned} \tag{2.7}$$

A and B are initial and final products whose concentration is maintained constant, and one furthermore has the conservation relation $E + C = \text{const}$. Plotting the steady-state value of X as a function of A with B constant, a curve of the form represented in Fig. 1 can be obtained. Clearly there exist values of A for which three steady-state solutions are possible. On the basis of the linear stability analysis, it is easily found that the lower branch of states (from $A = 0$ to P) and the upper branch (from P' to $A \to \infty$) are stable, while on the contrary the intermediate states between P and P' are unstable. An hysteresis loop can thus in principle be described by the system when A is varied, in such a way that the system is successively switched from P to P'. It is expected that the jumps between the lower and upper branch occur at P and P'.

In numerical calculations which take the fluctuations into account, it is however found that no such hysteresis loop seems to exist. In fact the system always jumps at some intermediate point ξ between P and P'. This is an excellent example of the effect of fluctuations on the macroscopic behavior. This result should also be related to the properties of non-equilibrium systems which have been described by Kobatake (1970), Schlögl (1972) and Kitahara (1975) and for which the existence of a construction analogous to the Maxwellian construction of vapor pressure of a Van der Waals gas is possible.

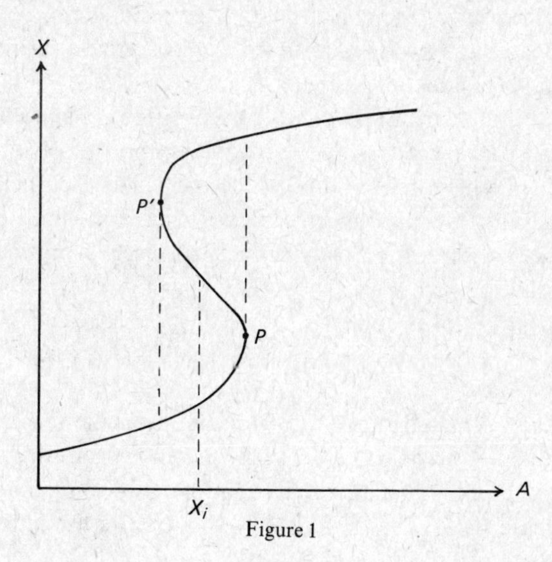

Figure 1

2.4. *Stability in respect to "small parameters"*

As we have already noted, stability problems are of crucial importance in all systems which involve many degrees of freedom and which as a result are continuously submitted to the effect of *fluctuations*, i.e., to random and spontaneous deviations from the average deterministic laws of evolution. Roughly speaking, the effect of fluctuations may manifest itself in two ways:

1) Fluctuations may simply correspond to random perturbations whose regression or amplification is described by the laws of evolution which govern the average motion of the system in the absence of fluctuations.

2) Fluctuations may also correspond to an alteration of the average deterministic laws of evolution of the system.

We have already considered an example of amplification of fluctuations on the macroscopic level. Another aspect corresponds to the possibility that new pathways are opened to the time evolution of the system as the result of fluctuations. *Small parameters* which would normally play no role in the macroscopic evolution become essential as the result of fluctuations. This situation may be considered as a special case of structural stability which will be studied in the next section. We will not discuss here particular models, which may be found elsewhere (see Prigogine et al., 1972), but remain as general as possible.

Suppose that we have a set of chemically interacting species X_i $(i = 1, \ldots, n)$ which all exist in relatively abundant quantities. The evolution in time of the $\{X_i\}$ is given by a set of differential equations

$$\dot{X}_i = F_i^e(X_1, \ldots, X_n) + \langle F_i \rangle (X_1, \ldots, X_n) \tag{2.8}$$

among the concentrations of the constituents X_i. F_i^e is a supply term related to the exchanges of substrates and products with the external environment; $\langle F_i \rangle$ describes the chemical processes inside the system.

It is assumed that the distribution of all concentrations in space inside the system is kept uniform and that there exists at least one asymptotically stable steady-state solution of (2.8), in the neighborhood of which the system is functioning. In agreement with our stability definitions, this means that any small deviation $\delta X_i = X_i(t) - X_i^0$ of the concentration of a chemical species i from its steady state value X_i^0 will regress in time in such a way that $\delta X_i \to 0$ as $t \to \infty$. Furthermore, in agreement with the linear stability analysis, the relation between δX_i and time is of the form $\delta X_i(t) \propto e^{\omega t}$ where ω is a typical normal mode whose real part here is negative.

Suppose now that we have to take into consideration the formation, by random fluctuation (mutations or error copies), of new products from the X_i, which we call $Y_j (j = 1, \ldots, m)$. The number of differential equations describing the joint evolution of X and Y will increase by m. Let us take $m = 1$ and write the kinetic equations for

Y and X in the following form:

$$\varepsilon \frac{dY}{dt} = G(\{X\}, Y, \varepsilon) \tag{2.9a}$$

$$\frac{dX_i}{dt} = F_i^e + F_i(\{X\}, Y, \varepsilon) \tag{2.9b}$$

F_i and G are functions of the old and new variables. ε is a parameter characteristic of the properties of the enlarged system, and such that in the limit $\varepsilon \to 0$ the original set of equations (2.8) is recovered. In other words, for $\varepsilon = 0$, there exists a value $\langle Y \rangle (\{X\})$ furnished by the condition $G(\{X\}, \langle Y \rangle, 0) = 0$ and such that it satisfies the identity

$$\langle F_i \rangle (\{X\}, Y(\{X\}, 0)) \equiv \langle F_i \rangle (\{X\}) \tag{2.10}$$

Mathematically, systems (2.9) can be constructed in such a way that the steady-state solution $\{X^0\}$ of (2.8) becomes unstable upon the addition of a new substance Y which follows a kinetics of the form (2.9a) and satisfies (2.10).[2]

In the limit $\varepsilon \to 0$ the effect of the new substance on the evolution of the $\{X\}$ disappears. For ε not equal to zero but small, the stability of the enlarged system will depend on the regression of $n + 1$ types of fluctuations in the concentrations characterized by $n + 1$ normal modes, n of which have values differing only by correction terms of order ε from the original normal modes of system (2.8). The supplementary mode introduced by the new substance Y is therefore the only one which can destabilize the old system. When this happens, it means that the solutions of equations (2.9) do not remain all the time in a neighborhood of order ε of the solution of (2.8). We will have evolution through unstable transitions of the original system upon addition of new substances. This evolution may then lead, for example, to a new state of high concentration of Y, dominated by the new substances.

Situations of this type have been analyzed within the context of prebiotic evolution for systems in which the new substances determine a drastic change in functional properties (Prigogine et al., 1972). ε is then the inverse of some typical kinetic constant k characterizing the new function and the new time scale of evolution imposed on the system. The essential property of the instability is that it occurs for $\varepsilon \to 0$ (i.e., $k \gg 1$); it corresponds to an increase in the interactions with the environment and in the specific dissipation (i.e., dissipation per unit mass). In other words, this instability, which is triggered by non-equilibrium environmental conditions ($F_i^e \neq 0$ and large), maintaining a continuous energy dissipation increases further the level of dissipation and creates therefore conditions which are favorable for the appearance of other instabilities. This behavior has therefore been called an "*evolutionary feedback*". We will consider this concept in more detail in Sec. 6.

[2] Notice also that (2.9a) has no flow-of-mass term from the surroundings: Y is produced solely from $\{X\}$ and the system is therefore essentially closed.

2.5. *Structural stability*

We have seen that fluctuations may lead to a different macroscopic evolution described by altered differential equations (see (2.9a) and (2.9b)). This leads to a new and more general formulation of stability theory. We may indeed ask if a motion is stable when new *small* terms are added to the differential equations independently if such terms may arise as the effect of fluctuations. This is the basic question studied in the theory of structural stability (see Andronov et al., 1966).

Let us consider a simple example. Suppose that we have a system whose motion in time is a solution of the following set of differential equations:

$$\frac{dx}{dt} = P(x, y), \qquad \frac{dy}{dt} = Q(x, y) \tag{2.11}$$

The question then is whether this motion is stable *when* $\varepsilon \to 0$ with respect to the motion described by the new equations

$$\frac{dx}{dt} = P(x, y) + \varepsilon p(x, y), \qquad \frac{dy}{dt} = Q(x, y) + \varepsilon q(x, y) \tag{2.12}$$

A simple example (Andronov et al., 1966) will show that qualitatively important stability changes might result from the transition (2.11) → (2.12) even in the limit $\varepsilon \to 0$.

Take the following equations

$$\frac{dx}{dt} = by, \qquad \frac{dy}{dt} = -bx \tag{2.13}$$

which are a linearized form of the Lotka–Voltera system. In the (x, y) phase space, (2.13) yields an infinite set of closed *orbitally* stable trajectories surrounding the point $(x = 0, y = 0)$ (Fig. 2).

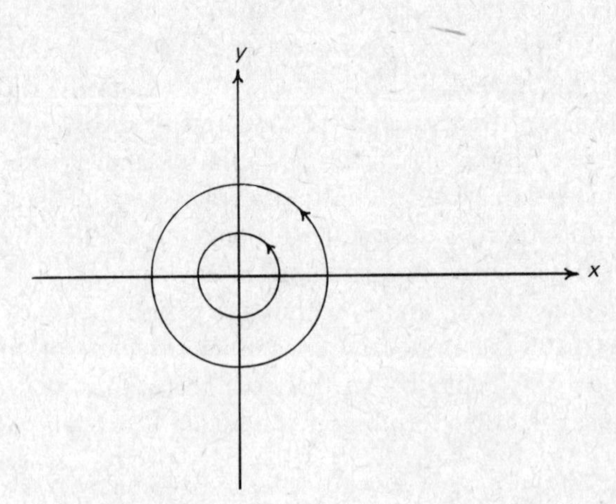

Figure 2

Compare now with the system

$$\frac{dx}{dt} = ax + by, \qquad \frac{dy}{dt} = -bx + ay \tag{2.14}$$

It is found that as soon as the parameter a is different from zero, however small, the point $(x = 0, y = 0)$ becomes *asymptotically* stable: it is then a focus to which all trajectories in phase space converge (Fig. 3).

In other words equations (2.11) are structurally unstable with respect to fluctuations which introduce additional terms like in (2.12). This exemplifies that one has to be very careful in using differential equations which might become structurally unstable, since very slight alterations will produce enormous effects. Indeed, with equations like (2.11), it is possible for the system to function at all times on a closed-trajectory situation even at an infinite distance from the point $x = y = 0$. On the contrary, when $a \neq 0$, all motions in time will converge toward the origin, which is the only stable state of the system.

As a second example, suppose that in a chemical system described by equations of the form (2.11) we now allow fluctuations which break spatial homogeneity. In order to take into account the response of the system to such disturbances of its average state, we have to add diffusion terms to (2.11) which in their simplest form are given by Fick's law. We thus have

$$\frac{\partial x}{\partial t} = P(x, y) + D_x \frac{\partial^2 x}{\partial r^2}, \qquad \frac{\partial y}{\partial t} = Q(x, y) + D_y \frac{\partial^2 y}{\partial r^2} \tag{2.15}$$

It is then essential to know if the solution of equations (2.11) is structurally stable or not in respect to the addition of diffusion terms. If this is not so, even in the limit $D_x, D_y \to 0$ spatial inhomogeneities due to fluctuations may be amplified and stabi-

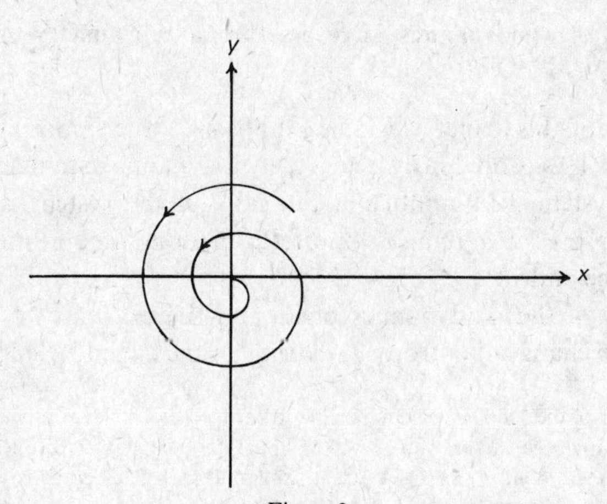

Figure 3

lized. A completely different distribution of matter and functional organization appears inside the system which is no longer predictable to first approximation on the basis of (2.11). In this respect, a broad class of partial differential equations of the second order have been studied by Fife (1972) who has been able to describe their behavior by analytic methods. On the other hand, several model systems have also been investigated in great detail. For example, the chemical scheme (Lefever, (1968)

$$A \rightarrow X, \quad 2X + Y \rightarrow 3X, \quad B + X \rightarrow Y + D, \quad X \rightarrow E$$

has a stability condition with respect to diffusion for the *homogeneous* steady-state regimes which can be expressed simply as

$$B < B_c = \left[1 + A \left(\frac{D_X}{D_Y} \right)^{1/2} \right]^2$$

where D_X and D_Y are the diffusion coefficients of X and Y.

When $B = B_c$, the system will develop a spatial pattern whose characteristic length λ is given by the relation[3]

$$\lambda^2 = \frac{(D_X D_Y)^{1/2}}{A}$$

Clearly, even for very small values of D_X and D_Y, B_c may remain finite and in an accessible range of concentration values. Similar results can be predicted in biological systems. In particular, the enzymatic reaction of phosphofructokinase with ATP and ADP has been studied in great detail (Higgins, 1964; Sel'kov, 1968; Goldbeter and Lefever, 1972; Goldbeter, 1973). In Sec. 5 we shall briefly summarize some of its particularly striking properties.

3. Non-equilibrium thermodynamics—irreversibility as a symmetry-breaking process

Already in Sec. 1 of this paper we have emphasized the importance of "distance from equilibrium". This is obviously a new thermodynamic parameter characterizing the state of the system. In equilibrium the state of the system is determined by quantities such as pressure, temperature, etc. In non-equilibrium supplementary variables have to be added.

The second law of thermodynamics postulates the existence of a state function S, the entropy. The change of entropy dS during a time dt can be split into two parts:

[3] Analytic calculations of the spatially or temporally ordered solutions which appear beond the bifurcation point have recently been carried out, as well as a study of their stability properties (G. Nicolis and J. F. G. Auchmuty, *Proc. Natl. Acad. Sci. USA* **71**, 2748 (1974); J. F. G. Auchmuty and G. Nicolis, to appear in *Bull. Math. Biol.* (1975).

the entropy production d_iS due to changes inside the system and the flow of entropy d_eS due to the interactions with the outside world:

$$dS = d_eS + d_iS \tag{3.1}$$

The entropy production d_iS due to changes inside the system is never negative

$$d_iS \geqq 0 \tag{3.2}$$

There are non-equilibrium situations where the macroscopic evaluation of entropy production and of the entropy flow is possible. This is so when there exists within each small mass element of the medium a state of *local equilibrium* for which the local entropy s is the same function of the local macroscopic variables as in the equilibrium state. As discussed in more detail elsewhere, this implies that dissipative processes are sufficiently dominant to exclude large deviations from statistical equilibrium. We may therefore apply the local-equilibrium assumption to transport processes described by linear laws as well as to not too fast chemical reactions (such that the activation energy is large in respect to thermal energy). When these conditions are satisfied we obtain for the entropy production per unit time

$$P = \frac{d_iS}{dt} = \sum J_\rho X_\rho \tag{3.3}$$

where J_ρ and X_ρ are the conjugate thermodynamic fluxes and forces. In the case of chemical reactions J_ρ and X_ρ are simply equal to v_ρ and $\mathscr{A}_\rho T^{-1}$ respectively the rate and affinity of chemical reaction. It is very recently that a general formulation of non-equilibrium statistical mechanics has been achieved which leads in turn to a microscopic definition of entropy of generality comparable to that of phenomenological entropy.

In order to do so, the dynamic nature of irreversibility had to be elucidated. We have now shown that irreversibility may be viewed as a symmetry-breaking process arising in well-defined classes of dynamic systems formed by many interacting units.

We have to limit ourselves here to a short summary of some of the ideas involved.

Let us strart from classical or quantum dynamics as expressed by Hamilton's equations of motion

$$\frac{dq}{dt} = \frac{\partial H}{\partial p}, \qquad \frac{dp}{dt} = -\frac{\partial H}{\partial q} \tag{3.4}$$

or Schrödinger's equation

$$i\frac{\partial \psi}{\partial t} = H\psi \tag{3.5}$$

Both descriptions may be unified through the Liouville–von Neumann equation for the distribution function (or density matrix):

$$i\frac{\partial \rho}{\partial t} = L\rho \tag{3.6}$$

with

$$L = \begin{cases} \{H, \rho\} & \text{Poisson bracket} \\ [H, \rho] & \text{commutator} \end{cases} \tag{3.7}$$

The solution of equation (3.6) requires a more precise specification of the problem in which we are interested. We suppose known the value of $\rho(t)$ at $t = 0$ and calculate $\rho(t)$ for $t \leqq 0$. This corresponds to an "*initial-value*" problem. If on the contrary we are interested in $\rho(t)$ for $t \leqq 0$, we have a "*final-value*" *problem*. Thermodynamic behavior is obtained when the initial-value problem leads to different solutions from the final-value problem. Entropy may be expressed in terms of the density matrix ρ and it may be shown that it increases for *all initial*-value problems. In other words, the second law of thermodynamics holds for a class (and not for all!) of solutions of the Liouville equation (3.6). We see that we can associate irreversibility with "symmetry breaking".

If the solutions for the initial-value problem coincide with those for the final-value problem, no entropy which satisfies the second law can be defined. The important point is that we have now established an exact general condition which a system has to satisfy to present thermodynamic behavior. We have called this condition the *dynamic condition of dissipativity* (it is related to the analytic continuation of the resolvent $1/(L - z)$ in the complex plane). It can be shown to be satisfied for systems, such as interacting gases or anharmonic lattices, to which we may expect the second law of thermodynamics to apply. We may say that thermodynamic description is the expression of the breaking of the original dynamic groups as described by equation (3.6) into two semigroups, one for the "*retarded*" solutions corresponding to the initial-value problem and the other for the "*advanced*" solutions corresponding to the final-value problem.

For *isolated* systems, entropy is a Lyapunov function in the sense defined in Sec. 2. Thermodynamic equilibrium is then stable with respect to arbitrary perturbations. However, this Lyapunov function has in general no simple macroscopic meaning (see Prigogine, George, Henin and Rosenfeld, 1973). It cannot be expressed in terms of observables such as temperature, etc. Using a somewhat paradoxical formulation we could say that the second law is in general a theorem in dynamics of large systems and not a theorem in thermodynamics considered as macroscopic physics. It is only *near* equilibrium that we can express this Lyapunov function in terms of macroscopic variables and that we are back to our conception of local equilibrium. But even in this restricted range we can introduce the distance from equilibrium as a parameter and study what happens when we impose stronger and stronger constraints on the system.

4. Thermodynamic stability theory

Let us start with the classical Gibbs–Duhem stability theory applicable to closed systems (which do not exchange matter with the outside world). The entropy flow (see (3.1)) is then given by

$$d_e S = \frac{dQ}{T} \tag{4.1}$$

where dQ is the heat received by the system and T the absolute temperature.

Using (3.1), (3.2) and (4.1) we obtain immediately from the first law of thermodynamics

$$T d_i S = T dS - dE - p\, dV \geqq 0 \tag{4.2}$$

where dS is the total entropy variation of the system. From (4.2) it is clear that an equilibrium state is stable provided any fluctuation of the variables satisfies the inequality

$$\delta E + p\, \delta V - T \delta S \geqq 0 \tag{4.3}$$

We thus have for isolated systems (E, V constant) the stability criterion $\delta S \leqq 0$. If any fluctuation around the equilibrium state of maximum entropy leads to a negative entropy variation, then by virtue of the second law, the equilibrium state is asymptotically stable in the sense of Lyapunov: all fluctuations will be damped. Alternatively, if V, T or p, T are maintained constant, (4.3) leads to an analogous stability condition for the thermodynamic potentials F or G.

Most stability problems of equilibrium states can appropriately be described in terms of the classical thermodynamic potentials E, F, G, H. Confronted with non-equilibrium situations this method encounters a major difficulty: the external properties of F or G indeed require a control of fluctuations in variables like p or T not only on the boundaries of the system but also at each point inside it. This is hardly realizable, particularly in open systems, where most of the time only the boundary conditions can be controlled. The results and methods of classical thermodynamics therefore cannot in general be transposed to the non-equilibrium domain. Such an extension needs a different type of approach, carried out in the recent years by Glansdorff and Prigogine (1971). It provides a new formulation of the equilibrium stability theory for problems with given *boundary values* and is also applicable to a very broad class of non-equilibrium situations. In fact, the range of validity of this theory extends to all situations for which the local equilibrium assumption holds. Let us simply sketch the main results.

4.1. *Stability condition of the equilibrium state*

For a system in a stable equilibrium state and subject to *fixed boundary conditions*, it can be deduced from the entropic balance that the excess entropy $\delta^2 S$, obtained by expanding the entropy around the equilibrium state,

$$S = S_e + (\delta S)_e + \frac{(\delta^2 S)_e}{2} \tag{4.4}$$

is a quadratic negative-definite expression[4]

$$T\delta^2 S = \left[-\frac{C_v}{T}(\delta T)^2 + \frac{\rho}{\chi}(\delta v)^2_{N_\gamma} + \sum_{\gamma\gamma'} \mu_{\gamma\gamma'} \delta N_\gamma \delta N_{\gamma'} \right] < 0 \tag{4.5}$$

and moreover that its time derivative is equal to the entropy production P of the system:

$$\frac{1}{2}\frac{\partial\, \delta^2 S}{\partial t} = P > 0 \tag{4.6}$$

In other words, $\delta^2 S$ is a Lyapunov function and its behavior around equilibrium states demonstrates the damping of all fluctuations inside the system.

This result is much more general than that of the classical Gibbs–Duhem theory as it is independent of the existence of any thermodynamic potential. It should however be noticed that it applies only to *small* fluctuations as higher order terms have been neglected in (4.5).

Let us now go over to systems subject to *non-equilibrium* boundary conditions and inquire under which conditions such systems still exhibit stability properties analogous to the equilibrium systems. Alternatively, the question we would like to investigate is whether there exists a Lyapunov function such that the behavior of systems out of equilibrium still can be understood in terms of an equilibrium type of physical chemistry.

4.2. *The linear range*

The central quantity of non-equilibrium thermodynamics is the entropy production P (cf. Sec. 3), which is a positive-definite quantity that can only vanish at equilibrium, i.e., when simultaneously $J_\rho = 0$ and $X_\rho = 0$. Therefore, we can define the *linear range* of non-equilibrium thermodynamics as the domain close to equilibrium in which linear relations hold between the fluxes and forces:

$$J_\rho = \sum_{\rho'} L_{\rho\rho'} X_{\rho'} \tag{4.7}$$

$L_{\rho\rho'}$ are phenomenological coefficients which satisfy the well known Onsager's reciprocity relations

$$L_{\rho\rho'} = L_{\rho'\rho} \tag{4.8}$$

[4] C_v, χ, ρ, N_γ are, respectively, the specific heat at constant volume, the isothermal compressibility, the density, and the molar fraction of γ.

Inserting (4.7) into (3.3) and using (4.8), it can be demonstrated that

$$\frac{dP}{dt} \leqq 0 \tag{4.9}$$

where the equality sign corresponds to the steady state.

We thus find that the steady states which belong to the linear range are characterized by an extremum principle, according to which entropy production is minimum for the steady state compatible with the constraints imposed on the boundaries. Furthermore P is a Lyapunov function, and we consequently see that not only around the equilibrium state, but also around the steady states immediately adjacent to it, the fluctuations will regress in time.

4.3. *The non-linear range*

The range of validity of linear thermodynamics for which (4.8) holds is more restrictive than the local equilibrium assumption. Even for situations corresponding to very large deviations from the equilibrium state, we may still admit most of the time that locally the equilibrium state is preserved at each point of the system, so that locally $\delta^2 s < 0$. The inequality sign is satisfied when the local equilibrium state is stable, as we always shall assume in what follows. This suggests an approach to stability problems based on the excess entropy $\delta^2 s$ as Lyapunov function.

In this way, whenever the excess entropy production satisfies the conditions

$$\frac{\partial\, \delta^2 s}{\partial t} = \sum_{\rho} \delta J_{\rho}\, \delta X_{\rho} > 0 \tag{4.10}$$

the steady state considered is stable. In (4.10), δJ_{ρ} and δX_{ρ} are the fluctuations of the fluxes and forces around the non-equilibrium steady state compatible with the boundary conditions.

It is interesting to note that far from equilibrium the stability properties are characterized by using the concept of the Lyapunov function in a reverse way compared with the equilibrium case. Indeed, at equilibrium the conditions of stability are deduced by stipulating the sign of $\delta^2 s$, while far from equilibrium, on the contrary, it is the time derivative of $\delta^2 s$ which furnishes a criterion of stability. We thus see that there exists a possibility even for purely dissipative systems (i.e., systems without hydrodynamic motions) to escape from the type of organization and functional properties which characterize the equilibrium regimes or the states close to equilibrium. *Non-equilibrium* might be the *source of order* producing new extremely coherent behaviors in space and in time provided condition (4.10) cannot be fulfilled for the steady states which extrapolate the properties of the equilibrium regime. In the case of chemical reactions, it can easily be verified that the kinetic properties underlying a breakdown of inequality (4.10) are of an autocatalytic or cross-catalytic nature (see the examples mentioned by Nicolis (1971)).

One of the most important characteristics of such instabilities is that the relation between order and fluctuations is much more complex than at equilibrium where everything is determined by the strict properties of the thermodynamic potentials. In the non-linear range, not only fluctuations are the trigger for the appearance of new structures, but moreover different fluctuations may correspond to different structures and several stable non-equilibrium regimes may coexist. Therefore in contrast with the ordering principles of equilibrium based on the classical potential, we may say that far from equilibrium a type of order prevails which we have called *order through fluctuations* and whose description would require a *new extension of chemical physics.*

5. Dissipative structures in an enzymatic reaction with positive feedback

Essentially we consider an enzymatic model involving an allosteric dimeric enzyme E which transforms a substrate S into a product P having a positive feedback effect on the enzymatic reaction. The reaction scheme thus has the form:

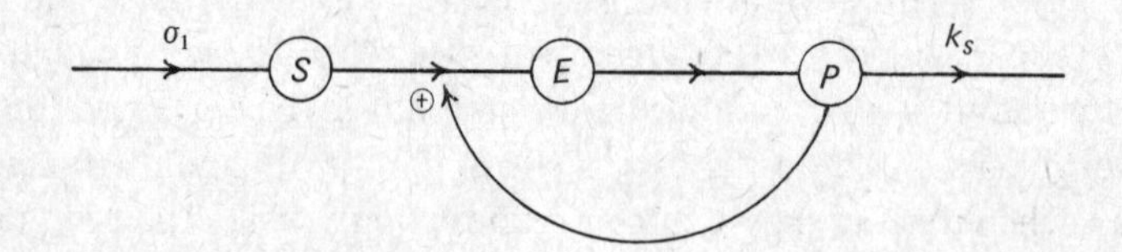

where σ_1 is the rate of injection of the substrate S and k_s a kinetic constant characteristic of the rate at which the product P leaves the system. The enzyme E exists in several conformations, some of which bind P preferentially. In doing so, they enhance the binding of the substrate S to the active conformation of the enzyme. A complete description of the model is given in Fig. 4 (Goldbeter et al., 1972).

This model has been applied to the phosphofructokinase reaction which is the principal oscillating step of the glycolytic chain. Here the substrate S and the product P are respectively ATP and ADP. Moreover, in several organisms, e.g., the yeast cells, PFK is an allosteric enzyme with regulatory properties similar to those of Fig. 4.

Now the functional property of this enzyme which is relevant for our discussion can be vizualized as in Fig. 5, where we have plotted the production rate $f(\alpha, \gamma)$ of P as a function of the concentration γ of P, the concentration α of S being kept constant. It is seen that $f(\alpha, \gamma)$ is a non-linear positive function of γ which *increases* in some interval until it reaches a plateau independent of α, γ.

Let us now consider the case where the boundary conditions imposed on the system have been chosen in such a way that it is functioning in the sigmoid part of the curve and let α_0 and γ_0 be the steady-state values of the concentrations of α and γ. The question we ask, is this *macroscopic* steady state stable with respect to infinitesimal

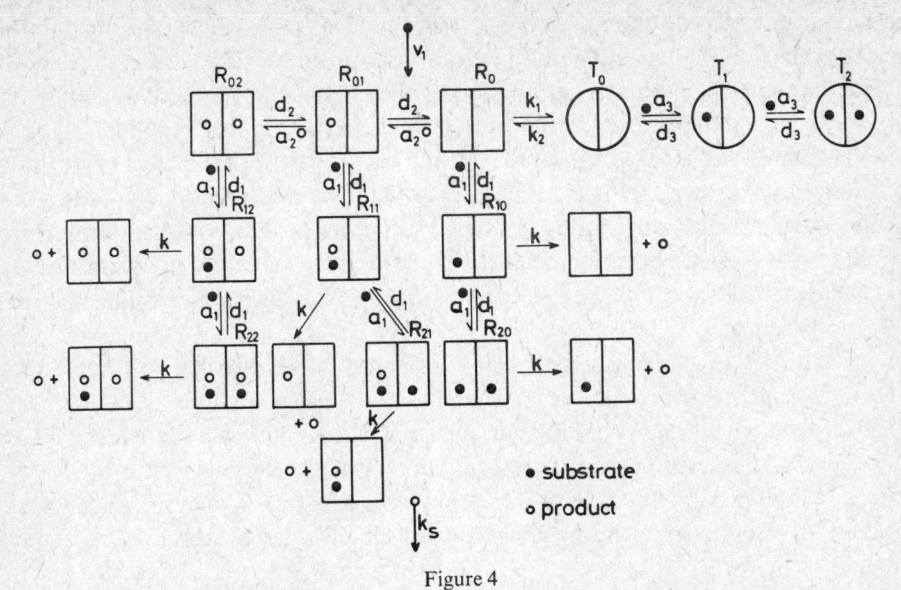

Figure 4
Model of an allosteric enzyme activated by its reaction product. ● substrate, ○ product.

fluctuations in the value of the concentrations due to random motion or thermal agitation? In other words, if at $t = 0$, $\gamma = \gamma_0 + \delta\gamma$, will we observe at $t \to \infty$, $\delta\gamma \to 0$ or on the contrary is there a possibility that $\delta\gamma$ amplifies in the course of time?

We can answer this question in a very simple way by considering the general stability criterion (4.10). If one considers in the sum (4.10) the terms related to the enzymatic reaction, one simply has for a fluctuation of γ

$$\delta f = (\partial f / \partial \gamma)_{\gamma = \gamma_0, \alpha = \alpha_0} \delta\gamma; \qquad \delta \mathscr{A} = \delta \log (\alpha/\gamma) = -\delta\gamma/\gamma_0$$

The product $\delta f \, \delta\mathscr{A}$ consequently contributes a negative term to (4.10), so that the system *may* become unstable when this term is dominant. Beyond such an in-

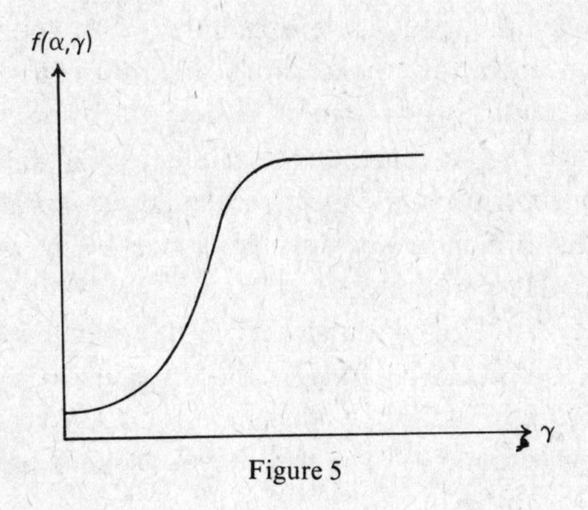

Figure 5

stability, it has been shown (Goldbeter, 1973) that in the absence of diffusion, the system exhibits sustained, stable oscillations in the concentrations of S, P and E in various forms. The amplitude and period of these oscillations are characteristic of the system and independent of the initial conditions; they correspond to a *limit cycle* (cf. Sec. 2). A typical result obtained with values of the parameters taken from the literature is shown in Fig. 6. We observe that the period in general is in good agreement with the periods of the glycolytic oscillations *in vitro*. Another point which can directly be tested concerns the effect of varying the substrate injection rate σ_1. Fig. 7 shows that the amplitude of oscillations goes through a maximum, whereas the period decreases.

From a thermodynamic point of view, the limit cycle shown before represents a *temporal dissipative structure*, i.e., a coherent state where all chemical constituents oscillate simultaneously with different relative phases, but with the same phases for each constituent within the reaction volume.

From a biological point of view, this behavior is usually called a biological clock. So let us now investigate the behavior of this clock, when diffusion is taken into account.

We consider a one dimensional medium which, to fix our ideas, can be viewed as a linear sequence of cells. Each cell is homogeneous, but in contact by diffusion with the neighboring cells. At each end, the concentration of all products is maintained constant. It is easily shown that under those conditions, keeping the value of the kinetic coefficients and rate parameters constant, the behavior of the system is simply related to a dimensionless parameter D/l^2 (D is the diffusion coefficient, and l the length of the system) which measures the coupling *between neighboring spatial regions* and might therefore be called the *diffusive coupling*.

Rougly speaking, one may say that as the intensity of the *diffusive coupling* decreases, the clock switches from a time-independent but space-dependent regime to a *concentration wave*. Keeping the same diffusive coupling but lowering the injection rate of the substrate, we can go from the time-independent regime to a *standing concentration wave*. In fact three principal types of regimes may appear: (1) spatial inhomogeneous steady-state structures, (2) propagating concentration waves, (3) standing concentration waves. As an illustration, Fig. 8 shows the behavior of one of the concentrations γ in regime (2) (see Goldbeter, 1973).

It is particularly striking to see that the characteristic space scale of behaviors (1)–(3), and of the concentration waves in particular, is supracellular. For values of the rate constants and diffusion coefficients in agreement with the experimental data in the literature (Hess and Boiteux, 1968; Hess, Boiteux and Krüger, 1969; Goldbeter and Lefever, 1972) it is found that an inhomogeneous time-independent regime will arise in a system of length l of the order of 10^{-2} to 10^{-1} cm, whereas propagating waves are observed for $l = 0.3$ cm.

It is also interesting to compare the time scale of the waves to the time scale of

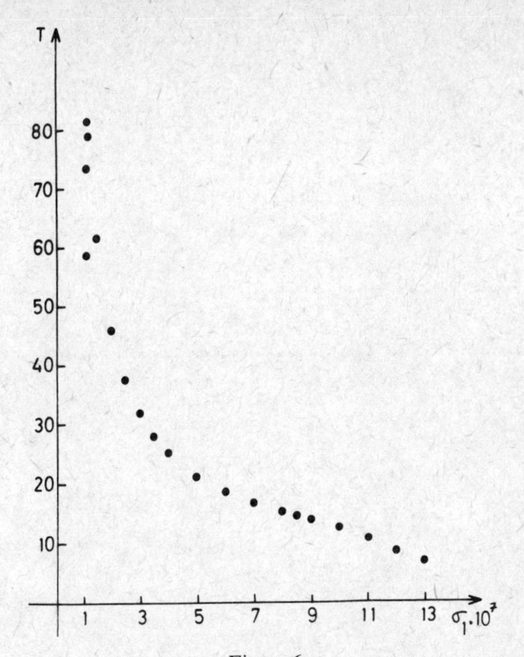

Figure 6
Oscillation period as a function of the substrate injection rate (for values of the parameters and details, see Goldbeter and Lefever, 1972a, b).

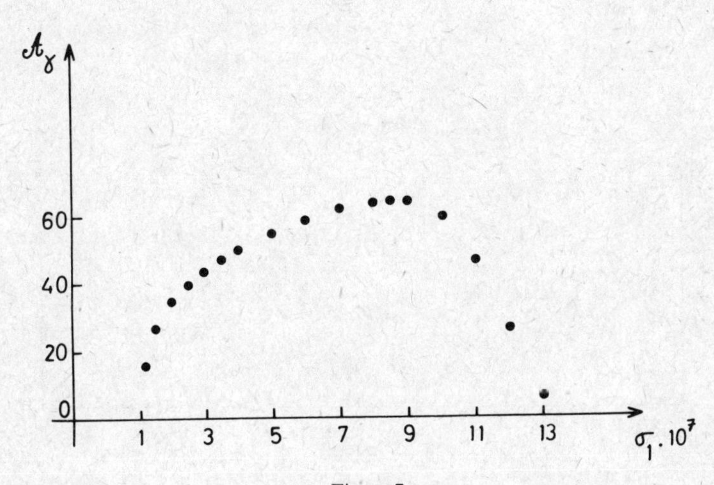

Figure 7
Amplitude of the product oscillations as a function of the substrate injection rate (see Goldbeter an Lefever, 1972a, b).

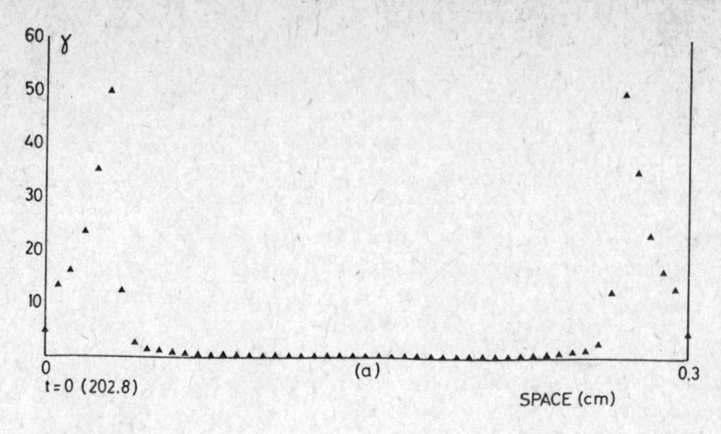
60
50
40
30
20
10
0
γ
(a)
0.3
t = 0 (202.8)
SPACE (cm)

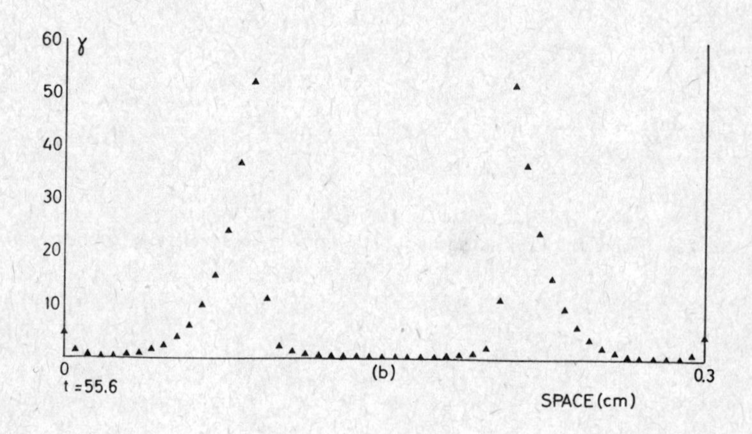
60
50
40
30
20
10
0
γ
(b)
0.3
t = 55.6
SPACE (cm)

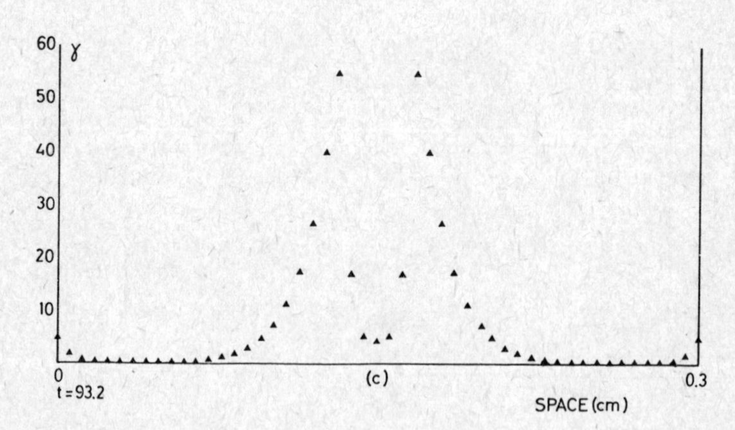
60
50
40
30
20
10
0
γ
(c)
0.3
t = 93.2
SPACE (cm)

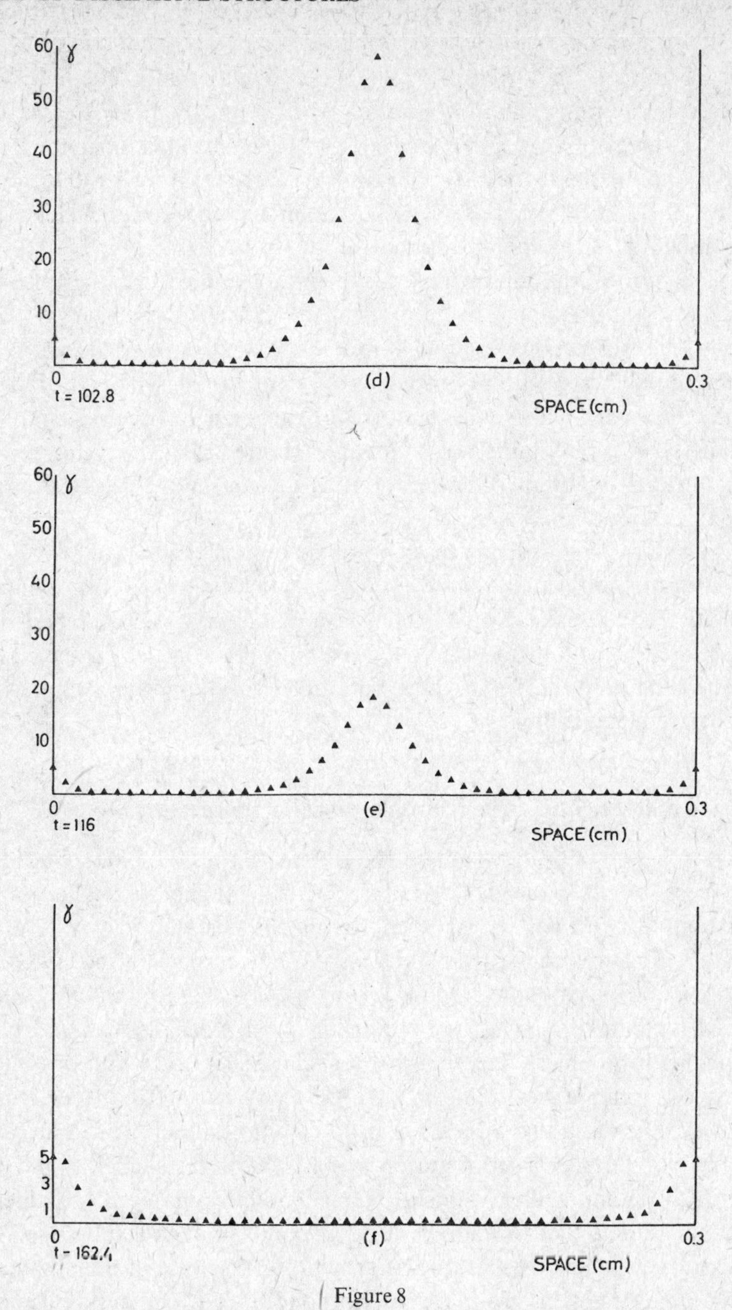

Figure 8

Distribution of concentration γ at successive time intervals ($T = 202.8$ sec). The time is set to zero in Fig. 8a, some twenty periods after the phenomenon started from the unstable homogeneous steady state. The dimension of the system is 0.3 cm. The steady-state concentrations maintained at the boundaries are $\alpha = 40.2$ and $\gamma = 5$.

diffusion. In the plateau region for the amplitude (see Fig. 7), the rate of propagation of the wave front is of the order of 10^{-3} cm/sec. The period of the individual limit cycles is then in the range of a few minutes. The time required by the wave front to travel at a constant rate over a distance of $6 \cdot 10^{-2}$ cm is about 1 minute. In contrast, the time required by the front to spread over the same distance by diffusion alone is of the order of 1 hour. (We thus see that cellular metabolism provides a means of *fast* transmission of sharp chemical signals.)

Finally, it is worth pointing out that all the important properties of the wave-like synchronization, the maintenance of well-defined phase shifts between various regions of space or the regulation of speed and amplitude, result from a single set of equations and assumptions. These assumptions tell us once and for all what is the type of molecular event considered, what is the geometric arrangement of the system units (cells) in space, and what is maintained constant at the boundaries. Everything else follows automatically and simultaneously, without any cause-and-effect relation between the different features of the wave.

These results thus demonstrate that, in principle, glycolysis could provide a means for signal transmission which in a morphogenetic field might control differentiation. These results also are expected to extend to a whole class of oscillatory metabolic reactions related to respiration and oxidative phosphorylation. In the next section we shall investigate a model of differentiation based on the coupling between pattern formation and growth.

6. Succession of instabilities. The Martinez model of morphogenesis

It has always been tempting to relate the differentiation of a morphogenetic field to the existence of a pre-pattern of concentration of some chemical called morphogen. The idea that this pre-pattern appears beyond an instability of a homogeneous steady-state regime was suggested some twenty years ago by Turing (1952) and later investigated in various systems (see for example Maynard-Smith (1961) and Sondhi (1963)). The fundamental advantage of this idea over older theories (morphogenetic gradient) evidently lies in the fact that the morphogenetic field is not postulated to be inhomogeneous from the beginning. Rather it explains how inhomogeneity may arise.

On the other hand, it also appears that differentiation and growth are associated with an evolution through a set of successive patterns, rather than with the transition in one step from a homogeneous state to a final definitely organized and differentiated pattern. If one admits that development corresponds to such a succession of states, one is lead to the conclusion that during growth the response of the organism to the distribution of morphogens gradually creates conditions which destabilize the initial pattern. At this point of instability, a new pattern appears which in turn controls the growth process until itself destabilized by growth.

An interesting example where a succession of instabilities is involved was recently discussed by Martinez (1972). This author shows how, in the course of development, a system of cells can evolve by a succession of instabilities through a set of distinct states. The important point to realize is that in order to regulate such a passage not only the system must be unstable, but the instabilities must in some way regulate the boundary conditions of the system. Indeed, if given *invariant* boundary conditions are imposed on a system, one expects that beyond a point of stability the system chooses, according to the fluctuations present initially, one stable regime from among the set of such states compatible with the boundary conditions. It then remains in this state for as long as this boundary is maintained. The interesting idea in the paper of Martinez is that the instabilities control their own boundaries, or more exactly the relation between boundary values and system dimensions, via *cellular division*:

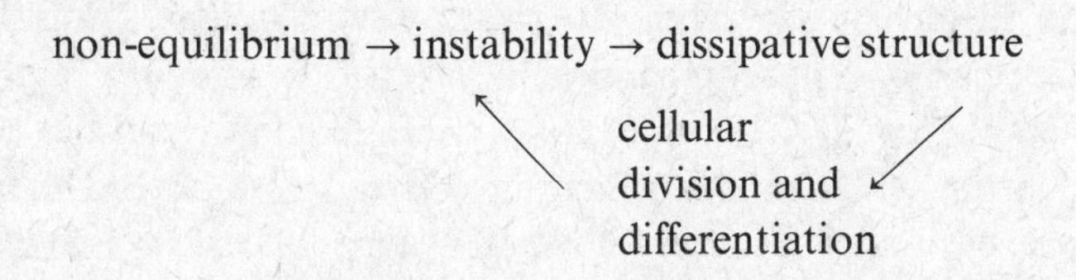

The incorporation of growth into a system of cells which undergo chemical reactions and diffusion requires a decision concerning the state which determines the cellular division. In its simplest version, the Martinez model involves two morphogens X and Y, and it is assumed that the realization of the inequality $X > Y$ between the concentrations constitutes the necessary prerequisite to cellular replication. On the other hand, since an important event should not be triggered by a random process like fluctuations, it is necessary to postulate that the inequality must hold over a period of time much longer than the characteristic regression time of random fluctuations. It is therefore assumed that the inequality must be realized in a steady state. Furthermore, in order to account for the fact that during the development of an organism, there is differential replication of the cells according to their position, one is lead to the idea that the steady state corresponds to an inhomogeneous pattern. The inequality should thus hold in some regions of the pattern and not in others.

These hypotheses can be summarized as follows:

1. One considers a linear array of cells.

2. A cell in the array divides if and only if the concentration Z of one of its constituents becomes bigger than some critical value Z_t.

3. In each cell, the synthesis of Z is controlled by two morphogens (activator X, inhibitor Y) according to the rate equation

$$\frac{dT}{dt} = \begin{cases} a(X - Y) - bZ & X > Y \\ -bZ & X \leqq Y \end{cases}$$

4. The concentrations of X and Y in each cell depend on another set of reactions and on the diffusion of Y between cells; a spatial dissipative structure may exist.

Now, as we mentioned in the preceding section, the properties of such a system depend on the ratio D/l^2, where D is the diffusion coefficient of the substance Y and l the length of the system. As a result, when such a pattern appears, neighboring cells will have different X and Y concentrations and accordingly different rates of synthesis of Z. The threshold Z_t will be reached in only some of the cells, which then will divide and consequently change the ratio D/l^2. This latter effect destabilizes the original pattern and a new distribution of matter among the cells appears, initiating further division of other cells. This process keeps going on; very complicated and organized situations may result if growth is considered in two or three spatial dimensions.

Although this is a highly idealized model involving chemical steps which are not likely to occur in cells, it nevertheless indicates very strikingly that "self-organizing automata" may exist on a purely chemical basis. There is no doubt, on the other hand, that similar results could be obtained with certain more realistic, but also more complicated, enzymatic reactions which have already established the possibility of the occurrence of spatial patterns of *supracellular dimensions* (see, e.g., Prigogine et al., 1969).

An important feature of self-organization and development viewed as a succession of instabilities is the fundamentally irreversible character of the whole process: each time the system reaches a point of instability, it spontaneously and *irreversibly* evolves towards a new structural and functional organization. Furthermore, these jumps can only occur at given times after the beginning of the whole process. In other words, the system has a natural time scale of irreversible aging associated with its own internal properties. In some sense, one could be tempted to think that we have here a first approximation of what biologists (see Jacob, 1970), in the evolution towards highly organized states, describe as the successive integration of levels of increasing structural and functional complexity.

7. Entropy production and evolutionary feedback

The Martinez model has attracted the attention to the fact that a self-organizing system cannot be described simply by the succession of states of increasing complexity that it presents. Moreover a typical time order is always associated with every process of self-organization and constitutes an intrinsic manifestation of the dynamic processes by which the system jumps from one state to another.

It is therefore interesting, when one considers a phenomenon like biological evolution, to look at the time intervals elapsing between the appearance of the different major stages of organization and to inquire whether this would not give some hints as to the mechanism of evolution itself. In this respect several authors (see, for example, de Duve (1973)) have recently pointed out that there is no proportionality between the increase in organization corresponding to an evolutionary step and the time it requires to occur. De Duve attributes to biological evolution a time scale of at least 3.2 billion years, half of which had elapsed until the appearance of the first protistes. Thereafter half of the remaining period had elapsed in each successive stage until the appearance of the first invertebrates, then the first vertebrates, and finally some 200 million years ago the first mammals. Man itself is probably "only" 2 million years old.

In other words, even if the organization and complexity of living system is difficult to evaluate quantitatively, these figures indicate an acceleration of evolution in the course of time. In its most *macroscopic* manifestations, evolution is an *autocatalytic* phenomenon: any progress is followed by another one which has a greater chance to occur. If one further considers that 1) the molecular mechanism of evolution most likely has not appreciably changed in going from the primitive to the evolved organisms, i.e., the frequency of random mutations in proteins and nucleic acids has remained the same and independent of the functional activity, and 2) the time interval between generations increases for higher organisms, one then realizes that this acceleration really is a paradoxical phenomenon: it cannot be explained by simple molecular considerations on reproduction and self-replication of macromolecules. Clearly we need something more than self-replication. We need something which plays the role of an *increasing selection pressure* in favor of the more organized states.

We do not know of course the exact nature of this driving force, but it certainly involves the interactions of living systems with the external world and suggests a relation between structure and energy dissipation which can be summarized by the following "*evolutionary feedback*" (for more details, see Prigogine et al., 1972)

non-equilibrium → dissipation → instability and new structure
↑ ↓
increase in dissipation

Each time that an instability is followed by a higher level of energy dissipation, one may assume that the driving force for the appearance of further new instabilities has been increased: some irreversible processes taking place inside the system are functioning more intensely and have accordingly increased their departure from the equilibrium state. The probability that there exists a class of fluctuations with respect to which these processes are unstable has become higher. On the contrary, if the result of the instabilities is to decrease the level of dissipation, one gets closer to the properties of an isolated system at equilibrium, i.e., closer to a regime where all fluctuations which cause a deviation from the state of maximum disorder are damped.

These considerations clearly indicate that in order to describe the behavior of such self-organizing systems we need some index related to the level of dissipation, whose variation would be characteristic of the organizing tendency manifested by the system. If we admit that the source of order is basically in the intensity of the purely dissipative irreversible processes, we are led to the suggestion that this role could be assumed by the specific entropy production of the system (s.e.p., entropy production per mole). Any change in mass, energy content, rate or affinity of reaction inside the system would be reflected by this quantity. Particularly, within the context of evolution, the behavior of this quantity has already been investigated on model systems which present instabilities due to the appearance of template effects in a polymerization reaction (see Prigogine et al., 1972). The characteristic result is that when the polymerization reaction is switched from a regime in which it occurs mainly by simple condensation of the monomers to a regime in which the monomers are assembled on a template, a discrete and important jump in s.e.p. is observed. In other words, it is found that an instability by which the system manifestly reaches a more complex stage of functional organization also creates conditions which favor a subsequent evolution of this state through non-equilibrium instabilities.

In the remaining part of this section, we shall briefly consider the relation between functional organization and s.e.p. on some very simple models.

7.1. *Simple linear chemical networks*

Let us first consider the transformation of some initial substrate A into K intermediate products X_k and a final product B, and investigate the behavior of the steady-state entropy production per mole σ as a function of the topology of the reaction scheme i.e., as a function of the reaction connections between the chemical species. It is assumed that the reactions follow linear kinetics and that all forward as well as all backward kinetic constants have been set equal. We first take a simple linear chain of reactions:

$$A \underset{l'}{\overset{l}{\rightleftarrows}} X_1 \rightleftarrows X_2 \rightleftarrows \cdots \rightleftarrows X_K \rightleftarrows B \tag{7.1}$$

where l and l' are respectively the forward and backward kinetic constants. When $l/l' = 1$, it is easily verified that σ as a function of the ratio $x = A/B$ and of the number K of intermediates X_k is given by the expression

$$\sigma = \frac{2(x-1)}{(K+1)(K+2)(x+1)} \ln x \tag{7.2}$$

We thus find that, independently of the value of the overall affinity of reaction (given by $\sim \ln(A/B)$), such a chain of reactions sees its s.e.p. decrease when its length increases, i.e., more intermediate steps have to be taken into account between the initial and final products A, B.

From a practical point of view, one could treat scheme (7.1) as a polymerization reaction: each step k of the chain corresponds to the addition of a monomer to the intermediate X_{k-1} in such a way that the concentration A_k of monomers remains constant in the medium and l represents the product of A_k with the kinetic constant of condensation ($l = k_a A_k$). It then can be verified that by equation (7.2), when the length of the final polymer B increases, its rate of production which is proportional to the steady-state value

$$X_K^0 = \frac{A + KB}{K + 1} \tag{7.3}$$

tends to diminish even if the overall affinity goes to infinity (cf. for example equation (7.3) with $A = $ const and $B = 0$).

Alternatively one could also say that when the chain synthesizes longer polymers, its consumption of monomers per unit time tends to diminish. One has

$$\frac{dA}{dt} \sim -k(A - X_1^0) \sim -k\left(A - \frac{KA}{K+1}\right) \to 0 \quad \text{as} \quad K \to \infty \tag{7.4}$$

Let us now compare these results with those obtained in the case of a linear scheme having a topology as different as possible from that of scheme (7.1). This would be the case when there are no reactions between the K intermediate products so that one has:

$$\begin{array}{ccccc} & \overset{l}{\nearrow}\!\!\underset{l'}{\swarrow} & X_1 & \rightleftarrows & \\ A & \rightleftarrows & X_2 & \rightleftarrows & B \\ & \searrow\!\!\nwarrow & \vdots & \nearrow\!\!\swarrow & \\ & & X_K & & \end{array} \tag{7.5}$$

It is readily shown that when $l = l'$, σ is given by

$$\sigma = \frac{K(x - 1)}{(K + 2)(x + 1)} \ln x \tag{7.6}$$

One sees that contrary to what happens with equation (7.2), the value of σ calculated with (7.6) increases with K and tends towards a finite constant value (for $x \gg 1$ one simply has $\sigma \sim \ln x$).

In summary, these results could be interpreted in the following way. Starting with a linear chain of reactions of the type (7.1) involving a number of intermediate compounds, any process of evolution by which the chain is progressively transformed into (7.5) is associated with an overall increase of σ. It must be noted that this variation of σ does not simply correspond to an increase in the number of loops which exist in parallel between A and B, or in the number of chemical reactions taking place, but actually reflects the nature of the functional changes introduced by the loops in the system.

One may thus consider that two competing antagonist effects are operative when a new loop appears: 1) the variation of the total mass of the system, 2) the change in the distribution of the chemical potential difference $\mu_A - \mu_B$ along the reaction paths. If the ratio A/B is sufficiently large, any new loop will lead to an increase of mass, which tends to decrease σ. This effect however is compensated if the distribution of the chemical potential difference among the intermediates, which corresponds to an overall *constant* affinity of reaction, can be associated with a higher rate of the overall transformation $A \to B$. For a given value of the boundary conditions, one may then say that the system interacts more strongly with the external world in such a way that a higher flux of matter passes through the system. It can easily be verified that we can distinguish on this basis between transformations occurring in reaction chains which, as in (7.7), may appear very similar at a first glance:

$$A \rightleftarrows X \rightleftarrows B \quad \text{(a)}$$

$$\text{(b)} \quad \begin{matrix} A \rightleftarrows X \rightleftarrows B \\ \searrow\!\!\!\nwarrow Y \nearrow\!\!\!\swarrow \end{matrix} \qquad \qquad \begin{matrix} A \rightleftarrows X \rightleftarrows B \\ \searrow\!\!\!\nwarrow Y \nearrow\!\!\!\swarrow \end{matrix} \quad \text{(c)} \tag{7.7}$$

In the case (a) $\to$ (b) one observes a decrease of σ, while with (a) $\to$ (c) one has the reverse effect.

7.2. *Steady-state* s.e.p. *of a simple catalytic model*

It is interesting to consider these results in relation to the effect of catalysis on a reaction chain as in (7.1). Some years ago, one of us (Prigogine, 1965) considered the following catalytic scheme:

$$A \rightleftarrows X_1 \rightleftarrows X_2 \rightleftarrows \cdots \rightleftarrows X_n \rightleftarrows B \tag{7.8}$$
$$\uparrow \quad \uparrow \quad \uparrow \quad \uparrow \quad \downarrow\!\uparrow\, M$$

It was verified that, provided $A/B \to \infty$, (7.8) of necessity has a higher level of dissipation than (7.1). It is interesting to observe the way in which this result is achieved. In the case of system (7.1), we find that between A and B there is a continuous degradation of the chemical potential $\mu_A > \mu_{X_1} > \mu_{X_2} > \cdots > \mu_{X_n} > \mu_B$. If $n \gg 1$, one has however $\mu_{X_i} - \mu_{X_{i+1}}$ tending to zero and as a result the overall rate of transformation of A into B is very low. In case (7.8), on the other hand, one finds that $\mu_A \simeq \mu_{X_1} \simeq \mu_{X_2} \simeq \cdots \simeq \mu_{X_n}$, but $\mu_{X_n}/\mu_B \gg 1$. In other words, the effect of catalysis is to propagate the chemical potential without degradation along the chain and to sustain between X_n and B a very large affinity of reaction. Thus by keeping one of the reactions very far from equilibrium, the system becomes more efficient. In fact, it behaves due to catalysis as if the chain were much shorter and involved only one intermediate compound.

8. Physical basis of self-organization

In any discussion of self-organization, one is confronted with the problem of defining and quantifying the notion of order. The description of a self-organizing process actually begins with the recognition of some gradient of order or hierarchy among the successive stages that a system presents in the course of its time evolution. In the next step, one then is confronted with the problem of finding a mechanism which accounts for the transition from one degree of organization in the hierarchy to the next one.

If one thinks of biological evolution on a very broad scale, some hierarchy among the variety of living forms can be recognized without great difficulty. When one goes from the molecular level to the level of cellular organites, cells, organs and apparatus, there is an obvious gradation which corresponds to a continuous increase in structural and functional complexity. This dualism between structural and functional aspects is in fact the intrinsic property of biological order. It also makes the notion of order in biology fundamentally distinct from the concept of order in physics, which is usually related with entropy.

To characterize biological order even in its molecular manifestations, entropy is clearly not an adequate function. An order criterion based on entropy differences between various systems would fail to distinguish between most arbitrary amino-acid sequences coming out of some type of a Monte Carlo game; it would be helpless in trying to determine those sequences which have some functional interest. Similarly the evaluation of the enormous information content of biological molecules is fairly useless as long as it does not give us any indication on the mechanism by which it has emerged. What must be taken into account is the fact that the purpose of biological structures is to accomplish certain functions with the best efficiency possible. In other words the degree of organization of a biological system is always evaluated with respect to what biologists would call its "project," i.e., seeing for an eye, hearing for an ear, etc. The better this project is realized, the more organized is the system. On the other hand, the transition toward the most elaborate forms of the project is then explained by the existence of a "selection pressure," which merely corresponds to the advantages gained by the organism over its environment when the project is realized with greater efficiency.

Recently Eigen (1971) has put forward a general theory of self-organization describing the evolution of macromolecules which formulates these considerations in a very elegant fashion. The "project" is self-replication of the molecules; the organization of a system is evaluated by a value function that expresses the fidelity with which the set of replication processes is performed; the gradient of organization defined in this way corresponds to the increasing probability of perpetuation of molecules which better minimize random mutations. The selection pressure is created by coupling this process with a constant influx of monomers and precursors.

In a somewhat different approach, Kuhn (1973) searched for a possible pathway of emergence of the genetic apparatus consisting of many little steps. Any one of these steps is obtained from the previous one by looking at the possible ways of evolution of the system in its environment, estimating the times required for each way, and then choosing the fastest possibility as the next step. The "project" here is to evolve faster and it tends to favor at the molecular level autocatalytic behavior of the type described by de Duve for biological evolution in general.

Results such as those reported in the preceding section have thus suggested to us a point of view which may be regarded as complementary to that of Eigen (1971) and Kuhn (1973). Within the perspective of evolution, one would have to look for instabilities or dissipative structures which increase the departure of open systems from equilibrium. Consequently a higher level of interactions with the outside world could be used as the source of order. Instabilities and kinetic properties would thus be classified into two classes according to whether or not they correspond to an increase of dissipation. The index or value function for such a *gradient of evolution* would be the entropy production measured intensively.

The results of course are very fragmentary, and yet they give some hope that it is not an unrealistic desire to try to associate with *la logique du vivant* a general physical quantity whose variation continuously follows the process of self-organization. Almost thirty years ago, one of us (Prigogine, 1947) wrote, "*ce n'est donc ni l'entropie, ni d'ailleurs aucun potentiel thermodynamique, qui permet de caractériser l'évolution irréversible d'un système, mais seulement la production d'entropie qui s'approche de sa valeur minimum.*

"*Dans certains cas cette valeur minimum de la production d'entropie ne peut être atteinte qu'en augmentant l'hétérogénéité, la complexité du système. Peut-être que cette évolution spontanée qui se manifeste alors vers des états à hétérogénéité plus grande pourra-t-elle donner une impulsion nouvelle à l'interprétation physico-chimique de l'évolution des êtres vivants*".

In this sentence we have the dualism between creation of structure and maintenance of structure. The situation at present seems to indicate that these two aspects can be characterized by a different behavior of entropy production: while the creation step leads to an increase of σ, the steps of maintenance on the other hand seem to follow the theorem of minimum entropy production.

References

ANDRONOV, A. A., VITT, A. A. and KHAIKIN, S. E. 1966. *Theory of Oscillators*. Oxford: Pergamon Press.

BLANGY, D., BUC, H. and MONOD, J. 1968. *J. Mol. Biol.* **31**: 13.

CESARI, L. 1962. "Asymptotic Behavior and Stability Problems in Ordinary Differential Equations," *Erg. Mathem. New Series*, **16**. Berlin: Springer Verlag.

DUVE, C., DE. 1973. "La Biologie au XXe siecle," Communication au Colloque *Connaissance Scientifique et Philosophie* organisé à l'occasion du bicentenaire de l'Académie Royale de Belgique.

EIGEN, M. 1971. *Naturwiss.* **58**:465.

FIFE, B. C. *Lecture Notes in Mathematics* **322**. Berlin: Springer Verlag.

GLANSDORFF, P. and PRIGOGINE, I. 1971. *Thermodynamic Theory of Structure, Stability and Fluctuations.* New York: Interscience, Wiley.

GMITRO, J. I. and SCRIVEN, L. E. 1966. In *Intracellular Transport,* K. K. WAREN (ed.). New York: Acad. Press.

GOLDBETER, A. and LEFEVER, R. 1972*a*. *Biophys. J.* **12**:1302.

GOLDBETER, A. and LEFEVER, R. 1972 *b*. In *Analysis and Simultation.* Amsterdam: North Holland.

GOLDBETER, A. and NICOLIS, G. 1972. *Biophysik* **8**:212.

GOLDBETER, A. 1973. *Proc. Nat. Acad. Sci. U. S. A.*

HERSCHKOWITZ-KAUFMAN, M. and NICOLIS, G. 1972. *J. Chem. Phys.* **56**:1890.

HESS, B. and BOITEUX, A. 1968. In *Regulatory Functions of Biological Membranes,* J. JÄRNEFELD (ed.). Amsterdam-London-New York: Elsevier Pub. Co.

HESS, B., BOITEUX, A. and KRÜGER, J. 1969. *Adv. Enzyme Regul.* **7**:149.

HIGGINS, J. 1964. *Proc. Natl. Acad. Sci. U. S.* **51**:989.

JACOB, F. 1970. *La logique du vivant.* Paris: Gallimard.

KITAHARA, K. to appear Adv. Chem. Phys. (1975).

KOBATAKE, Y. 1970. *Physica* **48**:301.

KUHN, H. 1973. In *Synergetics,* H. HAKEN and B. G. TEUBNER (eds.). Stuttgart.

LAVENDA, B., NICOLIS, G. and HERSCHKOWITZ-KAUFMAN, M. 1971. *J. Theor. Biol.* **32**:283.

LEFEVER, R. 1968. *J. Chem. Phys.* **49**:4977.

LEFEVER, R. and NICOLIS, G. 1971. *J. Theor. Biol.* **30**:267.

MARTINEZ, H. 1972. *J. Theor. Biol.* **36**:479.

MAYNARD SMITH, J. and SONDHI, K. C. 1961. *J. Embryol. exp. Morph.* **9**:661

MINORSKY, N. 1962. *Nonlinear Oscillations.* Princeton, N. J.: Van Nostrand Co.

NICOLIS, G. 1971. *Adv. Chem. Phys.* **19**:209.

NICOLIS, G. and PORTNOW, J. 1973. *Chem. Revs.* **73**:365.

OTHMER, H. G. and SCRIVEN. L. E. 1969. *Ind. and Eng. Chem.* **8**:302.

PRIGOGINE, I. and WIAME, J. M. 1946. *Experientia,* **II**:415.

PRIGOGINE, I. 1947. *Etude thermodynamique des Phénomènes Irréversibles.* Paris: Dunod, and Liège: Desoer.

PRIGOGINE, I. 1965. *Physica* **31**:719.

PRIGOGINE, I. and NICOLIS, G. 1967. *J. Chem. Phys.* **46**:3542.

PRIGOGINE, I. and LEFEVER, R. 1968. *J. Chem. Phys.* **48**:1695.

PRIGOGINE, I. 1969. In *Theoretical Physics and Biology,* M. MAROIS (ed.). Amsterdam: North Holland Publ. Co.

PRIGOGINE, I., LEFEVER, R., GOLDBETER, A. and HERSCHKOWITZ-KAUFMAN, M. 1969. *Nature* **223**:913.

PRIGOGINE, I. 1971. In *Theoretical Physics and Biology,* M. MAROIS (ed.). Amsterdam: North-Holland Publ. Co.

PRIGOGINE, I. and NICOLIS, G. 1971, *Quart. Rev. Biophys.* **4**:107.

PRIGOGINE, I., NICOLIS, G. and BABLOYANTZ, A. 1972. *Phys. Today* **25**:23.

PROGOGINE, I. GEORGE, C., HENIN, F. and ROSENFELD, L. 1973. *Chemica Scripta* **4**:5.

SCHLÖGL, F. 1972. *Z. Physik* **253**:147.

SCHRODINGER, E. 1945. *What is Life,* London: Cambridge University Press.

SEL'KOV, E. E. 1968. *Eur. J. Biochem.* **4**:79.

SONDHI, K. C. *Quarter. Rev. Biol.* **38**:289.

TURNER, J. *Phys. Letters,* **44A**, 395 (1973); *Bull. Math. Biol.* **36**, 205 (1974).

TURING, A. M. 1952. *Phil. Trans. Roy. Soc. London* B**237**:37.

WILSON, H. and COWAN, J. D. 1972. *Biophys. J.* **12**:1.

Hamilton–Jacobi approach to fluctuation phenomena

KAZUO KITAHARA
Chémie Physique II, Faculté des Sciences
Université Libre de Bruxelles, 1050 Bruxelles, Belgium

1. Introduction

In this short report, I would like to explain how to visualize evolution of fluctuations in chemical systems, or rather, in populations in a wide class of systems. The technical details have been published elsewhere (Kubo, 1971; Kubo, Matsuo and Kitahara, 1973; Kubo, Matsuo and Kitahara, 1973). Here I will give an outline of the theory and some interesting applications.

Phenomena of fluctuations can be described in terms of a probability distribution of physical quantities; for example, the number of particles of a certain species is realized with a probability $P(N)$. The average $\overline{N}$ is defined by

$$\overline{N} = \sum_{N=0}^{\infty} NP(N) \tag{1}$$

and it is supposed to be the macroscopically observed value. The probability distribution $P(N)$ may look like Fig. 1.

The fluctuation around the observed value $\overline{N}$ is estimated by the statistical variance, which is defined by

$$\Gamma = \sum_{N=0}^{\infty} (N - \overline{N})^2 P(N) \tag{2}$$

When the system is evolving, its behavior is described by the time-dependent probability distribution $P(N, t)$. The equation of motion for $P(N, t)$ may be constructed by considering the evolution of the system as a birth-and-death process

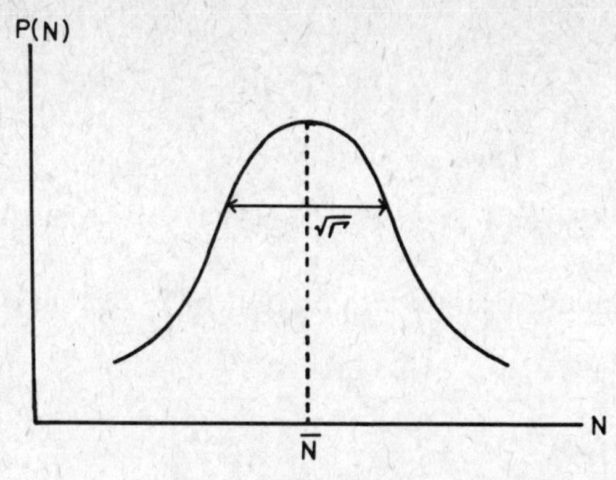

Figure 1
The probability distribution

(McQuarrie, 1967). Generally the equation takes the following form:

$$\frac{\partial}{\partial t} P(N, t) = -\sum_r W(N \to N + r)\, P(N, t) + \sum_r W(N - r \to N)\, P(N - r, t) \tag{3}$$

where $W(N \to N + r)$ stands for the frequency of the transition from the N-particle state to the $(N + r)$-particle state.

The amount of a transition r in a unit time is supposed to be small compared with the present number N, i.e.,

$$r \ll N \tag{4}$$

In order to describe the assumption (4) more explicitly, we introduce the volume (or the size) of the system, Ω, and scale the number of the particles N as

$$x = N/\Omega \tag{5}$$

Thus x is the concentration.

It is proved by Kubo, Matsuo and Kitahara (1973) that Eq. (3) has an asymptotic solution of the form

$$P(N, t) \sim \exp\left[\Omega \phi(x, t)\right] \tag{6}$$

in the limit of $\Omega \to \infty$. The function $\phi(x, t)$ can be expanded around the maximum,

$$\phi(x, t) = \phi(y(t), t) - \frac{1}{2\sigma(t)}(x - y(t))^2 + \cdots \tag{7}$$

This expansion results in the following approximate estimate of the average $\bar{N}$ and the variance Γ:

$$\bar{N} \cong \Omega y(t) \tag{8}$$

$$\Gamma \cong \Omega \sigma(t) \tag{9}$$

The deviation from the average is $\sim \sqrt{\Omega \sigma(t)}$ and it is small compared with the average $\bar{N}$, which is of order Ω.

Putting the asymptotic form (6) into Eq. (3), we obtain an equation for $\phi(x, t)$:

$$\frac{\partial \phi(x, t)}{\partial t} + H\left(x, \frac{\partial \phi(x, t)}{\partial x}\right) = 0 \tag{10}$$

where $H(x, p)$ is defined by

$$H(x, p) = \sum_r (1 - e^{-rp})\, w(x; r) \tag{11}$$

with the transition probability

$$W(N \to N + r) = \Omega w(x; r) \tag{12}$$

Eq. (10) is called *the Hamilton–Jacobi equation*, and it can be transformed into a set of ordinary differential equations (see Courant and Hilbert (1962) and the following section).

If we substitute Eq. (7) into Eq. (3), or into Eq. (10), we get equations for $y(t)$ and $\sigma(t)$:

$$\dot{y}(t) = C_1(y(t)), \tag{13}$$

$$\dot{\sigma}(t) = 2C_1'(y(t))\, \sigma(t) + C_2(y(t)) \tag{14}$$

where

$$C_1(x) = \sum_r r w(x; r) \tag{15}$$

$$C_2(x) = \sum_r r^2 w(x; r) \tag{16}$$

Eqs. (13) and (14) are sufficient to study the evolution of fluctuation.

2. Ehrenfest's model

Consider a random walk on a line shown in Fig. 2. The transition probability is given by (see Ming Chen Wang and Uhlenbeck (1945))

$$W(N \to N + r) = \begin{cases} (R + N)/2 & r = -1 \\ (R - N)/2 & r = +1 \end{cases} \tag{17}$$

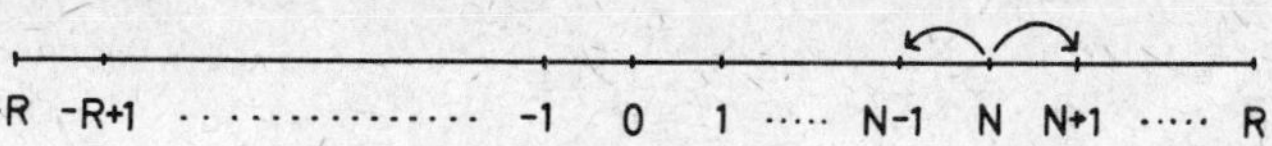

-R -R+1 .... -1 0 1 ..... N-1 N N+1 ..... R

Figure 2
Ehrenfest's model

Namely, when $N > 0$ ($N < 0$), the jump in the negative (positive) direction is more probable than that in the positive (negative) direction. Thus on the average the random walker is attracted to the origin $N = 0$.

When the boundaries $N = \pm R$ are far apart ($R \to \infty$), we may consider R as a large parameter Ω. We scale N as

$$x = N/R = N/\Omega \tag{18}$$

and rewrite the transition probability (17) as

$$w(x; r) = \begin{cases} (1+x)/2 & r = -1 \\ (1-x)/2 & r = +1 \end{cases} \tag{19}$$

Using Eqs. (13) and (14), the evolution equations for the most probable value $y(t)$ and the variance $\sigma(t)$ are

$$\dot{y}(t) = -y(t) \tag{20}$$

$$\dot{\sigma}(t) = -2\sigma(t) + 1 \tag{21}$$

These equations behave as shown in Fig. 3. The trajectory $(y(t), \sigma(t))$ tends to a unique limit. This asymptotic stability of the probability distribution is more clearly seen by studying the Hamilton–Jacobi equation (10), which in this case turns out to be

$$\frac{\partial \phi}{\partial t} + 1 - \cosh\left(\frac{\partial \phi}{\partial x}\right) - x \sinh\left(\frac{\partial \phi}{\partial x}\right) = 0 \tag{22}$$

i.e.

$$H(x, p) = 1 - \cosh p - x \sinh p \tag{23}$$

Suppose the initial condition $\phi(x, 0)$ is given. Following the standard method (Courant and Hilbert, 1962), we introduce a characteristic curve (x_t, p_t), which obeys the following set of equations:

$$\frac{dx_t}{dt} = \frac{\partial H}{\partial p}(x_t, p_t) \tag{24}$$

$$\frac{dp_t}{dt} = -\frac{\partial H}{\partial x}(x_t, p_t)$$

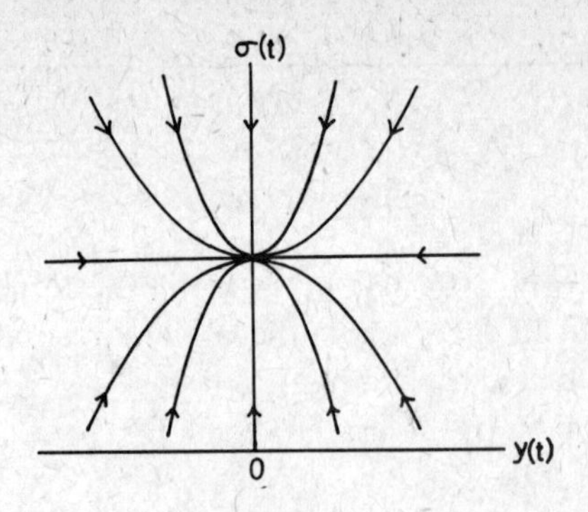

Figure 3
The evolution of $(y(t), \sigma(t))$

with the initial condition

$$p_0 = \frac{\partial \phi}{\partial x}(x_0, 0) \tag{25}$$

Thus the characteristic curve is parametrized by x_0,

$$\begin{aligned} x_t &= x(t; x_0) \\ p_t &= p(t; x_0) \end{aligned} \tag{26}$$

Eliminating x_0 from Eq. (26), we get p_t as a function of x_t,

$$p_t = \gamma_t(x_t) \tag{27}$$

It is known (Courant and Hilbert, 1962) that the solution of Eq. (22) is given by

$$\frac{\partial \phi}{\partial x}(x, t) = \gamma_t(x) \tag{28}$$

Geometrical interpretation of Eqs. (24) through (28) is as follows. For a given initial condition $\phi(x, 0)$, draw in the (x, p) plane a curve γ_0 defined by

$$\gamma_0: \quad p = \gamma_0(x) = \frac{\partial \phi}{\partial x}(x, 0) \tag{29}$$

From each point on the curve γ_0, we start the characteristic curve (24); thus the point on the curve γ_0 corresponds to the initial condition (25). The set of points which start from the points on the curve γ_0 forms a curve γ_t at time t as shown in Fig. 4. This curve $\gamma_t : p = \gamma_t(x)$ is in effect Eq. (27).

Since each characteristic curve has a constant of motion

$$H(x_t, p_t) = E(\text{const}) \tag{30}$$

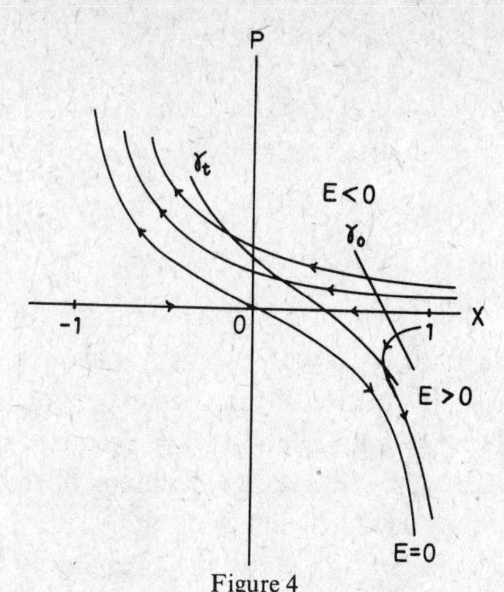

Figure 4
The evolution of characteristic curves

the characteristic curves are obtained by solving Eq. (30) for various values of E, as shown in Fig. 4.

From this pattern of the characteristic curves, we easily see that the curve γ_t tends to an asymptote γ_∞,

$$\lim_{t\to\infty} \gamma_t(x) = \gamma_\infty(x) = -2\tanh^{-1} x \tag{31}$$

which implies the asymptotic stability of the function $\phi(x, t)$, i.e.,

$$\lim_{t\to\infty} \phi(x, t) = \phi(x, \infty) = -2\int^x dx' \tanh^{-1} x' \tag{32}$$

Ehrenfest's model is very simple in that the evolution equations for the most probable value $y(t)$ and the variance $\sigma(t)$ have a unique steady solution. In the case of multiple steady states, the pattern of the characteristic curves becomes complicated and the function $\phi(x, t)$ is not asymptotically stable any more (see Kubo et al. (1973) and Kitahara (1974)).

3. Fluctuation in oscillating systems

Let us consider the following simple model of oscillation:

$$\begin{cases} \dot{x}_1 = F_1(x_1, x_2) \equiv x_1(a - r^2) - \omega x_2 \\ \dot{x}_2 = F_2(x_1, x_2) \equiv x_2(a - r^2) + \omega x_1 \end{cases} \tag{33}$$

where

$$r^2 = x_1^2 + x_2^2 \tag{34}$$

The coupled equations (33) have a steady state

$$x_1^{st} = x_2^{st} = 0$$

This steady state is

(i) a stable focus if $a < 0$ (Fig. 5a),
(ii) a marginal focus if $a = 0$ (Fig. 5b),
(iii) an unstable focus if $a > 0$ (Fig. 5c).

In case (iii), there exists a limit cycle with radius $\sqrt{a}$ around the steady state.

The transition from case (i) to case (iii) is similar to that found in some chemical reactions far from equilibrium (e.g., Zhabotinsky–Belousov reaction, etc.).

Eqs. (33) are supposed to describe average behavior of the two quantities (x_1, x_2) which are fluctuating around the deterministic path.

As in the previous sections, we will study the probability distribution $P(x_1, x_2, t)$ that the system takes the values (x_1, x_2) at time t.

A model is given in the form of a partial differential equation for the probability distribution $P(x_1, x_2, t)$:

$$\frac{\partial}{\partial t} P(x_1, x_2, t) = \left\{ -\frac{\partial}{\partial x_1} F_1(x_1, x_2) - \frac{\partial}{\partial x_2} F_2(x_1, x_2) + \frac{1}{2}\left(\frac{D}{\Omega}\right)\left(\frac{\partial^2}{\partial x_1^2} + \frac{\partial^2}{\partial x_2^2}\right)\right\} P(x_1, x_2, t) \quad (35)$$

Eq. (35) is obtained if we add random forces $f_1(t)$ and $f_2(t)$ to Eq. (33), namely,

$$\begin{cases} \dot{x}_1 = x_1(a - r^2) - \omega x_2 + f_1(t) \\ \dot{x}_2 = x_2(a - r^2) + \omega x_1 + f_2(t) \end{cases} \quad (36)$$

and assume

$$\langle f_i(t)\, f_j(t') \rangle = \frac{D}{\Omega}\, \delta_{ij}^{(Kr)}\, \delta(t - t') \quad (37)$$

(for detail, see Kitahara, 1974; Fuller, 1969).

Eq. (37) means that the random modulation of the evolution of (x_1, x_2) by the random forces is small ($\sim 1/\Omega$) and is of Markovian nature.

Eq. (35) has a solution of the form

$$P(x_1, x_2, t) \sim \exp[\Omega \phi(x_1, x_2, t)] \quad (38)$$

and the function $\phi(x_1, x_2, t)$ can be expanded around the maximum

$$\phi(x_1, x_2, t) = \phi(y_1(t), y_2(t), t) - \frac{1}{2}\sum_{i=1}^{2}\sum_{j=1}^{2}(x_i - y_i(t))\, Q_{ij}(t)\, (x_j - y_j(t)) + \cdots \quad (39)$$

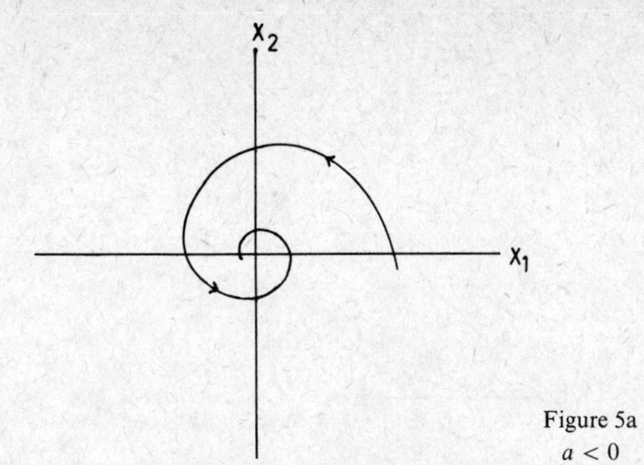

Figure 5a
$a < 0$

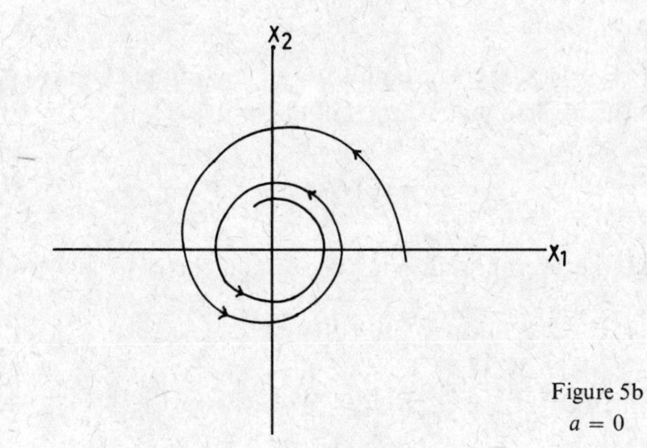

Figure 5b
$a = 0$

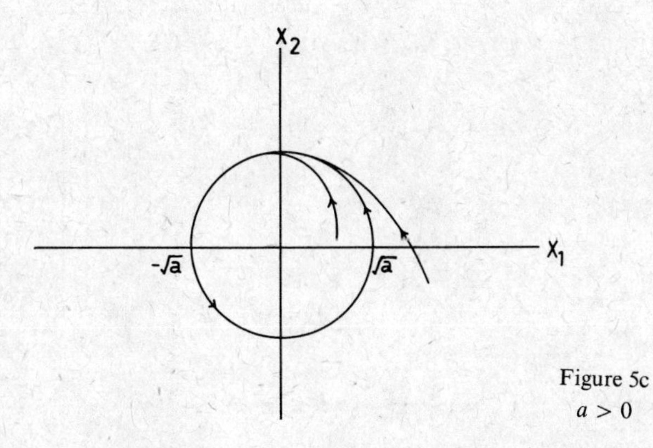

Figure 5c
$a > 0$

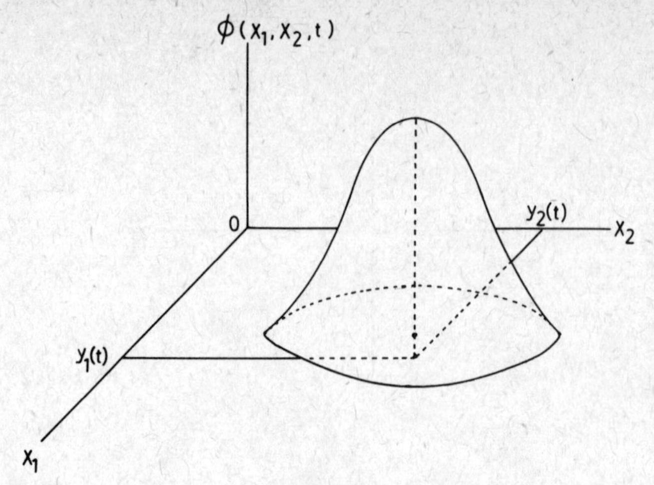

Figure 6
The form of $\phi(x_1, x_2, t)$

thus $(x_1 = y_1(t), x_2 = y_2(t))$ corresponds to the maximum of $\phi(x_1, x_2, t)$ (see Fig. 6).

We introduce the inverse matrix $\sigma_{ij}(t)$ of the matrix $Q_{ij}(t)$,

$$\sum_{k=1}^{2} \sigma_{ik}(t)\, Q_{kj}(t) = \delta_{ik}^{(Kr)} \tag{40}$$

Then for large Ω, the average values of x_1, x_2 and the covariances are estimated as

$$\langle x_i \rangle_t \cong y_i(t), \quad i = 1, 2 \tag{41}$$

$$\langle (x_i - \langle x_i \rangle_t)(x_j - \langle x_j \rangle_t) \rangle_t \cong \frac{1}{\Omega} \sigma_{ij}(t), \qquad i = 1, 2; \quad j = 1, 2. \tag{42}$$

Putting Eqs. (38) through (40) into Eq. (35), we obtain evolution equations for $y_i(t)\,(i = 1, 2)$ and $\sigma_{ij}(t)\,(i = 1, 2; j = 1, 2)$:

$$\dot{y}_i(t) = F_i(y_1(t), y_2(t)), \qquad i = 1, 2 \tag{43}$$

$$\frac{d}{dt}\begin{bmatrix} \sigma_{11} \\ \sigma_{12} \\ \sigma_{22} \end{bmatrix} = \begin{bmatrix} 2(a - 3y_1^2) & -2(2y_1y_2 + \omega) & 0 \\ -2y_1y_2 + \omega & -3r^2 + 2a & -2y_1y_2 - \omega \\ 0 & 2(-2y_1y_2 + \omega) & 2(a - 3y_2^2) \end{bmatrix} \begin{bmatrix} \sigma_{11} \\ \sigma_{12} \\ \sigma_{22} \end{bmatrix} + \begin{bmatrix} D \\ 0 \\ D \end{bmatrix} \tag{44}$$

where

$$y_1^2 + y_2^2 = r^2$$

When the system is macroscopically in the steady state, Eq. (44) becomes

$$\frac{d}{dt}\begin{bmatrix}\sigma_{11}\\ \sigma_{12}\\ \sigma_{22}\end{bmatrix} = \begin{bmatrix}2a & -2\omega & 0\\ \omega & 2a & -\omega\\ 0 & 2\omega & 2a\end{bmatrix}\begin{bmatrix}\sigma_{11}\\ \sigma_{12}\\ \sigma_{22}\end{bmatrix} + \begin{bmatrix}D\\ 0\\ D\end{bmatrix} \tag{45}$$

The eigenvalues of the matrix above are

$$\lambda_0 = 2a, \qquad \lambda_{\pm} = 2a \pm 2i\omega \tag{46}$$

Therefore, for $a < 0$, the covariance matrix approaches the steady solution,

$$\sigma_{11}^{st} = \sigma_{22}^{st} = -D/2a, \qquad \sigma_{12}^{st} = 0 \tag{47}$$

For $a > 0$, the covariances around the steady state diverge. This corresponds to the fact that the steady state is unstable and an oscillation with finite amplitude emerges.

In such a case, it is convenient to see the fluctuation around the oscillation. For simplicity, suppose the system is sufficiently old. Then the macroscopic motion $(y_1(t), y_2(t))$ is already on the limit cycle,

$$\begin{cases} y_1(t) = \sqrt{a}\cos\omega t \\ y_2(t) = \sqrt{a}\sin\omega t \end{cases} \tag{48}$$

Now we introduce a new frame of coordinates given by the orthogonal unit vectors $\mathbf{e}_L$ and $\mathbf{e}_T$, while the old one is given by the unit vectors $\mathbf{e}_1$ and $\mathbf{e}_2$ as shown in Fig. 7.

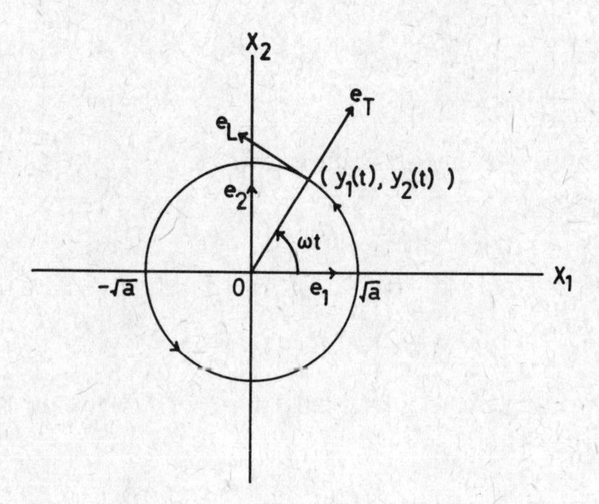

Figure 7
The transformation of coordinates

The deviations from the macroscopic path of evolution

$$\begin{cases} \Delta x_1 = x_1 - y_1(t) \\ \Delta x_2 = x_2 - y_2(t) \end{cases} \tag{49}$$

can be represented in the new frame as

$$\Delta x_L \mathbf{e}_L + \Delta x_T \mathbf{e}_T = \Delta x_1 \mathbf{e}_1 + \Delta x_2 \mathbf{e}_2, \tag{50}$$

thus

$$\begin{cases} \Delta x_L = \Delta x_1 \cos \omega t + \Delta x_2 \sin \omega t \\ \Delta x_T = -\Delta x_1 \sin \omega t + \Delta \acute{x}_2 \cos \omega t \end{cases} \tag{51}$$

Δx_L and Δx_T may be called the *longitudinal component* and the *transverse component* of the fluctuation, respectively. We introduce a new representation of the covariances,

$$\begin{cases} \langle \Delta x_L \, \Delta x_L \rangle_t \cong \dfrac{1}{\Omega} \sigma_{LL}(t) \\ \langle \Delta x_L \Delta x_T \rangle_t \cong \dfrac{1}{\Omega} \sigma_{LT}(t) \\ \langle \Delta x_T \, \Delta x_T \rangle_t \cong \dfrac{1}{\Omega} \sigma_{TT}(t) \end{cases} \tag{52}$$

$\sigma_{LL}(t)$ gives the variance in the direction of the macroscopic motion while $\sigma_{TT}(t)$ gives the variance in the perpendicular direction to the macroscopic motion.

Using Eq. (51), we derive the relation between the new and the old representations of the covariances,

$$\begin{bmatrix} \sigma_{LL} \\ \sigma_{LT} \\ \sigma_{TT} \end{bmatrix} = \begin{bmatrix} \cos^2 \omega t & \sin 2\omega t & \sin^2 \omega t \\ -\frac{1}{2}\sin 2\omega t & \cos 2\omega t & \frac{1}{2}\sin \omega t \\ \sin^2 \omega t & -\sin 2\omega t & \cos^2 \omega t \end{bmatrix} \begin{bmatrix} \sigma_{11} \\ \sigma_{12} \\ \sigma_{22} \end{bmatrix} \tag{53}$$

This enables us to construct closed equations for $\sigma_{LL}(t)$, $\sigma_{LT}(t)$ and $\sigma_{TT}(t)$:

$$\begin{aligned} \dot{\sigma}_{LL}(t) &= D \\ \dot{\sigma}_{LT}(t) &= -2a\sigma_{LT}(t) \\ \dot{\sigma}_{TT}(t) &= -2a\sigma_{TT}(t) + D \end{aligned} \tag{54}$$

The longitudinal variance $\sigma_{LL}(t)$ grows in time as $\sim Dt$, while the transverse variance $\sigma_{TT}(t)$ tends to a finite value,

$$\sigma_{TT}(t) = \left(\sigma_{TT}(0) - \frac{D}{2a} \right) e^{-2at} + \frac{D}{2a} \rightarrow \frac{D}{2a} \qquad (t \rightarrow \infty) \tag{55}$$

The divergence of $\sigma_{LL}(t)$ means that because of the random modulation by $f_1(t)$ and $f_2(t)$ in Eq. (36) the phase of the oscillation is randomized. $\sigma_{TT}(\infty)$ is finite because of the restoring force of the limit cycle in the radial direction.

In summary, we may say the followings:

(i) When the steady state is stable, the fluctuation is distributed around the steady state with a finite width.

(ii) When the steady state is unstable and there exists a limit cycle, the fluctuation is distributed around the limit cycle. The fluctuation around the unstable steady state grows to infinity.

For other models of chemical oscillations, such as the Lotka–Volterra model (Lotka, 1910) and the Nicolis–Lefever model (1971), etc., we may construct equations for the probability distribution of birth-and-death type and analyze the nature of fluctuation (Kitahara, in press). The Hamilton–Jacobi equation (10) can be extended to two-variable cases and we may calculate the function $\phi(x_1, x_2, t)$ itself.

References

COURANT, D. and HILBERT, R., 1962. *Methods of Mathematical Physics,* vol. 2. New York: Interscience, Wiley.

FULLER, A. T., 1969. *Int. J. Control.* **9**:603.

KITAHARA, K., 1974. Thesis, Université Libre de Bruxelles.

KITAHARA, K. *Adv. Chem. Phys.* (in press).

KUBO, R., 1971. Presented at the I.U.P.A.P. Conference on Statistical Mechanics, Chicago.

KUBO, R., MATSUO, K. and KITAHARA, K., 1973. Presented at the van der Waals Conference on Statistical Mechanics, Amsterdam.

KUBO, R. and KITAHARA, K., 1973. *J. Stat. Phys.* **9**:51.

LEFEVER, R. and NICOLIS, G., 1971. *J. Theor. Biol.* **30**:267.

LOTKA, A. J., 1910. *J. Phys. Chem.* **14**:271.

MCQUARRIE, D. A., 1967. *J. App. Prob.* **4**:413.

MING CHEN WANG and UHLENBECK, G. E., 1945. *Rev. Mod. Phys.* **17**:323.

The structure of reaction networks

George F. Oster
University of California, Berkeley

In 1948 Leo Szilard wrote a short story called "Report on Grand Central Terminal" wherein he relates the efforts of an alien race to make sense of the extinct human civilization. These beings can, of course, make little sense of the artifacts they find, but among the most confusing items they encounter are the pay toilets in Grand Central Station in New York. Their most closely reasoned argument eventually leads them to conclude that the pay toilets involve some sort of religious ceremonial act performed by humans in public places.

Of course, without knowledge of the physiological and social context of the pay toilets, their hypothesis, however fallacious, was certainly a rational one based on the evidence available to them. Moreover, their reasoning was frankly teleological since the toilets had obviously been constructed with some purpose in mind.

The reason I mention this story is that it recalls to me one of the many philosophical discussions I had with Aharon Katzir-Katchalsky. Frequently, late in the evening, when our working efficiency began to fall off, he liked to conclude the day's work with some light discourse relating our current research to some deeper philosophical considerations. This particular conversation centered about the role of teleological thinking in physical and biological science. Many of the familiar arguments were aired on this venerable philosophical topic. At some point Aharon recalled a remark made to us by Prof. Desoer, an electrical engineer at Berkeley, a remark whose significance Aharon felt we did not fully appreciate at the time it was made.

Desoer observed that teleology is a viewpoint deeply rooted in the bedrock of engineering thinking; the reason being simply that a principal goal of engineering science is the *design* rather than the analysis of systems. This view presupposes a "design purpose" for each piece of apparatus. However, the interesting point is that if we turn things around and ask a competent circuit designer, for example, to analyze an unfamiliar circuit of any complexity, he will likely find it difficult, if not impossible, to say much about it without prior knowledge of its design purpose.

He must be told, or guess, what the apparatus is supposed to do before he has much chance of understanding it. Thus, Aharon observed, if it is generally so difficult to ascertain the function and operation of apparatus built by men for a select purpose without first knowing ahead of time what task it was designed to perform, then what chance do we have of deciphering the functions of biological systems, infinitely more complex, and whose design function, if indeed there be any, is shaped by forces which we presently only dimly perceive?

One rational possibility is to emulate the engineer faced with the task of understanding an unfamiliar piece of apparatus: he simply hypothesizes a design purpose for the system, then checks to see if it does function acceptably according to those criteria. If not, he tries again with a different hypothesized design purpose. This procedure usually cannot resolve the question of the uniqueness of the design purpose: perhaps a color TV set functions just as well as an X-ray machine, as Ralph Nader claims. Nevertheless, it seems a reasonable and scientific way to proceed when faced by a system with unknown properties. Such blatant teleology is quite respectable in engineering (or cryptography for that matter) because we know ahead of time that the machine (or code) was indeed designed by men with a definite function in mind. However, can we claim that the same methodology is sound for approaching systems synthesized by nature rather than men? Well, of course, we didn't resolve that question that night, but Aharon concluded that it certainly would add to our comprehension of biochemical systems if we learned something about the synthesis of chemical networks.

However, even in engineering, design and synthesis of complex systems is still at a primitive stage compared with analysis—still much more of an art than a science. Nevertheless, an orderly search for unifying dynamic principles in biochemistry must begin somewhere, and the first requirement is a precise mathematical model for reaction systems. The reason for this is so that we can circumscribe the universe of possible behavior for reaction systems, and to construct model systems which we hope will provide a deeper insight into the possible design features of chemical systems constructed by evolutionary forces.

We found, however, much to our surprise, that a precise theory of reaction dynamics, comparable to, say, classical mechanics, did not exist. There are, in fact, a number of scientific meetings this year devoted to the mathematical foundations of reaction dynamics; the issue is still quite unsettled. So it was to this question of mathematical foundations that we clearly had to address ourselves before we could proceed further.

In this paper I would like to first sketch the broad strokes of the analytical theory of reaction systems that we have formulated, and indicate how some of the methods of abstract network theory can provide a concrete framework for modelling biochemical systems.

Although much of the presentation is unavoidably couched in mathematical language, I think it can be understood in a general way if the reader will focus on the qualitative, geometric notions and excuse my periodic recourse to arcane mathematical argot.

* * *

The methods of equilibrium thermodynamics have long played a central role in physical and biochemistry. One of the principal attractions of linear irreversible thermodynamics was the promise of extending the unifying force of the equilibrium theory to encompass irreversible phenomena. Unfortunately, the promise never quite lived up to expectations, because the formalism was largely restricted to linear, reciprocal systems operating very closely to thermodynamic equilibrium. This pretty much excluded a thermodynamic treatment of reaction dynamics, since in most cases of interest reactions proceed at a rather brisk rate and are neither linear nor reciprocal.

There were a number of early attempts to treat reaction dynamics by defining a driving "force" for reactions: the chemical affinity was constructed as a stoichiometric combination of chemical potentials. However, this also worked only close to equilibrium since, in general, the reaction rate is not a unique function of the thermodynamic affinity.

On the other hand, reaction dynamics has been treated rather satisfactorily, at least for gas phase and dilute solution reactions, by classical chemical kinetics, which rests rather heavily on the law of mass action. Now, chemical kinetics can be made *consistent* with equilibrium thermodynamics by introducing an independent postulate: detailed balance at equilibrium. However, chemical kinetics cannot be considered a real *extension* of thermodynamics to dynamical systems, since thermodynamic "forces" like chemical potentials play no role in generating the equations of motion. In a sense, chemical kinetics is a kinematic theory.

Quite recently, there has been a substantial effort by a number of workers to place chemical kinetics on a firm mathematical footing and provide a general view of the dynamics of chemical reactions, much in the tradition of classical mechanics.

However, being dyed-in-the-wool thermodynamicists, we never gave up hope of formulating a true thermodynamic theory of reaction rates. The clue was provided by a long neglected work by the physical chemist, Brönsted, in the 1930's, who noticed that all of the classical thermodynamic results could be obtained perfectly well without ever mentioning anything about free energy functions. Instead, he dealt with the work quantities. He published several papers on his method, which he called "energetics," although it had little to do with the energy methods of Gibbsian thermodynamics. But it was largely ignored by practically everyone, because for computing equilibria the free energy extremal principles of the Gibbs approach were not only more convenient but much more esthetic to the physicists, who lust eternally after extremal principles to characterize practically everything.

But, virtually unnoticed by the thermodynamic community, another profession had, in effect, adopted the methods of Brönsted and used them quite successfully for describing a particular nonequilibrium thermodynamic system. The electrical engineers probably never heard of Brönsted, nor ever thought of a circuit as a non-equilibrium thermodynamic system—and it is just as well that they didn't because they proceeded to develop the mathematical tools for treating quite complex systems without worrying too much about the energy and entropy methods that occupied the attention of chemists.

The first worker, to my knowledge, to exploit the mathematical connection between irreversible thermodynamics and network theory to any extent was Josef Meixner. He noticed that the mathematical property of passivity in electrical networks was just the concept required to eliminate the notion of entropy for nonequilibrium systems—a notion which had always been rather vague anyway. Meixner, however, used only the theory of linear, passive l-port circuits, and concerned himself primarily with the thermodynamics of materials, and did not worry too much about chemistry (he is a physicist, after all). But it was the problem of transport and chemical reaction in biological systems that most influenced our own work in this area.

Since our whole perspective has been to view a reaction system from a dynamic systems and control theory viewpoint, let me first briefly recall to you the formal definition of a dynamical systems model of a physical system. That is, a dynamical system consists of five mathematical objects (see Fig. 1):

1) A state space, Σ, whose points define what we mean by the system state.

2) An input space, U, which determines to what extent we can control the motion of the state point.

3)–4) An output space, Y, and a readout function, r, which filters our observations about the internal state.

5) A state transition function, φ, which specifies the internal dynamics by giving a prescription for how the state point moves in time under the influence of the inputs and the state itself.

With an appropriate choice of input space, output space and state transition function this framework encompasses all of the deterministic theories of physics. However, for the purposes of discussing chemical reaction dynamics we need only consider the finite-dimensional system representation:

$$\dot{\mathbf{x}} = \mathbf{f}(\mathbf{x}, \mathbf{u}), \qquad \mathbf{y} = \mathbf{g}(\mathbf{x}, \mathbf{u}).$$

In the following discussion I will focus mostly on the autonomous systems equations—the input-output aspects of the theory get involved in some technical problems of controllability and observability that would obscure the central idea, which is essentially quite simple.

The structural information for a reaction system is completely contained in the stoichiometric matrix, $\boldsymbol{\nu}$, which is of the order (species $\times$ reactions) (Fig. 2).

(1) State space, Σ
(2) Input space, U
(3) Output space, Y
(4) State transition function, φ
(5) Readout function, r

$\{\Sigma, U, Y, \varphi, r\} \triangleq$ Dynamical system

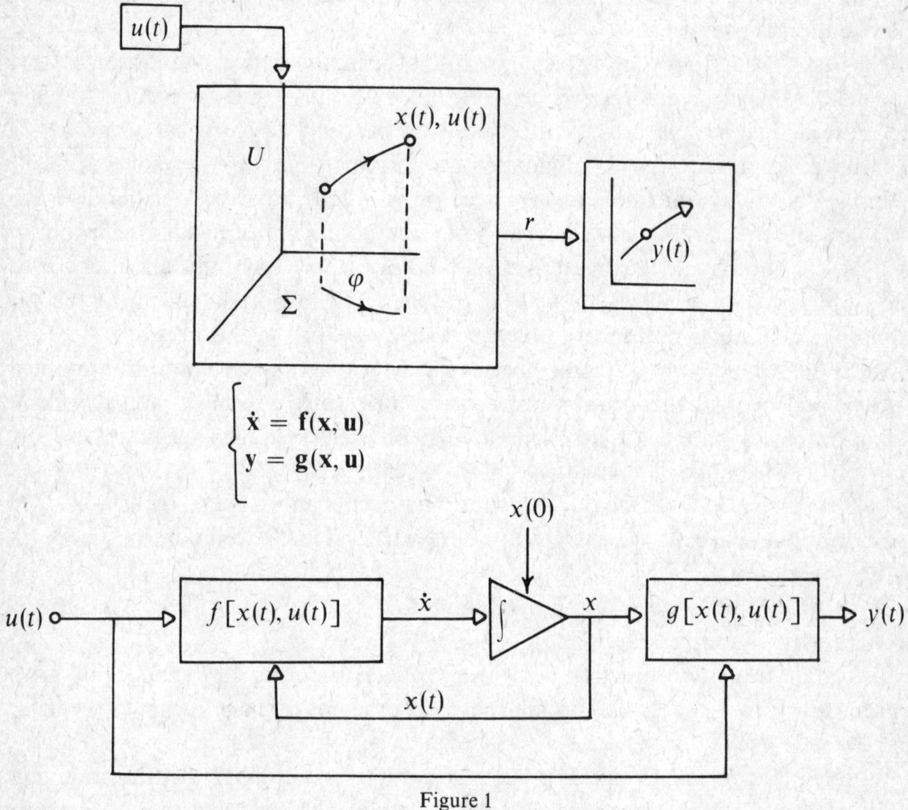

Figure 1

Now, for a homogeneous chemical system, conventional thermodynamics provides us with an adequate state space: that is, a set of N mole numbers if viewed as a point in $\mathbb{R}^N$ (species space) as in Fig. 3.

The problem is, first of all, to characterize the structure of the vector field that propels the state point about as reactions occur between the N chemical species. The law of mass action gives a prescription for this vector field on concentration space as a polynomial function in concentrations. But, as we shall see, this is not really the right space on which to view the dynamics, since there are a number of constraints on how the concentrations change in time. Moreover, as I mentioned before, the mass action law has to be patched up rather severely to deal with solution

Stoichiometric Matrix

$$\nu = \begin{array}{c} \\ 1 \\ 2 \\ 3 \end{array}\left[\begin{array}{cccccc} A & B & C & D & E & F \\ -1 & -1 & 1 & 0 & 0 & 0 \\ 0 & 0 & -2 & 1 & 0 & 0 \\ 0 & 1 & 0 & -1 & -1 & 2 \end{array}\right] \Big\} \text{ Reactions}$$

Species

$$A + B \rightleftharpoons C$$
$$2C \rightleftharpoons D$$
$$D + E \rightleftharpoons 2F + B$$

Figure 2

reactions. It is much better to view mass action as a special case of a more general scheme.

Chemical reaction dynamics lies, in a sense, "in between" two extreme classes of dynamical systems—both of which have been well studied (Fig. 4).

On the one hand, there are the potential flows, or gradient dynamical systems whose trajectories flow downhill on a potential surface, like molasses with no possibility of oscillatory or periodic behavior. At the other extreme are the Hamiltonian systems of classical mechanics which flow along constant energy surfaces. These systems are purely conservative and can, of course, exhibit a variety of periodic behaviors.

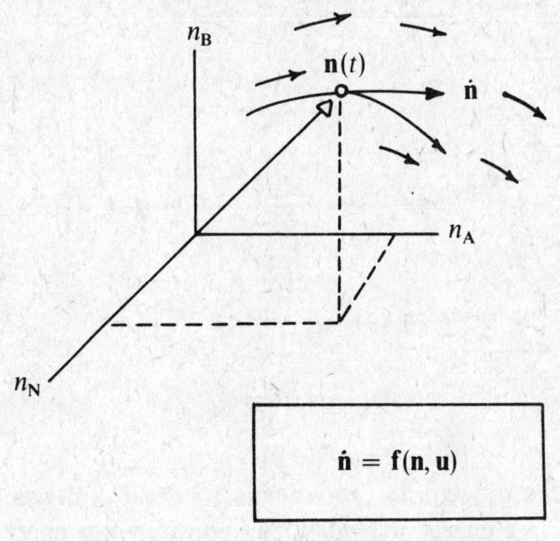

Mass action: $\dot{n}_i = k_f \prod n_i^{\beta_i} - k_r \prod n_j^{\beta_j}$

Figure 3

$\dot{\mathbf{x}} = \nabla\varphi(\mathbf{x})$
$\mathbf{x}(t)$
$\dot{\mathbf{x}} = \nabla\varphi(\mathbf{x})$
$\mathrm{curl}(\nabla\varphi) = 0$
(a)
φ = constant

Gradient flow

$\dot{\mathbf{x}} = \mathbf{J}\nabla H$
$\nabla H(x)$
$\dot{\mathbf{x}} = \mathbf{J}\nabla H(x)$

$$\mathbf{J} = \begin{bmatrix} \mathbf{0} & \mathbf{I} \\ -\mathbf{I} & \mathbf{0} \end{bmatrix}$$

$\mathrm{div}(\mathbf{J}\nabla H) = 0$
(b)
H = constant

Hamiltonian flow

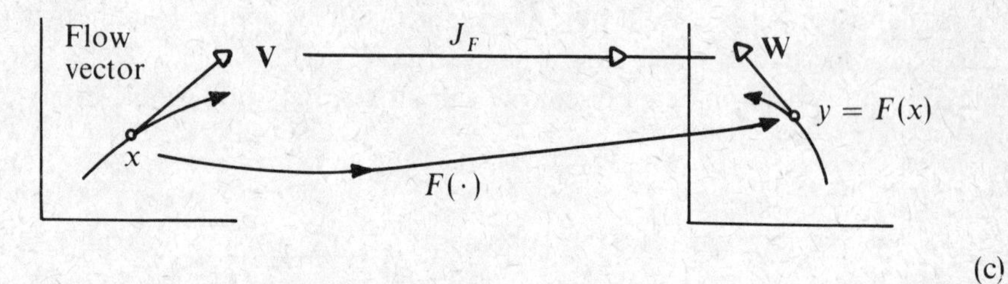

(c)

Force
covector
V
J_F^T
W
x
y
$F(\cdot)$

Figure 4

If we examine the trajectories of a chemical system, however, they flow neither directly down the free energy hill nor along constant free energy surfaces, that is, the chemical dynamic equations are "rotated" gradient flows, but not rotated so much as Hamiltonian flows.

Instead of treating the general nonlinear equations, which would enmesh us in a lot of technical details, I think I can give you something of the flavor by reformulating the familiar linear irreversible thermodynamic equations from a slightly different point of view. Suppose we have R reactions between N species. Then we know that each reaction can be specified by a reaction coordinate, i.e., the chemical advancements,

$$n_i(t) = n_i(0) + \sum_{j=1}^{R} \nu_{ij}\xi_j$$

So we can define a reaction space, $\mathbb{R}^R$, a point in which specifies the state of all R reactions. As the reactions proceed, this point traces out a trajectory in reaction space; the tangent to this curve is just the set of reaction rates (or flows) for the process. There are usually fewer reaction coordinates than species, so it is clear that the dynamics should be formulated on $\mathbb{R}^R$, not on the species space, $\mathbb{R}^N$.

Alternatively, the range of the affine map defined by the reaction stoichiometry defines an affine subspace of species space which is sometimes called the "reaction simplex." The choice of whether to use this space or the reaction space to describe the state of the system depends on the rank of $\boldsymbol{\nu}$, that is, the number of stoichiometrically independent reactions. For now, we shall consider only the case where $\boldsymbol{\nu}$ is full rank, so that the equations will be formulated on reaction space.

There is another difficulty in modelling complex reaction systems: the state space is frequently not Euclidean. This comes about because of two ubiquitous approximation techniques employed in chemistry. That is the quasi-equilibrium approximation and the pseudo-steady-state hypothesis. The quasi-equilibrium approximation assumes that certain species such as enzymes are present in only trace amounts, so that their concentrations relax to their equilibrium values almost immediately. The pseudo-steady-state hypothesis assumes that certain reactions proceed much more rapidly than others, so that those reactions go to their equilibrium extent almost immediately.

The result of either of these two physical assumptions is to confine the motion of the state point to a submanifold of the state space: a submanifold of species space in the case of the quasi-equilibrium assumption and a submanifold of reaction space in the case of the pseudo-steady-state hypothesis. The way this comes about can be seen in general on the following diagram (Fig. 6).

A dynamical system in $\mathbb{R}^3$, say three chemical reactions, can be written as

$$\dot{x}_i = f_i(\mathbf{x}), \qquad i = 1, 2, 3$$

If the rate of the first process is so rapid that $\dot{x}_1 \to 0$ on a time scale much shorter than x_2 or x_3, then $dt_1 = \varepsilon dt_{2,3}$, where $\varepsilon \ll 1$. The theory of a singular perturbation deals with the question of when it is permissible to simply set $\varepsilon \equiv 0$ and treat the

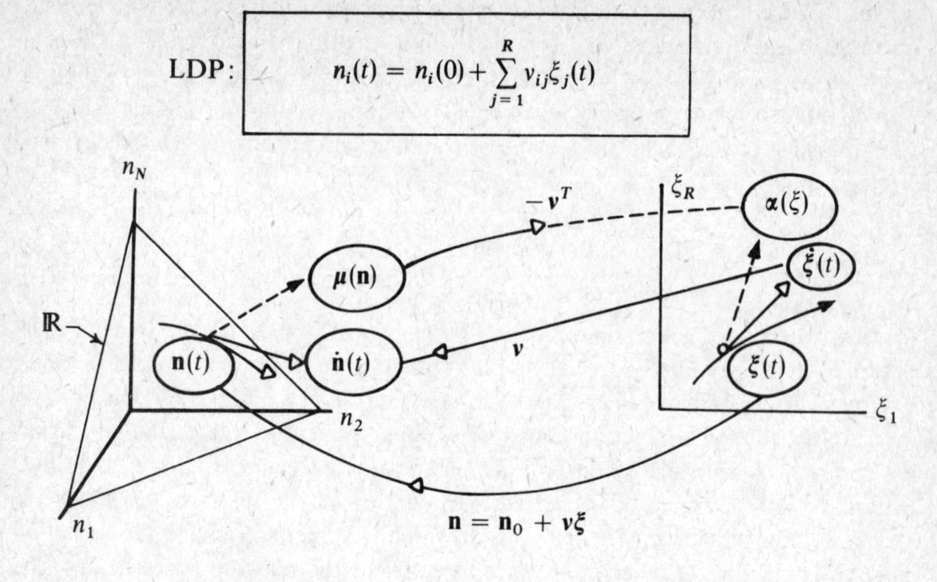

$$\text{LDP}\begin{cases} \mathbf{n}(t) = \mathbf{n}(0) + \mathbf{v}\boldsymbol{\xi}(t) \\ \dot{\mathbf{n}}(t) = \mathbf{v}\dot{\boldsymbol{\xi}}(t) \\ \boldsymbol{\alpha}(t) = -\mathbf{v}^T\boldsymbol{\mu}(t) \end{cases}$$

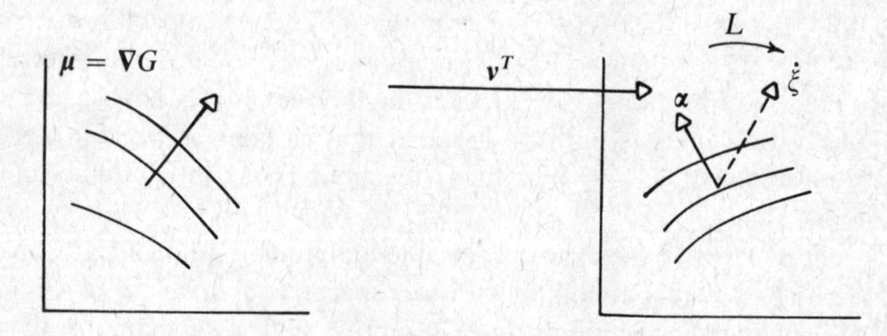

Figure 5

level surface $f_1(\mathbf{x}) = 0$ as a constraint surface to which $\dot{x}_2$ and $\dot{x}_3$ are confined. That is, the system relaxes to the "slow" manifold $f_1(\mathbf{x}) = 0$ quite quickly, and so we are principally interested in this constrained motion. The difficulty arises if this constraint surface is shaped peculiarly, for then it is not possible to write the differential equations globally; we have to be content with glueing together local solutions.

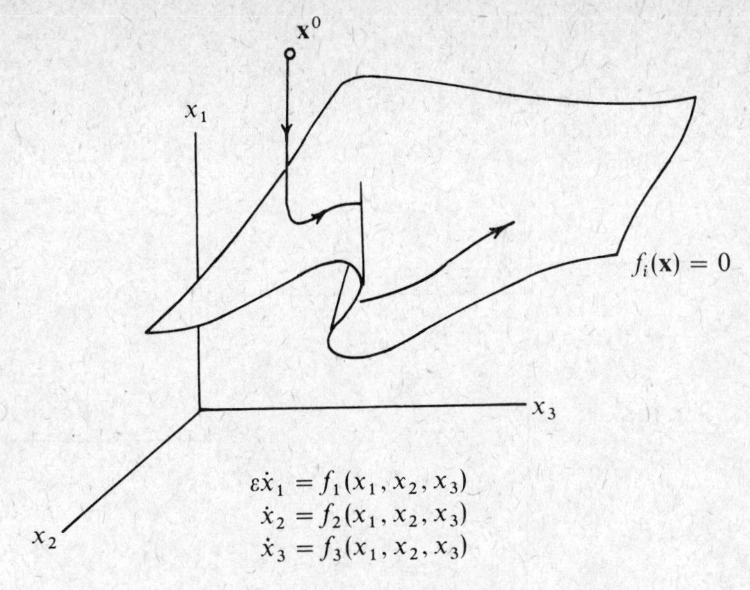

$$\begin{aligned} \varepsilon\dot{x}_1 &= f_1(x_1, x_2, x_3) \\ \dot{x}_2 &= f_2(x_1, x_2, x_3) \\ \dot{x}_3 &= f_3(x_1, x_2, x_3) \end{aligned}$$

Figure 6

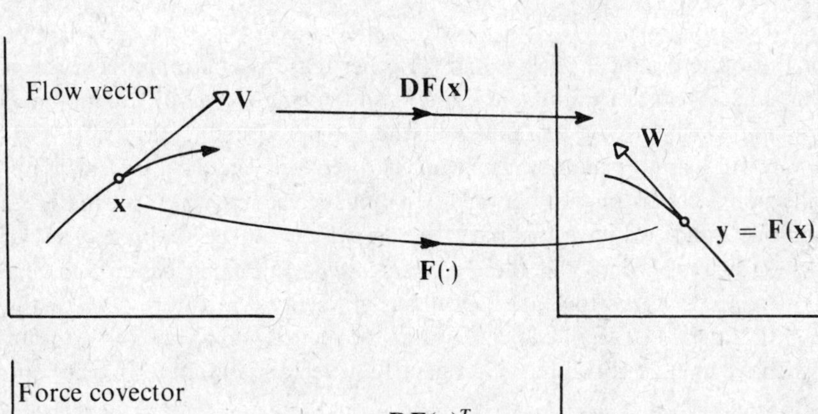

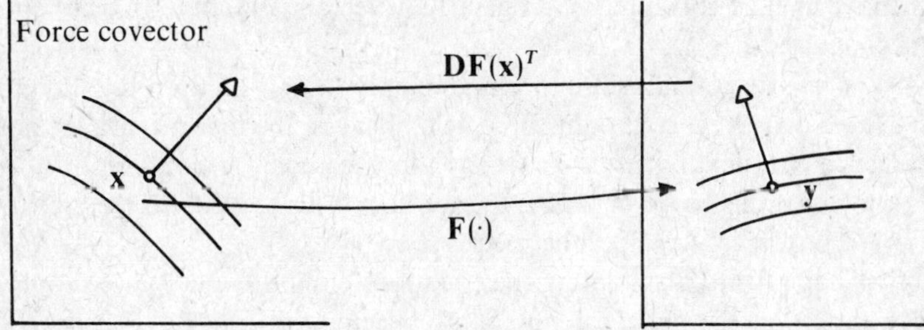

Figure 7

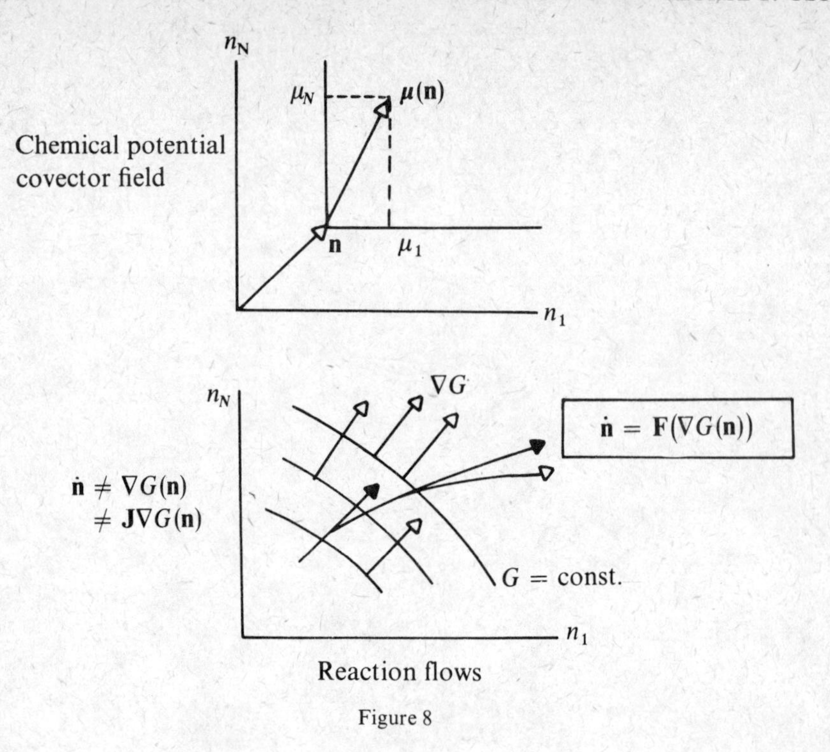

Figure 8

For this quite practical reason, and for other reasons of a more technical nature, it became necessary to formulate the theory in the language of differential geometry, but I shall tread lightly on these deep mathematical waters here (Fig. 7).

Now, it will help to make things clearer in discussing chemical systems if we agree to make a physical distinction between two kinds of vectors, which is usually ignored. On the one hand, we have ordinary vectors which can be thought of as velocity vectors to curves; I shall call these *tangent vectors*, or *flow vectors*. On the other hand, there are vectors which arise locally as gradient vectors normal to small pieces or potential surfaces. I shall call these kinds of vectors *force covectors* to remind us that when we change coordinates, tangent flow vectors and force covectors map in *opposite* directions (Fig. 8).

Now, it is true that equilibrium thermodynamics tells us that on the thermostatic state space we do have a covariant force field. That is, the thermostatic equation of state assigns a chemical potential covector $\boldsymbol{\mu}$ to each composition point. Moreover, we also know that this covector field is the gradient of a potential free energy function $G: \mathbb{R}^N \rightarrow \mathbb{R}$ whose level-sets fiber the species space $\mathbb{R}^N$.

As a first step toward formulating the equations of motion on $\mathbb{R}^R$ we should like a way of transferring the chemical potential covector field on $\mathbb{R}^N$ over to $\mathbb{R}^R$. The method for doing this is provided by the law of definite proportions which assures

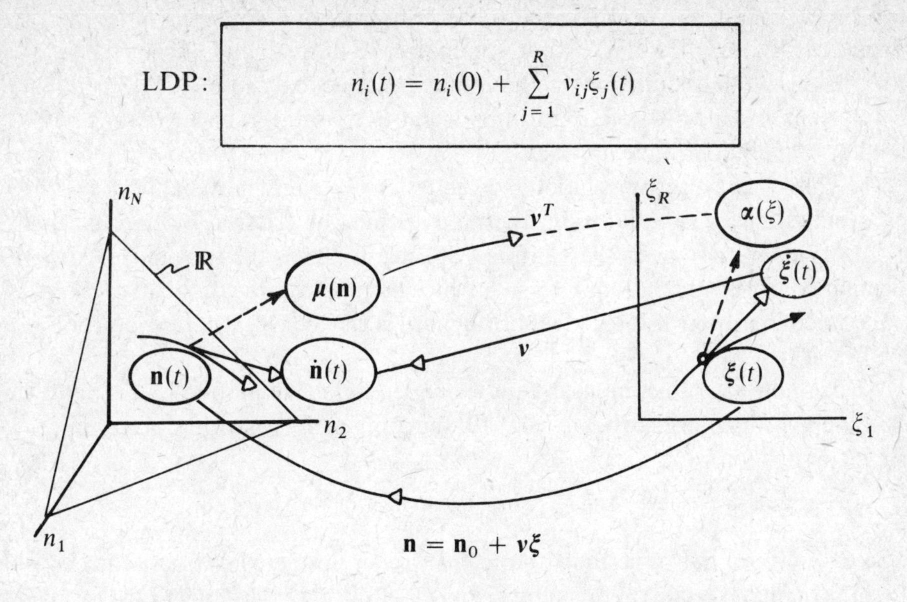

$$\text{LDP}\begin{cases} \mathbf{n}(t) = \mathbf{n}(0) + \boldsymbol{\nu}\boldsymbol{\xi}(t) \\ \dot{\mathbf{n}}(t) = \boldsymbol{\nu}\dot{\boldsymbol{\xi}}(t) \\ \boldsymbol{\alpha}(t) = -\boldsymbol{\nu}^T\boldsymbol{\mu}(t) \end{cases}$$

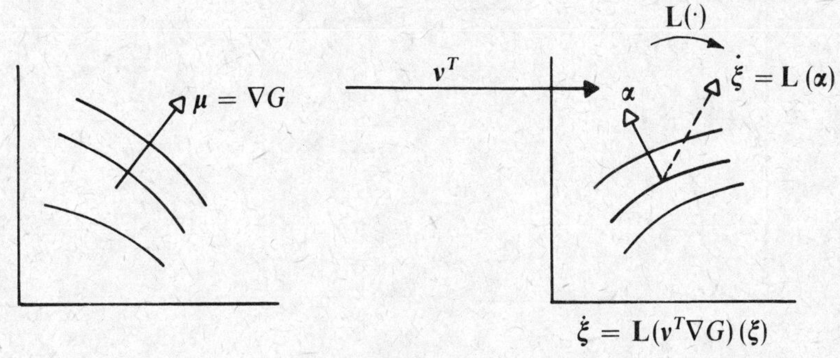

Figure 9

us that reactants and products disappear and appear in stoichiometric ratios. That is, reaction flow vectors are related to the rate of change of species by the stoichiometric matrix: $\dot{\mathbf{n}}(t) = \boldsymbol{\nu}\dot{\boldsymbol{\xi}}(t)$. Therefore, we know that the transpose of the stoichiometric matrix maps the chemical potential covectors on $\mathbb{R}^N$ over to cotangent vectors on $\mathbb{R}^R$. So we obtain a gradient field on reaction space which is just the usual chemical affinity (Fig. 9). All we need now is a way of converting the affinity covector field on reaction space to a vector field and we have our equations of motion. Unfortunately, though, the classic shortcoming of the linear theory is that the reaction rate is not a unique function of the affinity except very close to thermodynamic equilibrium. Let me hold off on this important fact, however, and push ahead to the familiar near equilibrium equations.

The role of constitutive relations, or equations of state, in dynamical systems is the same as it was in equilibrium thermodynamics, or classical mechanics; that is, to relate vector and covector quantities. So, clearly this is the spot to introduce the reaction constitutive relation, $\mathbf{L}: \boldsymbol{\alpha} \to \mathbf{j}$, which summarizes the dissipative properties of the reaction process. Close to equilibrium this is just the Onsager phenomenological coefficient $\mathbf{L}: \boldsymbol{\alpha} \to \xi$ (Fig. 9).

We can now assemble the near equilibrium equations of motion by just following the mapping diagram shown in Fig. 10 and obtain the following prescription for the vector field on $\mathbb{R}^R$:

$$\dot{\xi} = \mathbf{L}\big(-\boldsymbol{\nu}^T \boldsymbol{\mu}(\mathbf{n}_0 + \boldsymbol{\nu}\xi)\big) = \mathbf{F}(\xi).$$

So far we have not done much more than clean up the conventional irreversible thermodynamic treatment of reactions, and all of the old objections to the formalism remain. However, the notable aspect of this way of looking at the equations of motion is that by separating the components of the dynamics into contributions from equilibrium properties ($\boldsymbol{\mu}$), dissipative properties ($\mathbf{L}$) and stoichiometry ($\boldsymbol{\nu}$), which connects them, we can remedy all of the old defects, and get a great deal more besides.

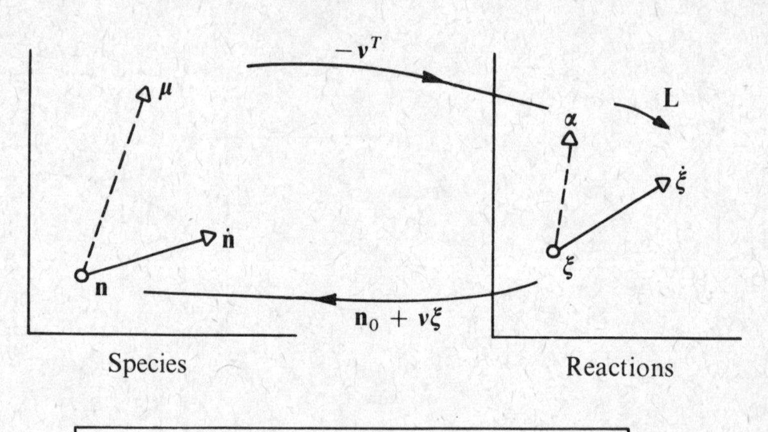

$$\dot{\xi} = \mathbf{L}\big(-\boldsymbol{\nu}^T \boldsymbol{\mu}(\mathbf{n}_0 + \boldsymbol{\nu}\xi)\big) = \mathbf{F}(\xi)$$

Figure 10

Now let us return to the problem of generalizing the equations of motion to the nonlinear regime. This amounts to more than just replacing the linear functions by nonlinear ones, since, as I mentioned before, the thermodynamic affinity does not uniquely characterize the reaction rate. Another central feature of reactions is that they are not reciprocal. That is, perturbations of reactant and product do not have symmetrical effects on one another, so that there is nothing comparable to the Onsager theorem for reaction constitutive relations. These facts deal a serious blow to the simple formulation I just outlined.

I shall not go into the details because of space limitations, and let me just point out one or two facts that the generalization hinges on.

$$R \rightleftharpoons P$$

$$\hat{\mathbf{v}} = [\mathbf{v}_R \,\vdots\, \mathbf{v}_P]$$

$$\mathbf{n}(t) = \mathbf{n}_0 + \hat{\mathbf{v}}\begin{bmatrix} \xi^f \\ \xi^r \end{bmatrix} \qquad \dot{\mathbf{n}}(t) = \hat{\mathbf{v}}\begin{bmatrix} \mathbf{j}^f \\ \mathbf{j}^r \end{bmatrix}$$

$$\boldsymbol{\alpha}^f = \mathbf{v}_R^r \boldsymbol{\mu} \qquad \boldsymbol{\alpha}^r = \mathbf{v}_P^T \boldsymbol{\mu}$$

$$(\boldsymbol{\alpha}^f, \boldsymbol{\alpha}^r) \longrightarrow (\mathbf{j}^f, \mathbf{j}^r) = (-\mathbf{j}, \mathbf{j})$$

$$\begin{aligned} \dot{\xi} &= \mathbf{F}(\xi) \\ &= \boldsymbol{\Lambda}(\hat{\mathbf{v}}\boldsymbol{\mu}(\mathbf{n}_0 + \hat{\mathbf{v}}\hat{\xi})) \end{aligned}$$

$$\boxed{\dot{\xi} = \boldsymbol{\Lambda}(\nabla\varphi(\hat{\xi}))}$$

$$[\dot{\mathbf{x}} = \mathbf{J}\nabla H(\mathbf{x})]$$

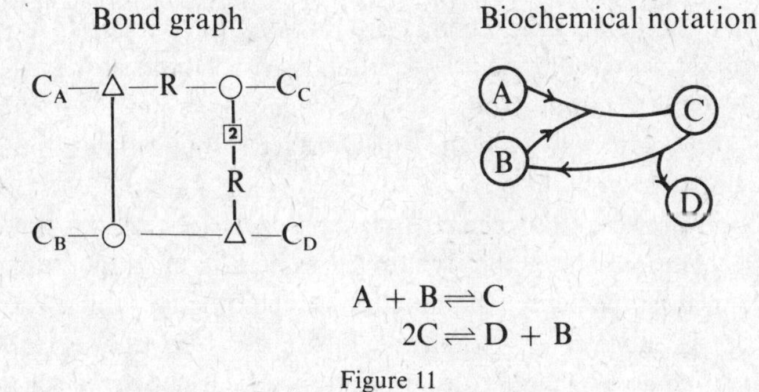

$$A + B \rightleftharpoons C$$

$$2C \rightleftharpoons D + B$$

Figure 11

In the first place, we notice that the set of reactions form a natural partitioning of the chemical species into reactants and products. These are not necessarily disjoint sets, since a given species can participate as a product in one reaction and a reactant in another. This leads us to define the stoichiometric matrix in a slightly different way, by partitioning it into reactants and products:

$$\hat{\nu} = [\nu^R, \nu^P]$$

This induces us to define a forward and reverse reaction rate $\mathbf{j}^f$, $\mathbf{j}^r$. That is, a rate measured by the disappearance of reactant or the appearance of product. Then, we can define, as de Donder did back in 1927, forward and reverse reaction affinities as stoichiometric combinations of reactants and products, respectively: $\boldsymbol{\alpha}^f = \nu^R \boldsymbol{\mu}$, $\boldsymbol{\alpha}^r = \nu^P \boldsymbol{\mu}$.

Now the law of definite proportions requires that the forward and reverse reaction rates be equal, but the forward and reverse affinities may be completely independent; so the reaction constitutive relation must have the form: $(\boldsymbol{\alpha}^f, \boldsymbol{\alpha}^r) \rightarrow (\mathbf{j}, -\mathbf{j})$. This is just the additional degree of freedom we need to obtain a thermodynamic theory of reaction rates that is not tied to reciprocity or the law of mass action. It reduces as a special case to both the linear irreversible thermodynamic formalism and to mass action equations, but permits us to generalize to a much wider class of open reaction systems.

One extremely interesting consequence of these physical constraints on our mathematics is that one can easily show that the law of definite proportions is, in fact, contradictory with reciprocity except at thermodynamic equilibrium. This surprising conflict between the basic rule of chemistry and the principle cornerstone of linear irreversible thermodynamics highlights the impossibility of treating reactions within the conventional framework.

What we finally end up with is a set of canonical equations of motion which look very much like the near equilibrium case, but the meaning of the symbols is quite different (Fig. 11). As I mentioned in the beginning, the vector field for reaction dynamics is sort of a rotated gradient flow, like a dissipative, nonholonomic constraint in mechanics. However, they are quite different in structure from the equations of classical mechanics. In fact, the chemical equations have a lot in common with certain equations from circuit theory—certain types of transistor circuits—whose properties are also poorly understood.

Let me now show you how even the old Onsager formalism is intimately related to abstract circuit theory.

The connection between the species and reaction spaces, that is the equilibrium and irreversible components of the dynamical system is the stoichiometric matrix, ν. It is clear that, in some sense, the reaction stoichiometry, which shows how the reaction network is "hooked up" (that is, which species are connected via reaction pathways), is similar to Kirchoff's Laws in an electrical network, which also show

how the electrical parts are interconnected. The common denominator is that both can be formulated as *topological* constraints. It was the possibility of using topological methods to examine reaction networks that first drew our attention to this particular way of looking at chemical systems.

To see the connection more precisely, we need one or two results from algebraic topology.

As shown in Fig. 12 a *linear graph* is a purely topological object consisting of nodes and branches. The topology of a linear graph can be coded in its *connection matrix*, which by choosing a *tree* can be partitioned in the form

$$\mathbf{B} = [\mathbf{I}, \mathbf{D}^T], \quad \mathbf{Q} = [-\mathbf{D}, \mathbf{I}].$$

If we now view the graph as a network with voltages on and currents flowing through each branch in any manner whatsoever, then we can form a vector of all the branch

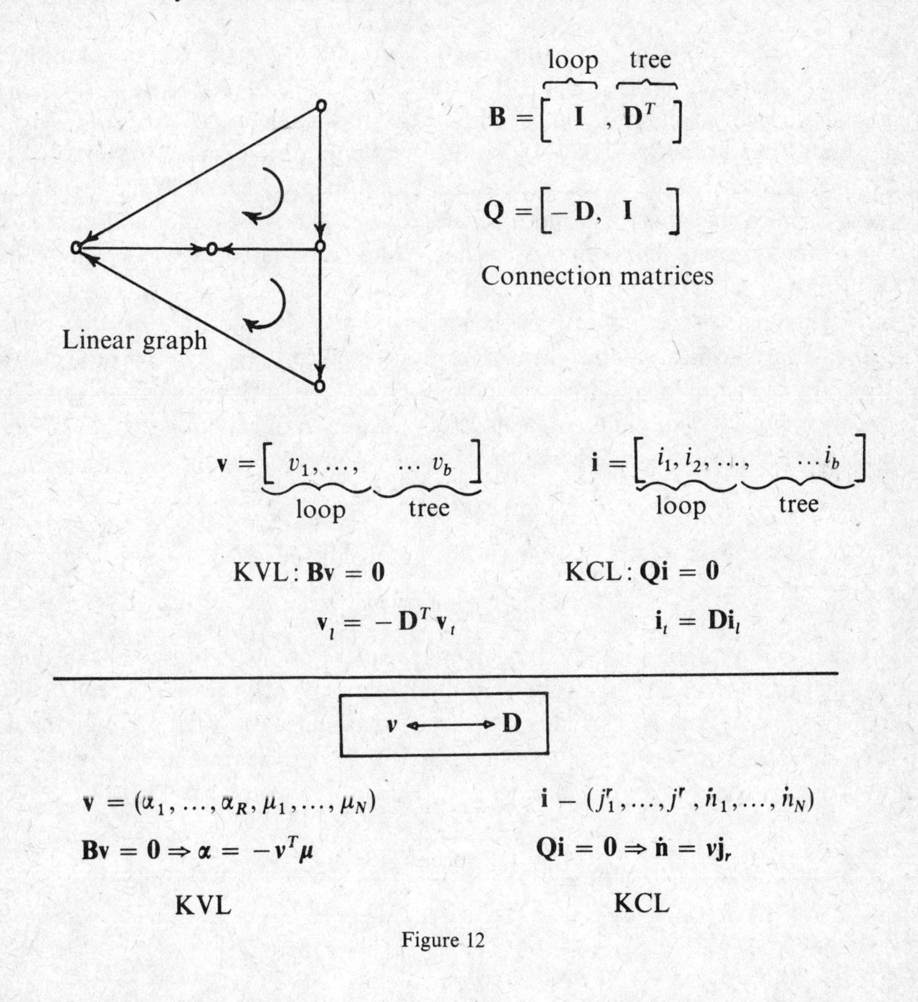

Figure 12

voltages and currents (Fig. 12). If we multiply these vectors by the connection matrices we obtain the equations $\mathbf{Bv} = \mathbf{0}$ or $\mathbf{v}_l = -\mathbf{D}^T\mathbf{v}_t$ and $\mathbf{Qi} = \mathbf{0}$ or $\mathbf{i}_t = \mathbf{Di}_l$, where the subscripts l and t refer to the link and tree branches of the graph. These are just Kirchhoff's Laws for the graph, which are the topological restatements of flow conservation and continuity of potentials.

On the other hand, if we consider for simplicity a reaction with only unit stoichiometry, its stoichiometric matrix can be represented as linear graph if we simply associate each species with a *tree branch* and each reaction with a loop (*link*). The reaction graph precisely codes the topology of the reaction network: $\boldsymbol{\nu} = \mathbf{D}$. Moreover, if we define a vector of all chemical potentials and affinities $(\boldsymbol{\mu}, \boldsymbol{\alpha})$ and a vector of both molar and reaction flows $(\dot{\mathbf{n}}, \mathbf{j})$, and then multiply them by the connection matrices, we regain the defining equations for the reaction quantities:

$$\boldsymbol{\alpha} = -\boldsymbol{\nu}^T\boldsymbol{\mu}, \quad \dot{\mathbf{n}} = \boldsymbol{\nu}\mathbf{j}.$$

So we see, in this simple case, that reactions do indeed obey Kirchhoff's Laws, and can be coded in conventional network form.

What does this transfer of old wine into new bottles gain us? A great deal, I would say, both from a practical and a theoretical viewpoint. First, and most importantly, it points the way to the correct generalization to open systems with nonlinear reaction dynamics, and thereby a more general view of the possibilities and limitations of reaction dynamics. On the practical side, it should permit us to apply some rather sophisticated methods of network analysis to the analysis, and as I mentioned at the beginning, even the synthesis of chemical networks, a fact of considerable interest to industrial chemical engineers. Before proceeding further, let me give a glimpse of the power and unifying force of topological methods in the network formulation.

A central result of graph theory can be obtained by simply noting that the product of the connection matrices for the reaction graph always vanishes:

$$\mathbf{BQ}^T = \mathbf{0}.$$

This means that the vector of flows is always orthogonal to the vector of forces:

$$\mathbf{V}^T\mathbf{i} = (\boldsymbol{\mu}, \boldsymbol{\alpha})^T(\dot{\mathbf{n}}, \mathbf{j}) = 0.$$

This is a purely topological fact and is independent of any property of the system save that the flows and forces obey Kirchoff's Laws, which, as we saw, they always do for chemical systems. This identity is true for nonlinear, nonstationary dynamics. Then, if $\mathbf{V}$ and $\mathbf{i}$ are orthogonal, so must their time derivatives and their variations about any stationary state:

$$\mathbf{V}^T\mathbf{i} = 0 = \delta\mathbf{V}^T\,\delta\mathbf{i}.$$

If we now project this sum onto the species and reaction spaces, we have:

$$\sum_R \dot{\alpha}_k j_k = -\sum_S \mu_s \dot{n}_s \quad \text{and} \quad \sum_r \delta\alpha_r\,\delta j_r = -\sum_s \delta\mu_s\,\delta\dot{n}_s.$$

$$\mathbf{BQ}^T = \mathbf{0}$$

$$\Rightarrow \mathbf{v}^T\mathbf{i} = \mathbf{0} \qquad \text{Tellegen's Theorem}$$

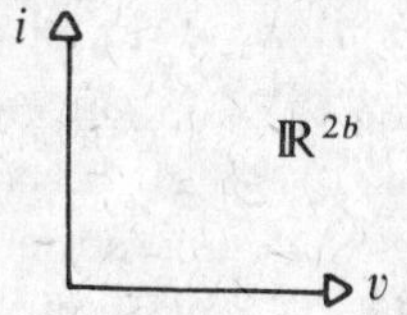

$$\begin{array}{rl|rl}
\dot{\mathbf{v}}^T\mathbf{i} = \mathbf{0} & & \delta\mathbf{v}^T\delta\mathbf{i} = \mathbf{0} & \\
(\dot{\mathbf{v}}^T\mathbf{i})_{rxn} & = -(\mathbf{v}^T\dot{\mathbf{i}})_{SP} & (\delta\mathbf{v}^T\delta\mathbf{i})_{rxn} & = -(\delta\mathbf{v}^T\delta\mathbf{i})_{SP} \\
 & = -\boldsymbol{\mu}^T\dot{\mathbf{n}} & & = -\frac{1}{2}\frac{d}{dt}[\delta\mathbf{n}^T D^2 G \delta\mathbf{n}] \\
 & \leqq 0 & & \geq 0 \\
\text{i.e., } \Sigma J_i\dot{X}_i & \leqq 0 & \text{i.e., } \Sigma\delta J_i \delta X_i & \geq 0
\end{array}$$

Figure 13

But it is easy to show that, if there are no phase changes in the system, and if the boundary inputs are held constant, then both right hand sides are sign definite quadratic forms, and we have the result

$$\sum_r \dot{\alpha}_r j_r \leq 0 \quad \text{and} \quad \sum_r \delta\alpha_r\, \delta j_r \geq 0.$$

These two inequalities were first proposed by Professors Prigogine and Glansdorff, who used them to characterize the local stability properties of nonequilibrium stationary states. Professor Eigen has assigned them a central role in his theories of chemical evolution.

My only purpose in rederiving them here is to point out the subtle, but pivotal role played by topological concepts in reaction systems, and to indicate how network concepts can unify and provide a deeper insight into the structure of chemical dynamics.

The fact that topological constraints plays such a central role in chemical systems is not so surprising if we simply reflect for a moment on any piece of electronic apparatus. The normal functioning of such a system depends crucially not so much on the geometrical relationship of its components as on how it is hooked up, that is, by the topological relationship of its components. The effects of topology should be no less pervasive for biochemical networks. Moreover, as a system becomes more complex wholly new properties emerge which are not perceived on the subunit level, i.e., collective properties.

I might digress briefly to comment on another observation which Katchalsky found disturbing; and that is, far from comprehending the functioning of biological systems, we hardly understand the operation of many of the machines which we ourselves have built and know quite well their design purpose. By "understand" I mean here that we frequently possess no "reduced description" or model of their operating characteristics. For example, there are no decent models for the telephone exchange, no reduced description—only the blueprints for the whole system—so that, in a sense, the machine is its own minimal representation. I suppose it is simply an article of faith that biological systems, like many physical systems, will have reduced mathematical descriptions—but, so far as I can determine, such assertions are, indeed, acts of faith.

Now, from the network viewpoint, we can see that, like a TV set, a reaction network acquires its behavioral complexity from the topological constraints imposed by the reaction stoichiometry. These constraints can be viewed as confining the system reaction rates and affinities to a linear submanifold of the state space—but their peculiar property is that they are *neutral* constraints. That is, like the reaction dissipation the stoichiometry imposes algebraic constraints amongst the reaction flows and affinities but unlike the reaction constitutive relation, these constraints are lossless. Professor Prigogine has emphasized the role of dissipation in creating ordered thermodynamic *spatial* structures; but for reaction systems, we see that the central temporal organizational constraints are, in fact, neutral—they exact no price in entropy production—they create *functional* dynamic order without dissipation.

In fact, by examining the equations of motion, we see that the two types of constraints, neutral and dissipative, act in concert to produce the system trajectory. It is also interesting to note that by smoothly varying the neutral constraints, we can possibly alter the system's qualitative dynamic behavior, even making the system go unstable, or oscillate, without altering the dissipative constraints at all. The resulting functional structures so created, although they are maintained by dissipation, are as much a consequence of the dissipationless neutral Kirchhoff constraints (and so might be called Kirchhoff structures vs. the spatial ordering called dissipative structures). I might point out that, for a given set of reactants, the dissipation is fixed and not subject to manipulation. But, by introducing enzymes to select amongst alternative reaction pathways, it is the neutral constraints that are being manipulated. From this viewpoint, one could argue that chemical evolution invokes the selection of reaction pathways, and thus there is information growth in the increasing complexity of the reaction network—but dissipation plays a secondary role.

Now, let me return for a moment and sketch the central ideal behind the use of abstract circuit theory as a natural mathematical setting to discuss chemical reactions.

First of all, we see that a reaction can be considered as a certain subclass of general dynamical systems—one that is formed by the interconnection of simpler subsystems by neutral constraints. The subsystems that are interconnected fall into two classes:

1) Each chemical species is a dynamical system governed by the equations

$$\dot{\mathbf{x}}(t) = \mathbf{u}(t) \leftrightarrow \dot{\mathbf{n}} = \mathbf{u}(t)\,(= \text{system inputs})$$

$$\mathbf{y}(t) = \mathbf{g}(\mathbf{x}(t)) \leftrightarrow \boldsymbol{\mu}(t) = \mathbf{g}(\mathbf{n}(t))$$

The study of these equations is the concern of equilibrium thermodynamics. They describe an essentially trivial dynamical system, whose distinguishing characteristic is that the readout function $g(\cdot)$ is a gradient vector field $\mathbf{g}(\cdot) = \nabla G(\cdot)$ (i.e., the system is *reciprocal*). Dynamical systems of this type are called "capacitive," since they are mathematically isomorphic to the equations of an electrical capacitor.

2) On the other hand, the reaction process itself can be viewed as an algebraic dynamical system:

$$\mathbf{0} = \mathbf{x}(t) - \mathbf{u}(t) = \mathbf{f}(\mathbf{x}, \mathbf{u}) \leftrightarrow \boldsymbol{\alpha}(t) = \mathbf{u}(t)$$

$$\mathbf{y}(t) = \mathbf{g}(\mathbf{x}, \mathbf{u}) \leftrightarrow \mathbf{j}(t) = \boldsymbol{\Lambda}(\hat{\boldsymbol{\alpha}}(t))$$

The study of these equations, when $\boldsymbol{\Lambda}$ is linear, is the subject of irreversible thermodynamics.

The distinguishing feature about this simple dynamical system is that $\mathbf{u}^T\mathbf{y} = \boldsymbol{\alpha}^T\boldsymbol{\Lambda}(\hat{\boldsymbol{\alpha}}) \geq 0$, i.e., the reaction is a *passive* dynamical system. This is the appropriate mathematical form of the Second Law. Dynamical systems of this type are called "resistive," since they are of the same form as an electrical resistor.

Now let us return to the linear graph representation of the reaction stoichiometry, because it shows us how to hook up reversible and irreversible thermodynamics to form a general theory.

The graph represents the neutral interconnections of the reaction stoichiometry. These interconnections can be decomposed into 3 (really two) basic types: a gen-

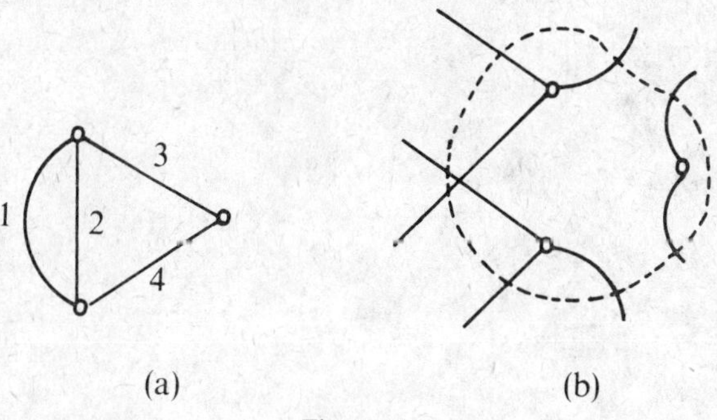

Figure 14

eralized "parallel connection" and a generalized "series connection." That is, by performing a few mental gymnastics, we can take the graph representing the reaction stoichiometry and extract all of the branches leaving only a "terminal" box of the interconnection nodes. This object is called a "connection *n*-port." The connection *n*-port can then be further decomposed by collecting together all of the parallel and all of the series connections, and thus reticulate the formal structure of reaction systems into the schematic form shown in Fig. 15. By these machinations we can effect a decomposition of the stoichiometric matrix in such a way that we can introduce a graphical notation for reaction systems which can be considered a mathematically rigorous extension of the heuristic graphical notation employed by biochemists for diagramming metabolic reaction chains. The principal purpose for introducing such a notation for reaction systems (aside from the fact that it comes free having done the mathematics) is that it provides an easy and intuitive graphical notation which can be interfaced nicely with computer network analysis and synthesis programs developed by the electrical engineers to treat reaction and reaction-transport systems in a convenient and efficient fashion.

However, much remains to be done before one can synthesize reaction networks to order the way one would design an electronic circuit—although this was one of our original goals.

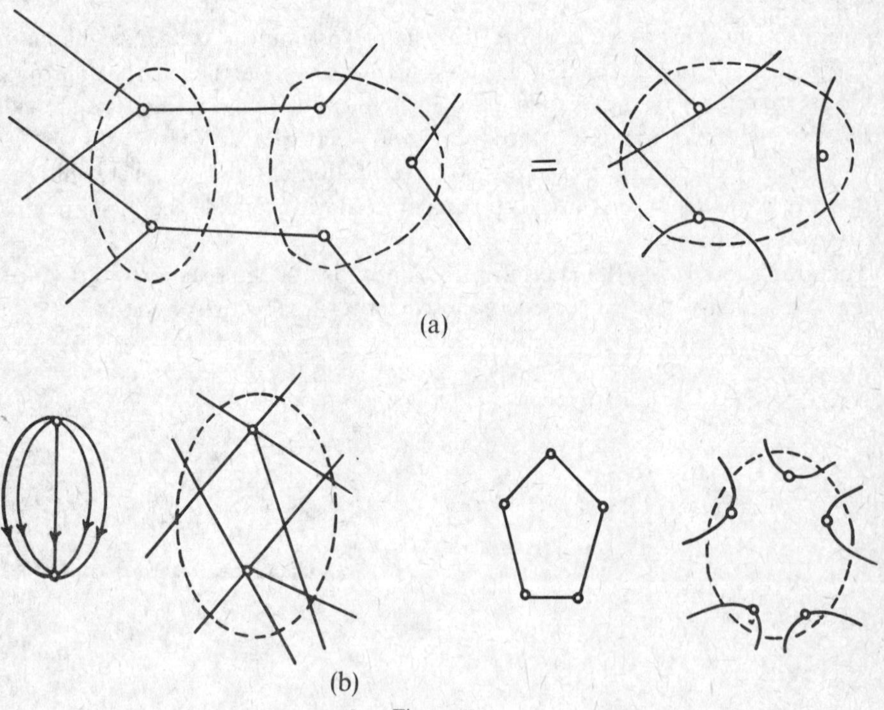

Figure 15

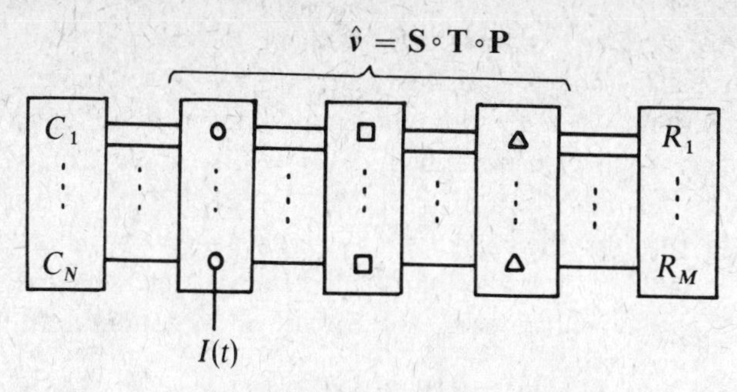

Figure 16

In conclusion, let me just sketch how the network formalism can be extended to treat transport coupled reaction systems. However, the systems I want to address myself to are not the continuum diffusion-reaction systems described by Prof. Prigogine which are so suggestive of pre-biotic chemical patterns. Rather, I want to move on down evolution's trail and consider how arrays of cells interacting through membranes might behave. That is, we consider an array of cells of any geometry, each containing a set of reactions and communicating across their boundary surfaces. Of course, if all intercellular transport is a linear process, and each cell contains an identical set of reacting species, then in the limit of zero cell size we would regain the continuum diffusion-reaction equations. But there is no need to make any

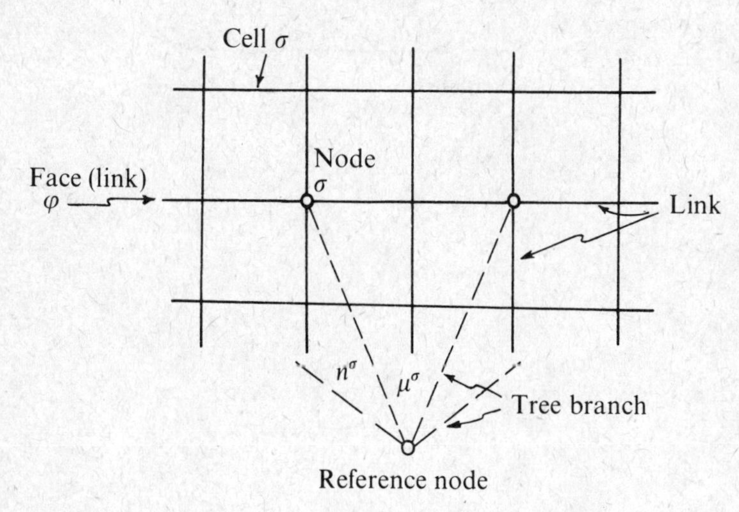

Figure 17

such restrictions in the network approach. First of all, the connectivity properties of any cellular array can be coded as a linear graph by a very simple algorithm: simply place a node within each cell and connect the nodes by one branch for every cell face across which transport occurs. Then, the flow of a single chemical species in the cellular array corresponds to a distribution of flows on the network.

If one then carries through the analysis, the mathematical structure for pure diffusion in a cellular array is formally almost identical to that encountered in the case of pure reaction (within a single cell). The equations of motion are:

$$\dot{\mathbf{n}} = \mathbf{D}\Upsilon_i(-\mathbf{D}^{\mathrm{T}}\mu_i(\mathbf{n}_i)) = \mathbf{F}(\mathbf{n})$$

where $\mathbf{D}$ is the same connectivity matrix for the graph and Υ_i is the constitutive relation describing the transport properties of the cellular membranes.

The equations for N simultaneously diffusing species can be constructed by assigning vector rather than scalar quantities to the branches of the graph and introducing some tensor product notations (for example, the connection matrix for the graph representing N diffusing species in a cellular array is $\tilde{\mathbf{D}} = \mathbf{D} \otimes \mathbf{I}_N$, i.e. $\tilde{\mathbf{D}}$ is obtained from $\mathbf{D}$ by substituting $\mathbf{I}_N$ for 1. The final equations look exactly the same.

Next, if we introduce a set of chemical reactions into each cell, the diffusion and reaction networks can be interconnected in a straightforward way to generate a

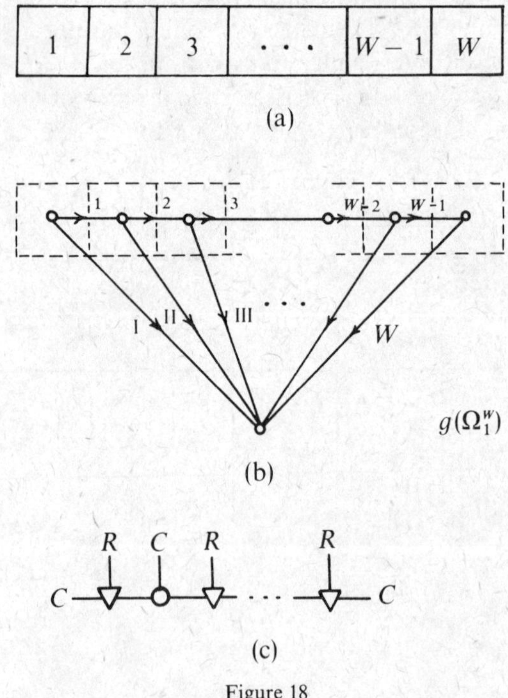

Figure 18

M reactions
N species
L faces
W cells

$$\Omega_{NM}^{WL} \triangleq \boxed{\begin{matrix} \vdots \\ A+B \rightleftharpoons C \\ C \rightleftharpoons D \\ \vdots \end{matrix}}$$

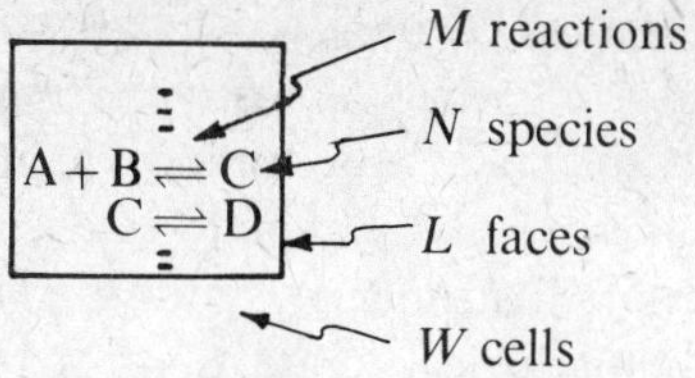

$$\tilde{\mathbf{n}} = \begin{bmatrix} \mathbf{n}^1 \\ \vdots \\ \mathbf{n}^w \end{bmatrix} = \sum_{\sigma=1}^{W} \mathbf{e}^\sigma \otimes \mathbf{n}^\sigma$$

$$\tilde{\zeta} = \begin{bmatrix} \zeta^1 \\ \vdots \\ \zeta^L \end{bmatrix} = \sum_{\varphi=1}^{L} \mathbf{e}^\varphi \otimes \zeta^\varphi$$

$$\tilde{\mathbf{D}} = \mathbf{D} \otimes \mathbf{I}_N$$
$$\dot{\tilde{\mathbf{n}}} = \tilde{\mathbf{D}} \circ \tilde{\Upsilon} \circ - \tilde{\mathbf{D}}^T \tilde{\mu}(\tilde{\mathbf{n}}) = \tilde{\mathbf{F}}(\tilde{\mathbf{n}})$$

Cellular diffusion equations

$$\frac{d\tilde{n}}{dt} = \tilde{\nu}\tilde{\Lambda}\left(-\tilde{\nu}^T\tilde{\mu}(\tilde{n})\right) + \tilde{D}_1\tilde{\Upsilon}'\left(-\tilde{D}_1^T\tilde{\mu}(\tilde{n})\right) + \tilde{D}_2\tilde{\Upsilon}''\left(-D_2^T\tilde{\mu}(\tilde{n}) - \tilde{D}_3^T\tilde{E}(t)\right)$$
$$= \tilde{F}\left(\tilde{n}, \tilde{u}(t)\right)$$

Linearized diffusion-reaction equations(simplified):

$$\delta\dot{\tilde{\mathbf{n}}} = \left[(\mathbf{I}_w \otimes \tilde{\mathbf{n}}) + (\mathbf{\Delta}' \otimes \mathbf{T}') + (\mathbf{\Delta}'' \otimes \mathbf{T}'')\right]\delta\tilde{n}$$

(Othmer and Scriven)

Figure 19

set of canonical equations for cellular transport reaction. The general form of these equations is rather complicated. The crucial aspect is that the effects of the cellular topology and the reaction topology on one another are explicitly displayed. Thus, as Scriven and Othmer have shown, the configuration of the cellular array profoundly influences the reaction dynamics. But beyond that, the reaction may itself regulate the topological relationships of the cells, causing certain configurations to become unstable, the cells resorting themselves into different arrangements. Once again, the structural order thus created from the reaction dynamics is generated by the evolution of the neutral constraints, and the dissipative constraints, although acting in concert, play a secondary role. In a sense we have an informational, or nonentropic restructuring, driven by the dissipative process of chemical reactions.

In conclusion, let me note that underlying all the fancy mathematics is a theme of great simplicity—that the logical foundations of reaction systems are formally identical with a certain class of network problems and that this isomorphism can be exploited to gain a deeper insight into the theory of reaction dynamics and to unify equilibrium and irreversible thermodynamics with chemical kinetics. Beyond that, it was Aharon Katzir-Katchalsky's hope to provide a practical tool to study the synthesis of reaction networks with an eye to inferring the design purpose nature had in mind when she constructed them. To the extent that we have succeeded in carrying out this program will always be a tribute to the incomparable insight and inspiration Aharon provided to all of us fortunate enough to have worked with him.

Source and transmission of information in biological networks

H. ATLAN

Polymer Department, Weizmann Institute of Science, Rehovot, Israel

Network thermodynamics, as developed by Aharon Katzir-Katchalsky in collaboration with G. Oster and A. Perelson (1971, 1973), provides a method which should be suitable for analyzing complex systems composed of a large number of coupled chemical reactions with various kinds of transports, both in the homogeneous and the non-homogeneous phase. It is worth emphasizing that these are dynamic networks whose elements are not the compounds participating in the reactions, but the reactions themselves. In electrical circuits the elements are resistors, capacitors, inductors, etc., while in these networks the elements are the chemical reactions and the transport processes. In this paper I will analyze the conditions under which this method can be applied to real biological systems.

First, however, I would like to give what seems to me a good example of the different questions which can arise in biology and which require different methods of analysis. There are some questions which can be answered by the classical methods of molecular biology, but there are other questions which cannot, and which require new methods of analysis, of the same kind as network theories—new for biologists, not for physicists and engineers.

This example is a problem in cell differentiation. Spore formation in *Bacillus subtilis* has been studied extensively as a model for the morphologic, biochemical and immunogenic modifications which accompany cell differentiation (Hansen, Spiegelman and Halvorson, 1970; Sonenshein and Losick, 1970; Losick, Shorenstein and Sonenshein, 1970; Shapiro, Azabian-Keshishian and Bendis, 1971). It was found that new biosyntheses were triggered by the modifications of the medium responsible for spore formation. These new syntheses were brought about by new messenger RNA, which in turn was caused by modifications in the RNA polymerase. Now, these modifications were caused by either synthesis of new RNA polymerase or denaturation of the existing RNA polymerase. Whatever the case, new causal factors or phenomena will be looked for until the initial cause—i.e., the modification in the medium which triggers the sporulation—is reached. So far the classical methods of biology, searching for molecular modifications in a step-by-step sequential manner, are perfectly appropriate.

More recently (Shapiro, Azabian-Keshishian and Bendis, 1971) the same approach was extended to another bacterial system, namely, bacteria of the genus *Caulobacter*. This system seems to be a better model for cell differentiation since the morphologic, biochemical and immunogenic modifications are not induced by changes in the medium, but occur spontaneously as obligatory steps in the cell cycle. When these modifications are prevented, the cells cannot divide and die. In *Caulobacter*, as in *B. subtilis*, new biosyntheses have been found responsible for these modifications, new messenger RNA are responsible for the new biosynthesis, and changes in RNA polymerase were looked for as causes for the synthesis of new mRNA, and so on, along the lines of the method used to study spore formation. But here, in *Caulobacter*, there is no hope for an end to this search for a linear sequence of causes, and for finding the initial cause, since there is no inducer from the external medium, and therefore no *one* initial cause.

The method required here is one which would allow the cell to be treated as a system in a given state at a given time, determined by the value of a set of coupled variables, which is responsible for the next state of the cell. The whole set of variables is subject to variations. Since it is hoped that network thermodynamics can provide such a method, it is important to understand the conditions for its uses.

The first condition is that it should be possible to find biological systems to which the method can be applied and tested. In Sec. 1, I shall present some recent experimental findings on the existence of complex couplings between transport across biological membranes and biosynthetic reactions; the possible role these couplings play in cell control and regulation is discussed in Sec. 2.

Sec. 3 deals with the conditions under which network theories can be applied, as they are, to biological systems, and the conditions under which they need basic improvements, in view of the special kind of problems to be solved. In particular, I intend to discuss the problem of evolving networks starting from the analysis of a somewhat simplified form of the Zhabotinsky reaction as an example for a dissipative structure.

Sec. 4 discusses briefly in a more formal manner the problem of the sources of information responsible for the organization of so-called self-organizing systems.

1. Coupling between ion transport and protein synthesis

Trying to answer the question of the practical utility of network thermodynamics for real biological systems, we were looking for experimental systems of medium complexity in which couplings between biochemical reactions and transport could be studied.

Medium complexity means something between molecular cell-free systems and whole cellular systems. For this purpose we first tried to build semi-artificial—what we called cell-like—systems, where well-known biochemical reactions were confined within a vesicle limited by a cell membrane. Thus interactions between transport across the membrane and reactions within the vesicle could be studied systematically, after the kinetics of transport on the one hand, and the reactions on the other hand, had been studied separately.

These attempts are still in process.

In the meantime, in collaboration with M. Herzberg and H. Breitbart, I have turned towards a kind of cell-like system which is provided by nature, i.e., the reticulocyte. Reticulocytes are almost mature red blood cells which have lost their nucleus and almost all their mitochondria. Their cytoplasm, like that of mature red blood cells, is homogeneous, i.e., structureless as far as can be judged from electron microscope pictures. But there are still active ribosomes and messenger RNA so that translational phases of protein synthesis take place actively, leading to synthesis of hemoblogin. In other words, we have a system made of an active biological membrane, the cell membrane, and a well-known biosynthesizing machinery, which can be studied comparatively *in vitro* and within the environment limited and controlled by the membrane.

I shall present some results of the experiments showing that a functional coupling does indeed exist between the transport of potassium through the membrane and the rate of protein synthesis at the level of the ribosomes within the cells. Contrary to what we expected, this coupling does not take place directly via K^+ concentration within the cell.

Figure 1 shows the schematic structure of K^+ carriers which were used to modify drastically the transport properties of the membrane by increasing very specifically its permeability to K^+.

These carriers are Valinomycin (a well-known antibiotic) and dicyclohexyl-18-crown-6 (to which we shall refer as DC), a synthetic macrocyclic compound. Valinomycin and DC are known to make both artificial lipidic membranes and biological membranes highly and specifically permeable to potassium, by binding specifically K^+ ions in a one-to-one reaction within the ring of their molecule, and carrying K^+ ions across the membranes.

The main results of the experiments, repeated many times under various conditions (Herzberg, Breitbart and Atlan, 1974), can be summarized as follows:

Adding the K^+ carriers to the incubation medium of reticulocytes produces a *rapid* inhibition of up to 90% of protein synthesis, measured by incorporation of a radioactive amino acid (Fig. 2), whereas no inhibitory effect whatsoever can be found when these compounds are added to an *in vitro* cell-free preparation of active

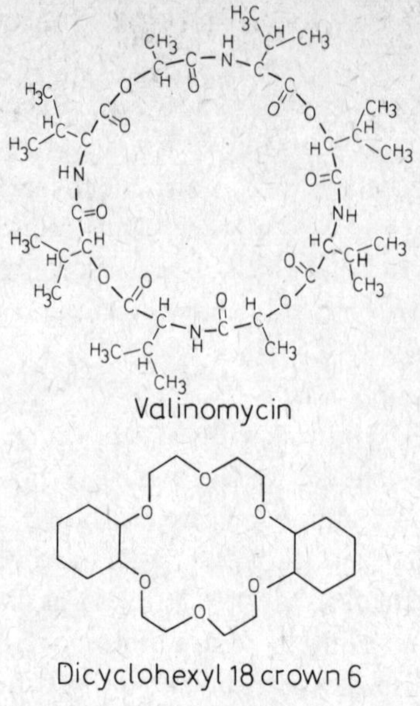

Valinomycin

Dicyclohexyl 18 crown 6

Figure 1
Schematic structure of two Potassium carriers.

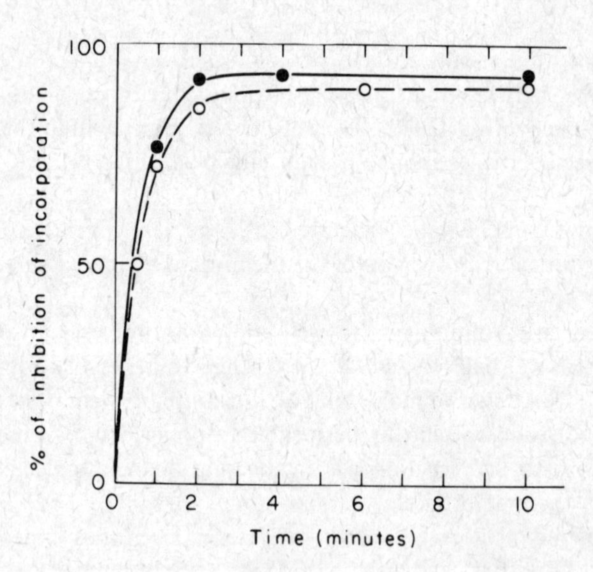

Figure 2
Time course of valinomycin (●—●) and dicyclohexyl-18-crown-6 (○ - - ○) action. ^{14}C leucine incorporation was measured in reticulocytes and compared with controls incubated without drugs (Herzberg et al., 1974).

ribosomes extracted from reticulocytes of the same population. This inhibitory effect has been localized at the level of the peptide chain elongation, since no modification of the monosome–polysome distribution has been found by analytical centrifugation.

The kinetics of this inhibitory effect for the two carriers is represented in Fig. 3. It can be seen that they act within very different ranges of concentration. Since these effects were obtained only on intact cells, and not on *in vitro* systems, we concluded that this inhibitory effect on ribosomal activity within the cell is *membrane mediated*. Moreover, within these concentration ranges, the effect is completely reversible simply by washing out the drug; this shows that *no* permanent structural alteration of the ribosomes takes place.

For Valinomycin at *higher concentrations* a new phenomenon occurs, in which an irreversible modification with some inhibitory factor bound to the ribosomes seems to be involved. At lower concentrations of Valinomycin, and for all possible concentrations of DC, the effect is reversible and so far no inhibitory factor could be detected. As to the possible mechanisms responsible for this inhibition, we presume that changes in K^+ concentration within the cell can be eliminated since i) in the *in vitro* system they do not produce any effect; ii) in the cell system they do not have enough time to take place.

Amino acid and ATP deprivation as indirect effects of changes in permeability of the membrane have also been eliminated by control experiments.

What remains at this state is a correlation between the flux of K^+ through the membrane and the rate of protein synthesis. There can be a number of different intermediate factors responsible for this coupling, namely:

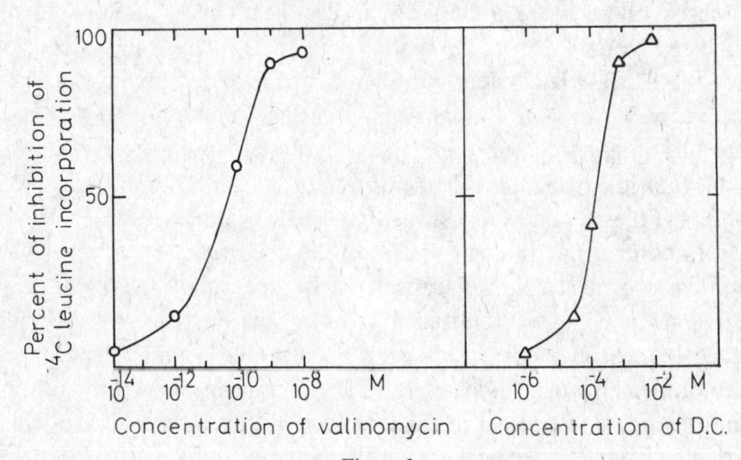

Figure 3

Concentration curve of valinomycin and dicyclohexyl-18-crown-6 action.

1) The membrane potential, which may be a good candidate but must be measured.

2) The permeability of the membrane to other compounds, the transport of which would be coupled to that of K^+. This includes H^+ ions, which would mean a modification of the pH, and possibly electrons, which would mean a modification of the redox potential.

3) The value of the ADP/ATP ratio, which can be modified even without any gross modification of the ATP content, and which was shown by Freudenberg and Mager (1971) to be effective in controlling protein synthesis in reticulocytes, also at the elongation step.

Anyway, whatever the detailed precise mechanisms are, a functional coupling between ion-transport properties of the cell membrane and polysome-elongation activity within the cellular environment is involved.

A phenomenon which could have a similar physiological meaning has been reported (McDonald, Sachs, Orr and Ebert, 1972; Orr, Yoshikawa-Fukada and Ebert, 1972) in experiments where the rate of DNA replication in baby hamster kidney cells appears to be under the control of the cell membrane potential. Again, in these results, as in our experiments, changes in K^+ concentration and ATP level within the cell do not seem to be involved.

This shows that functional couplings between transport and reactions can exist not only in the direction from the reaction to the transport, as in active transport, but also in the reverse direction, from transport to reaction. In fact, such couplings have been known for a long time as taking place in mitochondria according to Mitchell's description of the oxydative phosphorylation coupled to proton concentration gradient, in chloroplasts and in bacterial vesicles as well; but in all these cases, they involve only metabolic reactions such as oxydative phosphorylation, photosynthesis, or oxydoreductions. Whereas here, the reactions involved are biosynthetic reactions, i.e., protein synthesis in the reticulocyte system, and DNA replication in the B.H.K. system of Ebert's group (McDonald, Sachs, Orr and Ebert, 1972; Orr, Yoshikawa-Fukada and Ebert, 1972).

These results imply that this kind of coupling can serve as a possible way to transmit information from the cell membrane to the interior of the cell, in order to control biosynthesis, and even, as was suggested by Cone (1971), Ebert (1972a, b), and others, to control the different phases of the cell cycle.

This point is worth stressing because usually one thinks about information at the cellular level as being transmitted only by means of structural specificities, i.e., signals reach the cell in the form of specific molecules bound to specific receptors on the membrane. This triggers changes in the membrane structure and properties, so that another signal is emitted in the form of another molecule released from the membrane, which binds specifically to its target molecule, and so on. Clearly, information is transmitted via a set of equilibrium reactions of specific binding of structurally adapted molecules.

Here, however, we have another kind of transmission of information which takes place by means of functional couplings between transport and reactions, not via a set of subsequent equilibrium bindings but via kinetic couplings between non-equilibrium reactions. Hence the changes in the *rate* of one process—the transport—trigger changes in the rate of the other, coupled, process—the reaction.

There are some advantages and disadvantages in the transmission of information via changes in rates, over the one which works via structural specificities. It is probably less specific to start with, although, by combination of couplings, a functional specificity can be retained at the expense of increase in the number of couplings, i.e., in complexity (in other words, non-structural specificity can be achieved by combinatorial arrangements so that specific functional patterns are generated, as was suggested by Dr. Jerne in his paper, but this is at the expense of increase in complexity). However, coding by rate is probably more flexible and richer in coding possibilities, whereas specific bindings act more like switches, namely, all-or-none processes. In fact, we find here the same advantages and disadvantages of controls by means of continuous versus discrete coding, or analog versus digital.

2. Control and regulation in biological networks

The question of what is control, and how control and regulation are achieved in biological networks as compared to engineering networks, was amongst the problems which haunted Aharon Katzir-Katchalsky in the last part of his work on network thermodynamics. In his work with Oster and Perelson (1973) it is shown that the same kind of control mechanism by parametric modulation which exists in engineering systems can also be found in chemico-diffusional networks.

The main idea is that control and regulation act through modulation of parameters characterizing the elements of the network, say values of resistances, capacitances, transducer moduli, etc. We have seen from what exists in cell systems, where not only structural specificities but also kinetic couplings seem to play a role, that this general idea of control by parametric modulation, i.e., by changes in rates, applies to biological systems as well. In fact, however, it is not so simple, and one must be careful in transposing the engineering concepts of control to entire biological systems.

Parametric control means that the value of the parameters which characterize the elements of the network is subject to changes according to some function. If the function is imposed from the outside, it is a simple control mechanism. If the function is an output from some part of the network itself, and there is some feedback or feed forward, it is a regulatory control, or a regulation mechanism. The action of this controlling function on the network is invariably represented as a phenomenon which is physically separated from the network itself. The effective agent in control mechanisms is signals or information transmission. The energy necessary for signals—

and in general for transmission of information—has little to do with the energy exchanges involved in the functioning of the network. A particular case of information flow is the topology itself, i.e., the blueprint to build the network. For example, in bond graph notation, if a resistance R is controlled by some function of a controlling variable r, this can be drawn as in Fig. 4, with an arrow representing the energy flow in the resistor, and a different (dotted) arrow representing the signal flow.

In other words, energy flows and information flows are always separated and in usual network analysis, can therefore be analyzed separately, as if transmission of information did not cost any energy. The reason for this is well-known today following the work of Brillouin (1956): if one wants to transform a certain amount of information units into energy units, one multiplies by a factor $kT \ln 2$, which is equal to 10^{-16} and in most cases makes the energy cost of information completely negligible.

This applies particularly well to *enzymatic* control where the resistance of a reaction—the relation between the rate of the reaction and its affinity—is a function of the catalytic activity. This activity depends on the conformational state of the enzyme molecule, which itself depends on properties of the medium, for example concentrations of given ions. Therefore changes in ionic concentrations can act as controlling signals on the reaction via changes in the conformational state of the enzyme molecule. Here again the separation is obvious between the energy flow, i.e., the flow of chemical energy of the reaction, and the information flow, which utilizes changes in conformation, in entropy and free energy of the enzyme molecule that have nothing to do quantitatively with the free energy of the reaction. So far it would seem that the engineering concept of control applies very well to chemico-diffusional and biological networks, and our experiments on reticulocytes seem to confirm it even more.

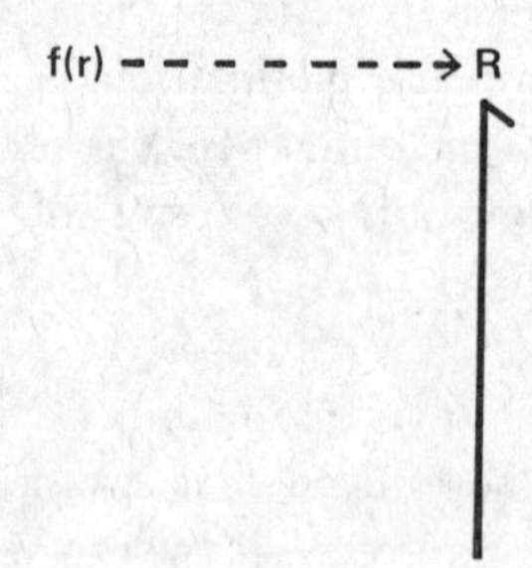

Figure 4

Parametric control of a resistance R by some function of a controlling variable (r) in bond graph notation. Energy and signal flows are separated and represented in two different ways.

However, all this is true only if we focus our attention on a *part* of a chemico-diffusional network; for example, if we isolate the enzymatic reaction and the signals in the form of the ionic concentrations changes, without being interested in what produces these concentration changes. If we were interested in the entire network, then we would have to take into account the fact that these concentration changes are the result of other reactions or transport phenomena, or both. In fact, it appears that in the mechanisms producing these changes, their cost in energy is of the same order of magnitude as that involved in the controlled reaction. In other words, if one looks at the entire network, the information flows and energy flows are not quite as well separated as they appear to be if one looks at a part of the network. In addition, it is clear from network thermodynamics (Oster, Perelson and Katchalsky, 1971, 1973) that all the *chemical* resistances and capacitances (and inductances (Atlan and Weisbuch, 1973)) are functions of the concentrations. For example, resistances, defined as $\partial A/\partial J$, are of the form $RT/k_f \prod_i c_i^{-\nu_i}$, capacitances are of the form c_i/RT, etc., and this is the main source of non-linearities in these networks. Of course this feature is, in principle, rich in possibilities of control and regulation, since the feedbacks are, so to speak, built in from the beginning into the elements of the network. This, however, constitutes a qualitative jump in the nature of the control mechanisms because of the intricacies between information flows and energy flows.

To my mind this is one of the differences between the analysis of partial networks and entire networks. It stems, partly, from the special nature of chemical networks, where changes in concentration act as both signals and energy flows, and partly from the well-known difference in the position of the engineer facing the problem of design and the position of a person facing the induction problem of determining the laws and properties of a black box. From a quantitative point of view, this could be an indirect way of coming back to an old prediction of Brillouin (1956) that biological complexities might lead to such situations where the 10^{-16} factor would not be enough to make information quantities negligible when expressed in energy units.

3. Zhabotinsky reactions and evolving networks

There is another, perhaps the most obvious, difference which emerges between partial and entire networks. It has to do with the notion of *changing* or *evolving networks*. As we know, couplings between reactions and transport are interesting not only because they can provide mechanisms of control, but also because under appropriate conditions they can give rise to instabilities, with the appearance of structures, which have already been discussed.

In principle, of course, network theory can be used to analyze such systems and to tell us about the conditions of oscillations, the appearance of concentration waves leading to a rearrangement of matter in space and causing transition from a homogeneous mixture to a non-homogeneous distribution. Within the framework of the network approach, the problem of subsequent instability, with ensuing new structures, can be stated—if not yet solved—as the problem of networks with changing topology.

3.1. *Zhabotinsky reaction as an example of chemico-diffusional networks*

As an example, I would like to show what can be done with the Zhabotinsky reaction, which has stimulated a lot of work in the last few years being a nice and relatively simple system giving rise to both temporal and spatial oscillations. This work is being carried out in collaboration with G. Weisbuch and J. Salomon, 1975.

Figure 5 shows a graph of the Zhabotinsky reaction in what appears as its most likely form today (Noyes, Field and Körös, 1972; Field, Körös and Noyes, 1972). There is a chain of progressive reductions of bromate to bromous acid and hypobromous acid until malonic acid is oxydized with formation of bromomalonic acid, followed by release of bromide ions. On the one hand, the bromide ions released at the end are necessary for the initial steps of the reduction of the bromate, so that a feedback loop exists in the network. On the other hand, this chain is coupled to an autocatalytic reaction of bromous acid production, with bromous acid itself participating as a reactant. This autocatalytic reaction is coupled to oxydation

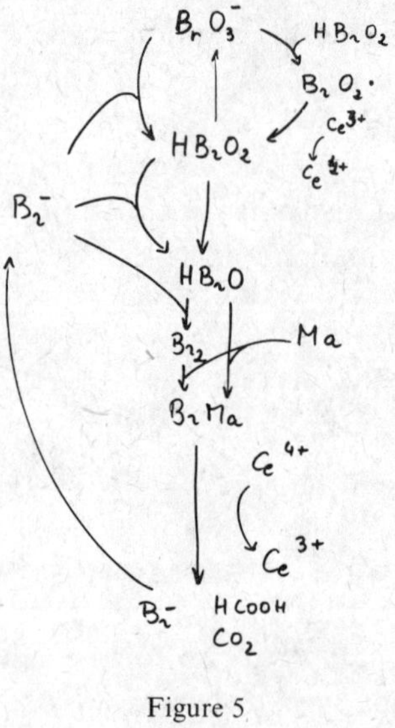

Figure 5

Schematic representation of the Zhabotinsky reaction system (see text, and Noyes et al., 1972; Field et al., 1972).

of Ce^{3+} to Ce^{4+}. At the end, the release of bromide ions from bromomalonic acid is coupled to the reduction of Ce^{4+} to Ce^{3+}.

What makes the system oscillate is the existence of two reaction pathways, one which requires high concentrations of bromide ions and the other through the autocatalytic reaction. When the concentration of bromide is high, the first pathway (on the left side of the figure) is operative until the bromide has been used up (more bromide is used than produced). Ce^{4+}, which is necessary for the last reaction, is used up at the same time. This pathway cannot operate anymore, and the system switches to the autocatalytic pathway. This regenerates Ce^{4+}, and bromide ions accumulate without being used until they again reach a high concentration, so that the first pathway can be reactivated, and so on.

An analysis of this network, using the rate constants found in the literature and determining the conditions of oscillations, is still under study (Weisbuch, Salomon and Atlan, 1975), and I would like to show a somewhat simplified version, in which the existence of temporal oscillations can be established by classical methods of kinetic analysis with the help of an analog computer.

In Fig. 6 the two pathways are represented in a simplified manner as a set of five reactions with only three variables: X for bromide ions, U for the bromous acid which participates in the autocatalytic reaction, and V for hypobromous acid and bromomalonic acid together:

$$X \xrightarrow{k_1} U$$

$$U + X \xrightarrow{k_2} V$$

$$U \xrightarrow{k_3} 2U$$

$$U \xrightarrow{k_4} V$$

$$V \xrightarrow{k_5} X$$

All the other compounds are assumed to be constant and included in the rate constants.

A set of three differential equations can be written to describe the kinetics of the system:

$$\dot{X} = -k_1 X - k_2 UX + k_5 V$$

$$\dot{U} = k_3 U - k_2 UX - k_4 U + k_1 X$$

$$\dot{V} = k_2 UX + k_4 U - k_5 V$$

As is well known, an instability of small fluctuations around a stationary state can always be studied by linearizing the equations around the stationary state and by solving the characteristic equation of the linearized system for roots with a

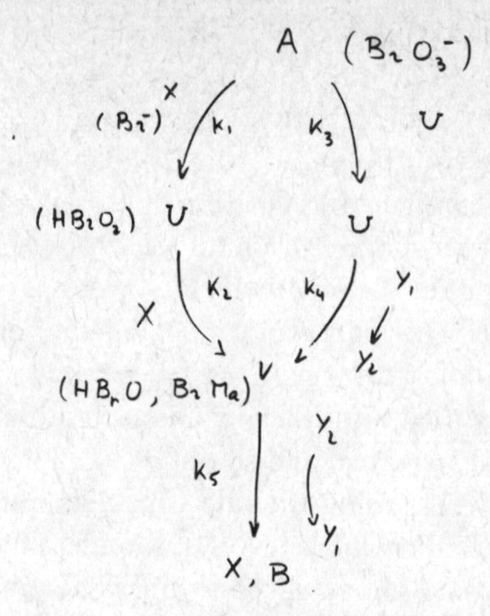

Figure 6
Representation of a simplified Zhabotinsky system reduced to three variables and 5 reactions (see text).

positive real part. That is to say, we assume exponential solutions of the form $e^{\omega t}$ and look for conditions such that $\mathrm{Re}\,\omega > 0$. Under these conditions, the existence of a limit cycle beyond the instability can be established; the shape and properties of the limit cycle are shown in Fig. 7 for some values of the kinetic constants.

This limit cycle, plotted as V vs. X, corresponds to the existence of stable oscillations after the onset and beyond the instability of the stationary state. The stationary state is represented by a point within the cycle and, using different initial conditions close to the stationary state, we would trace a spiral moving out from the interior and stabilize on the same limit cycle. Figure 8 represents the temporal oscillations of the three variables; this corresponds to an experimental situation where the reaction takes place in a stirred solution so that different diffusion rates are not allowed for the different compounds.

Now, if we allow diffusion in one direction with different diffusion coefficients D_X, D_U, D_V for the three variables, a straightforward application of Fick's law leads to a new set of equations:

$$\frac{\delta X}{\delta t} = -k_1 X - k_2 UX + k_5 V + D_X \frac{\delta^2 X}{\delta r^2}$$

$$\frac{\delta U}{\delta t} = k_3 U - k_2 UX - k_4 U + k_1 X + D_U \frac{\delta^2 U}{\delta r^2}$$

$$\frac{\delta V}{\delta t} = k_2 UX + k_4 U - k_5 V + D_V \frac{\delta^2 V}{\delta r^2}$$

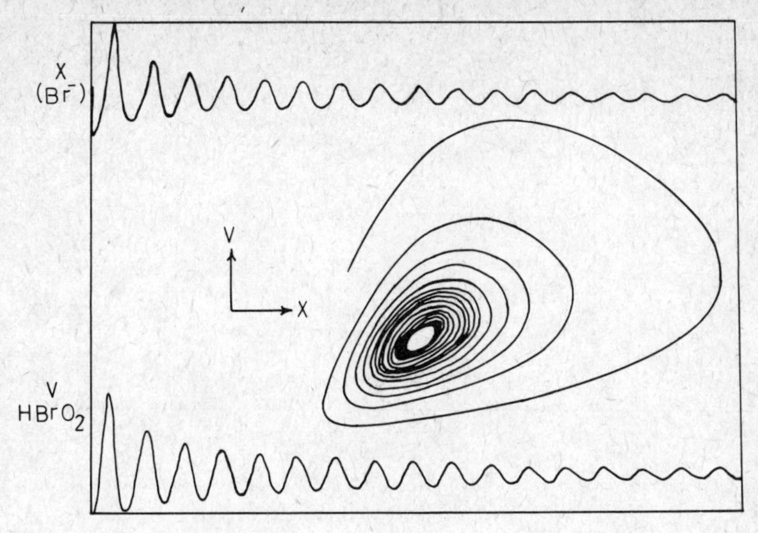

Figure 7

Limit cycle showing the existence of stable ascillations beyond the instability of the stationary state in the simplified three variables system (see text).

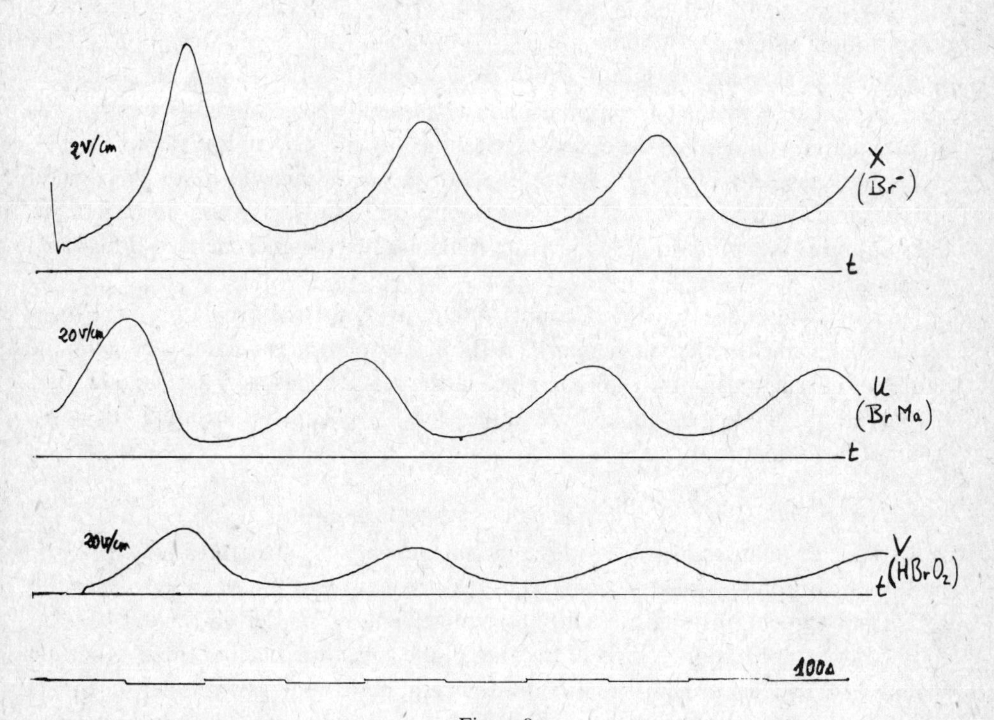

Figure 8

Temporal oscillations of the three variables in the simplified system.

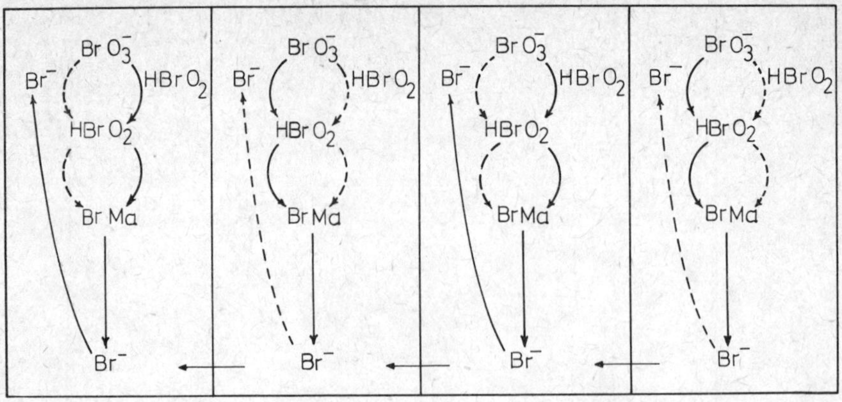

Figure 9

Discrete model of diffusion from box to box for the simplified system, leading to concentration waves and oscillations in space.

where r is the space variable. It is possible to find conditions (i.e., sets of values for the D's) such that oscillations in space will occur. This was done by numerical computation using a C.S.M.P. modelling program, after transformation of the continuous space variable r into a discrete variable. In other words, the reactions were assumed to take place homogeneously in small boxes and the compounds (one compound in the simplest case) were assumed to diffuse from box to box (Fig. 9).

For critical values of the D's, concentration waves appear, which are propagated and become stationary after a sufficiently long time, so that a non-homogeneous distribution of the compounds in the boxes is maintained through chemico-diffusional coupling.

This ultra-simplified scheme accounts qualitatively for the oscillating properties of the Zhabotinsky reaction network. Although unrealistic as compared with the entire set of reactions known to take place in the real Zhabotinsky system, it is thus of interest in helping us to analyze three-variable chemico-diffusional networks leading to dissipative structures.

3.2. *The concept of chemico-diffusional evolving networks*

If we want to take seriously the idea that such dissipative structures play a role in biological morphogenesis, we must consider the existence of not only one, but a set of subsequent instabilities with rearrangement of matter in space, possibly followed by stabilization, a kind of freezing of the structure. Each of these structures would serve as a set of new initial conditions and new topology for a new chemico-diffusional network, which in its turn will evolve toward another instability leading to a new structure, and so on.

In the case of the Zhabotinsky reaction network, for example, we are dealing initially with a reaction network in a homogeneous phase. If we ignore the differences in diffusion coefficients, we obtain temporal oscillations. If we introduce different diffusion coefficients, we obtain oscillations in space, i.e., a redistribution of the reactants and products in space in a non-homogeneous pattern, through the creation of a concentration wave.

In terms of a network representation, this is similar to what happens in the theory of lines with localized constants, which can be treated in a discrete way, as represented in Fig. 9: the chemical network is realized homogeneously in a sequence of small boxes distributed in space, and the reactants and products are able to diffuse from box to box. The continuous space dimension r is replaced by a discrete box number. Under conditions producing oscillations in space, we end up with our set of boxes in a new state, with bromide ions and other variable reactants not distributed equally among the boxes.

Now, this is of course the basis for morphogenesis according to the original idea of Turing (1952), as developed by Prigogine and his school, and recently by Katzir-Katchalsky, Scriven and others.

If we are to try to apply this idea to real biological morphogenesis, we must allow two additional assumptions:

1) The new structure that arose beyond the instability somehow becomes frozen, so that it is less dependent upon the rate of dissipation and is perhaps transformed into a crystal-like, minimum-free energy structure, rather than a dissipative one. In other words, the process of structuring of the instability would be responsible for the *appearance* of the structure, i.e., for the redistribution of matter, but a subsequent reaction would be responsible for its relative fixation.

2) This structure, after it has appeared and has been relatively stabilized (whether as a dissipative structure or as an equilibrium structure), will act as a new non-homogeneous chemico-diffusional network. For example, the regions of space in which some compound has accumulated may behave as membranes, simply because transport across the accumulation introduces new parameters. In network terms, new network elements and new connections arise as a result of the appearance of the structure.

In other words, we start with a chemico-diffusional network in a homogeneous phase. If the conditions are such that a structural instability occurs, we end up with heterogeneous phases, which means a new network as far as the network elements and connections are concerned. The new network can, in its turn, lead to a new structural instability, i.e., another new network, and so on. Ideally, one should be able to predict from the initial network the subsequent ones, and this is how we were led to the idea of changing or evolving networks.

For a long time the engineers have restricted themselves to linear networks with constant coefficients. Then they started to analyze time-varying and eventually non-linear networks. The next step, which seems to emerge from the analysis of biological networks, is the analysis of changing or evolving networks, whose functioning generates new *elements* and new *connections*, so that the network as a whole, and not only its coefficients, is changing.

I shall now briefly sketch some preliminary ideas as to how to treat such networks. Since the *elements* are assumed to be time-variable, a way of dealing with the appearance of new elements is to assign zero initial values to the elements which exist in the final form and do not exist in the initial form. Of course this is not very satisfactory, because it implies knowledge of the final form of the network.

To deal with the changing *connections*, the bond graph technique of Paynter, as proposed by Oster, Perelson and Katchalsky (1971, 1973) for network thermodynamics, appears highly suitable. In bond graphs the topology of the network, i.e., the connections, is represented in fact by special kind of *elements* called junctions (parallel and series). These elements are defined, as other elements, by what can be considered as a special case of constitutive relations, which in this case are the two Kirchhoff laws (Fig. 10, where 0 stands for parallel junction and 1 for series junction).

Within this formalism, a variable topology is reduced to variable junctions: allowing the network to change its connections amounts to allowing changes in the junctions. Thus, since the junctions are basically network elements, networks with variable junctions are not basically different from networks with variable elements.

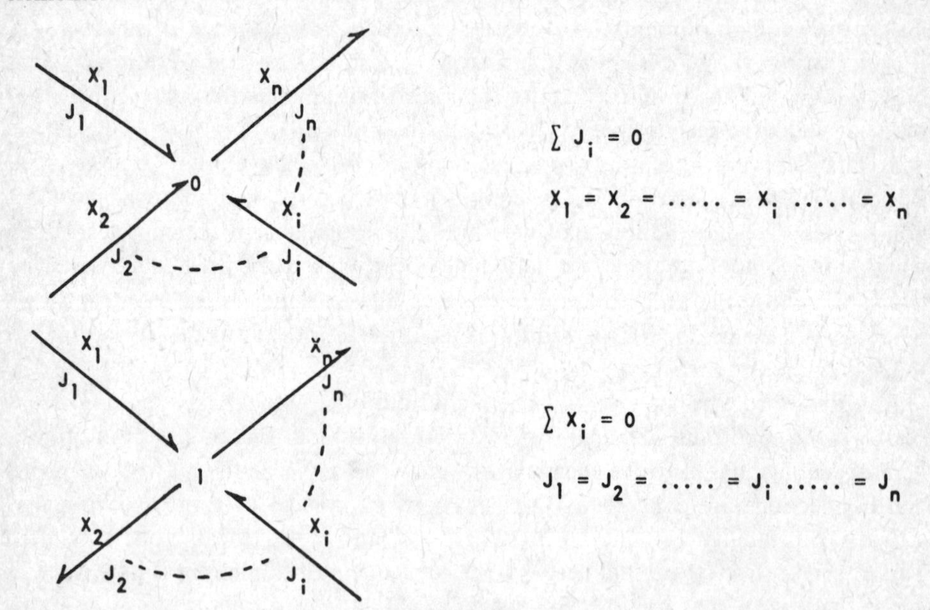

Figure 10
Junctions and corresponding Kirchoff laws in bond graph notation. (0 stands for parallel and 1 for series junction).

3.3. *The Tellegen theorem in chemico-diffusional networks*

We have demonstrated, with the help of the bond graph formalism, that a basic theorem of network analysis, the *Tellegen theorem* valid for time-variable non-linear networks, also applies to changing or evolving networks. This theorem states that, in a closed network, the sum of the products of all the forces X and the flows J through all the elements of the network is zero:

$$\sum_i X_i J_i = 0$$

This is also valid in an open network if it is transformed into a closed one by including the reservoirs in the analysis. The Tellegen theorem is valid for any network, with no restriction on the elements, which can be time-variable and non-linear, the only condition being that the Kirchhoff laws are satisfied. It can be shown, however, that the Kirchhoff laws are quite general, being in fact equivalent to the laws of conservation and assumptions of local equilibrium. By demonstrating this theorem within the formalism of bond graphs (Atlan and Katzir-Katchalsky, 1973), we showed that it remains valid also if the network topology is allowed to change, i.e., if parallel junctions are transformed into series junctions and even if new junctions appear, as long as the total number of one-port elements remains the same.

Thus the Tellegen theorem emerges as an invariance relation which can be applied not only to time-varying and non-linear networks, but also to evolving networks.

The special interest of this theorem, apart from its generality, lies in the fact that it is more than a trivial statement of conservation of power. This was originally shown by Tellegen himself, who called it a quasi-power theorem (Penfield, Spence and Duinker, 1970), since the conserved quantity is not merely the sum of products of conjugated forces and flows. In fact, it is not necessary that the forces be defined in the thermodynamic sense as differences in potential conjugated to the flows; the forces need only be differences in arbitrary values assigned to adjacent nodes of the network in space. In other words, we only require time derivatives, i.e., flows, and space derivatives, i.e., gradients; it is not stipulated that the gradients be the conjugated forces in the thermodynamic sense. This point is relevant here, since it can be applied to chemico-diffusional networks.

In a network represented in the form of a line, as in Fig. 9, we can write this theorem as follows. The diffusion flows between the boxes are taken as flows, and the values of the *chemical* flows at a given time in each box are taken as the potentials, so that in place of the differences in potentials conjugated to the diffusion flows we take the differences in chemical flows. The theorem is thus written as

$$\sum_r J^D \Delta J^R = 0,$$

where r is the box index. I am proposing this relation for any given diffusing and reacting compound in a chemico-diffusional network. Although it may appear a little strange, it is a direct consequence of the Tellegen theorem in its quasi-power form.

The problem of how to treat evolving networks is worth exploring if one wants to bridge between artificial chemico-diffusional networks and real organisms, even unicellular, viewed as natural, entire, evolving chemico-diffusional networks. Whether this approach gives an answer to the question of the origin of biological information—the origin of both intracellular and intercellular diversification—still remains to be seen.

4. The formal nature of self-organization

An analysis of the evolution of changing networks should make it possible to determine the conditions under which this evolution goes toward progressively increasing diversification, which implies progressively increasing structural information content in Shannon's sense, and, conversely, the conditions under which the evolution goes toward progressively increasing homogeneity or uniformity.

Since we are still far from such an analysis, I intend to devote the last section to some observations of a more formal nature on the possible origin of diversification in a non-externally-programmed system, which can be called a "self-organizing" system. However, before discussing self-organization, I will recapitulate how Ashby (1962) demonstrated that self-organization cannot exist.

It was shown that self-organization, in the strict sense of the organization of a system changing itself without being affected by the environment, is impossible. Indeed, in the most general way, organization can be defined as what makes a system in a state S_1, given inputs I, go to a next state S_2. In other words, organization acts as a mapping f of the set product $\{I\} \times \{S\}$ into $\{S\}$.

$$\{I\} \times \{S\} \xrightarrow{f} \{S\}$$

Now, changing organization would mean changing f. But it is impossible for the organization to change by itself, because if so the law of change of f would be in fact the organization of the system, and this law would be constant.

Therefore changes in organization can come only from the outside, so that self-organization would seem to be self-contradictory. But there are two possibilities for the changes to come from the outside. One is a set of instructions which constitute a program; the other is simply random perturbations (so-called "noise"). It is in the second case, when changes in organization (other than mere destruction) occur in response to random perturbations, that we speak of *self-organizing systems.* We thus come back, by a different route, to what is implicit (and sometimes explicit) in the ideas of M. Eigen on self-organization of matter.

Now, classically, it is known that in a noisy channel, noise can only decrease the amount of information transmitted in the channel. It therefore seems difficult to understand how noise-producing factors can be a source of increase in information. However, we showed a few years ago (Atlan, 1968, 1972) that in an organized system *two* kinds of channels should be distinguished, where the effects of noise are different as far as the rate of change in information content in the system is concerned.

The first kind of channel is between one state of the system at time t and the subsequent state of the system at time $t + \Delta t$. This channel ends up at the observer, and the information content of the system, which measures its degree of diversification (in terms of the degree of improbability that the system would be assembled in its precise form quite by chance), is in fact the output of this channel to the measuring observer. In this channel noise acts as a disorganizing agent by decreasing the transmitted information.

Channels of the second kind transmit information between the various parts of the system. Noise acting on these channels decreases the information transmitted from one part to another but, by the same token, increases the relative autonomy of the parts within the system. This results in an increased diversification or variety, i.e., an increased information content of the entire system as transmitted to the observer.

This idea has been applied to a simple system in which a communication channel exists between information carriers using different alphabets, like nucleic acids and proteins, with their four-letter and twenty-letter alphabets, respectively. This change in the alphabet with increase in the number of letters has proved to be a necessary condition for the positive effects of noise (what I have called the autonomy-producing ambiguity) to compensate and possibly overcompensate the negative effects of noise (what I have called the destructive ambiguity), so that the overall rate of change in the information content (i.e., in the diversity of the organized system) can be positive.

The general procedure to estimate the effects of noise on the information transmitted in a channel between input messages x and output y is as follows. The transmitted information H is equal to the difference

$$H = H_0 - H_n$$

Here

$$H_0 = H(y) = \sum_j p(j) \log_2 p(j)$$

is the information content of the output messages, given by Shannon's function of the probabilities $p(j)$ of the letters in the message; and

$$H_n = H(y|x) = -\sum_{ij} p(i)\, p(j|i) \log_2 p(j|i)$$

is the ambiguity produced by the noise, a function of the conditional probabilities $p(j|i)$ of the output letters given the input.

The following equation (Yockey, 1958)

$$\frac{dp(j|i)}{d\lambda} = -J(\lambda)\,p(j|i) + \frac{1}{N_j}J(\lambda)$$

relates these conditional probabilities—and thereby the ambiguity $H(y|x)$—to a "dose" of noise λ. N_j is the number of different letters in the alphabet of the output y, and $J(\lambda)$ an appropriate function. This equation permits writing a differential equation for the information transmitted in a channel as a function of time, where the effects of time are assumed to be those of all possible noise-producing factors.

This general procedure has been applied (Atlan, 1968) to the case with the two kinds of ambiguities, one with a *minus* and the other with a *plus* sign, acting on the information transmitted from the system to the observer, with two subsystems y_1, y_2 using alphabets with different numbers of letters N_j and $N_{j'}$. The basic equation is then

$$H = H(y_1) - H(y_1|x) + H(y_2|y_1)$$

By making use of Yockey's equations for $H(y_1|x)$ and $H(y_2|y_1)$ we end up, after integration, with a function of the form

$$H(t) = H_0 + \left(1 - \frac{1}{N_j}\right)\log_2(N_{j'}/N_j)(1 - e^{-J_0 t})$$

where the function $J(\lambda)$ is assumed to be a constant J_0, and the "dose" of error-producing factors λ is replaced by time t (see Atlan, 1968, 1972).

As a result, it appears that the direction of variation of $H(t)$ depends upon the ratio $N_{j'}/N_j$, so that H can be an increasing function only when $N_{j'} > N_j$, i.e., the number of letters of the proteins, say, is larger than that of the nucleic acids. In other words, this may give a possible explanation for the fact that has led to the discovery of the genetic code, but whose necessity has never been explained, namely the change with the increase in the number of letters of the alphabet which is observed when one goes from the four bases of the nucleic acids to the 20 amino acids of the proteins.

Later (Atlan, 1972, 1974) this was generalized to provide a theory of self-organization, i.e., changes in information content of systems under the effect of noise, with the possibility of initial increase followed by the more classical, although inevitable, decrease (Fig. 11). The period of self-organization, i.e., the initial increasing period, is accompanied by a decrease in redundancy up to a time t_M. Then, when the initial redundancy has been used up, the only possible effects of noise are destructive.

On this basis it has been proposed that the degree of organization of a system should be measured not by its information content, which provides a very poor

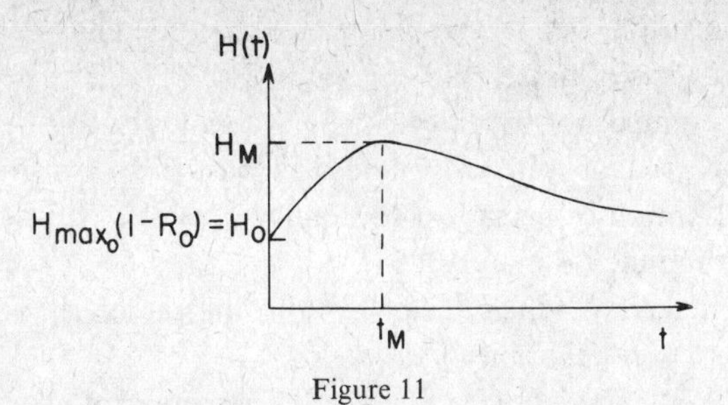

Figure 11

Variation of information content versus time with an initial increase figuring "self-organization." $H_{\max 0}$, R_0 and t_M are the parameters necessary to define the organization (text, and Atlan, 1972; Atlan, 1974).

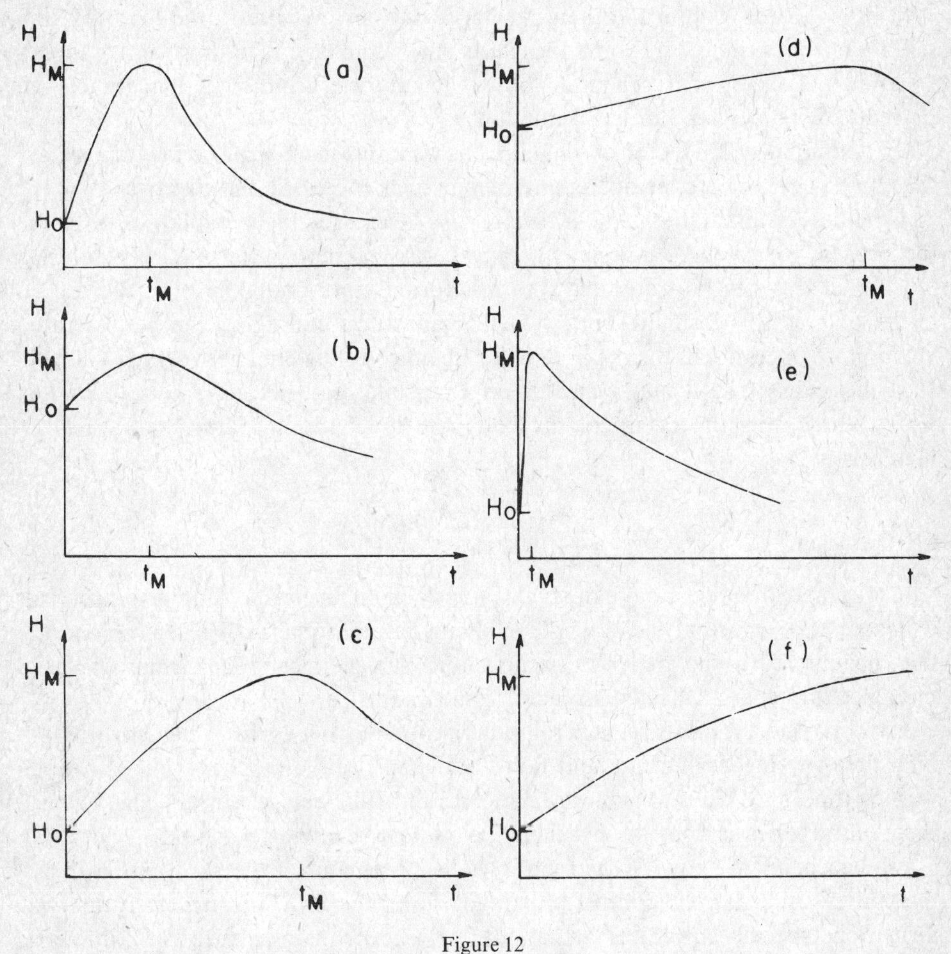

Figure 12

Different kinetics representing different kinds of organization (all of the self organizing type).

definition of organization, but by the kinetics of the time-change of the information content under the effects of noise. As is shown in Fig. 12, different kinds of organization are to be represented not by different numbers of bits, but by different kinetics. To characterize a particular kinetics, we need at least three parameters (see Fig. 11):

1) the initial maximum information content $H_{\max 0}$
2) the initial redundancy R_0, and
3) a time parameter t_M which measures the inertia of the system to changes produced by noise, i.e., a reliability factor.

The first of these parameters ($H_{\max 0}$) has the meaning of structural complexity; the last parameter, the reliability factor t_M, has the meaning of functional stability; the second parameter, the redundancy R_0, has a mixed structural and functional meaning, as is well known from automata studies.

In other words, one of the main results is that an increase in the diversity of a self-organizing system, e.g., under the effects of random factors feeding a spontaneous "learning" of the system, can occur only when starting from something which can be reduced, namely the initial redundancy.

But redundancy is a factor of reliability, as we know from error-connecting coding theories. Therefore, reducing redundancy, which increases the diversity and the possibilities of adaptation and learning, also decreases the reliability and brings the process to an end—unless a jump generates a new initial high redundancy.

How all this can be translated into less formal, more tangible physicochemical terms, or how this formal theory of self-organization and aging can be translated into a more operational theory of evolving chemico-diffusional networks is a subject of further research. If I may conclude on a personal note, the ambition to conduct this research was the origin of my collaboration with the late Aharon Katzir-Katchalsky.

5. Conclusion

Applying network thermodynamics to biology helps in formulating new problems encountered in the analysis of whole evolving natural systems. These problems are not encountered in the analysis of artificial physicochemical systems, nor of partial biological systems artificially isolated from organisms.

Network theory must be generalized to evolving networks which involve not only time-varying coefficients and non-linear constitutive relations, but also time-varying topology. Ways to achieve such generalization were discussed. The Tellegen theorem was shown to be an invariance relation for evolving networks.

Physicochemical network theory allows the treatment of the problems of transmission of information (control and regulation) not only through structural information (specific structures acting like switches), but also through parametric control, i.e., modulation of network parameters. The systems thus become more flexible and more efficient, since transmission of information via changes in *rates* (of reactions and/or transports) is not limited to all-or-none processes. Examples of such transmission of information by changes in rates can be found in recent works showing that cell membrane transport properties can control specific steps in DNA and protein synthesis. However, whereas parametric control is sufficient to treat artificial networks, or portions of natural networks artificially isolated, difficulties are encountered when it is applied to whole biochemical networks. This is due to confusion between usually separated signal flows and energy flows in closed chemical networks, where both flows utilize changes in concentrations as physical supports.

The problem of sources of information (i.e. blueprints and programs), which is trivial in artificial—man-made and programmed—networks, is much less straightforward in natural, informationally closed, networks, where one is forced to use such strange concepts as internal program and program identified with the machine. The "order-from-noise principle" of McKay, Bateson and Von Foerster, which has forseen (albeit in a vague way) the role of randomness in organization and the fact that the only source of new information is noise was established more precisely. This provides a clue for a formal understanding of how new information (in the sense of Shannon) or diversification can be created. The role of random fluctuations as a source of information through reduction of redundancy was stressed in the logic of (apparently) spontaneous diversification processes.

References

ASHBY, W. R., 1962. "Principles of the Self-organizing System." In *Principles of Self Organization* (H. VON FOERSTER and G. W. ZOPF, eds.), Pergamon Press, pp. 255–278.

ATLAN, H. and WEISBUCH, G., 1973. "Resistance and inductance-like effects in chemical reactions: influence of time delays." *Israel J. Chem.* **11**:479–488.

ATLAN, H. and KATZIR-KATCHALSKY, A., 1973. "Tellegen's theorem for bond graphs. Its relevance to chemical reactions." *Curr. Mod. Biol.* **5**:55–65.

ATLAN, H., 1968. *J. Theoret. Biol.* **21**:45.

ATLAN, H., 1972. *L'organisation biologique et la theorie de l'information*. Paris: Hermann.

ATLAN, H., 1974. *J. Theoret. Biol.* **45**:295–304.

BRILLOUIN, L., 1956. *Science and Information Theory*. New York: Academic Press.

CONE, C.D., 1971. *J. Theoret. Biol.* **30**:151–181.

FIELD, R. F., KÖRÖS, E. and NOYES, R. M., 1972. "Oscillations in chemical systems II." *J. Amer. Chem. Soc.* **94**, 25:8649–8664.

FREUDENBERG, H. and MAGER, J., 1971. "Studies of the mechanism of the inhibition of protein synthesis induced by intracellular ATP depletion." *Biochim. Biophys. Acta* **232**:537–555.

HANSEN, J. N., SPIEGELMAN, G. and HALVORSON, H. O., 1970. "Bacterial spore outgrowth: its regulation." *Science* **168**, 3937:1291–1298.

HERZBERG, M., BREITBART, H. and ATLAN, H., 1974. *European J. Biochemistry* **45**:161–170.

LOSICK, R., SHORENSTEIN, R. G. and SONENSHEIN, A. L., 1970. "Structural alteration of RNA polymerase during sporulation." *Nature* **227**:910–913.

McDonald, T. F., Sachs, H. G., Orr, G. W. and Ebert, J. D., 1972. *Developmental Biology* **28**:290–303.

Noyes, R. M., Field, R. F. and Körös, E., 1972. "Oscillations in chemical systems I." *J. Amer. Chem. Soc.* **94**:4, 1394–1395.

Orr, C. W., Yoshikawa-Fukada, M. and Ebert, J. D., 1972. *Proc. Nat. Acad. Sci. USA* **69**:243–267.

Oster, G., Perelson, A. and Katchalsky, A., 1971. "Network thermodynamics." *Nature* **234**, 5329: 393–399.

Oster, G., Perelson, A. and Katchalsky, A., 1973. "Network thermodynamics: dynamic modelling of biophysical systems." *Quart. Rev. Biophysics* **6**:1–134.

Penfield, P., Jr., Spence, R. and Duinker, S., 1970. *Tellegen's Theorem and Electrical Networks.* Cambridge, Mass.: The MIT Press.

Shapiro, L., Azabian-Keshishian, N. and Bendis, I., 1971. "Bacterial differentiation." *Science* **173**: 884–892.

Sonenshein, A. L. and Losick, R., 1970. "RNA polymerase mutants blocked in sporulation." *Nature* **227**:906–909.

Turing, A. M., 1952. "The chemical basis of morphogenesis." *Phil. Trans. Roy. Soc.* **B 237**:37–72.

Noyes, R. M., Field, R. F. and Körös, E., 1972. "Oscillations in chemical systems I." *J. Amer. Chem.*

Weisbuch, G., Salomon, J. and Atlan, H., 1975. "Analyse algébrique de la stabilité d'un système à trois composants tiré de la réaction du Zhabotinsky." *J. Chim. Phys.*

Yockey, H. P., 1958. In *Symposium on Information Theory in Biology* (H. P. Yockey, R. L. Platzman and H. Quastler, eds.). New York: Pergamon Press, pp. 50–59 and 297–316.

Evolutionary games

M. EIGEN

Max Planck Institute of Biophysical Chemistry, D-3400 Göttingen-Nikolausberg, Germany

The origin of life is tautologous with the origin of biological information.

How can information in general originate?

The term information has two aspects: the probabilistic aspect of absolute information, and the semantic aspect of the meaning or value of information. The "semantics" of biological information must refer to those properties of matter which guarantee the maintenance of the functional organization of the particular state of matter we call "alive."

Classical information or communication theory (Shannon and Weaver, 1949), a branch of applied mathematics, deals with the first aspect. It relates the amount of information of a given set of symbols or digits to the probability distribution for all their possible combinations. More specifically—according to Renyi (1961)—it provides a quantitative measure of the "gain of information" associated with any change of the probability distribution conditioned by certain clues of observation.

Changes of probability distributions may also occur within a stationary system of matter as a consequence of instabilities brought about by random fluctuations (Eigen, 1971). Such a system must possess filtering or selection properties which are inherent in the material and controlled by certain environmental conditions involving a finite flow of free energy (Shrödinger, 1944).

Which are the prerequisites of such a material of selforganization, generating the information for its own maintenance and further evolution?

The systems may be categorized according to their dynamic response to random fluctuations occurring in the population variable. Three prototypes are discussed:

1) Stable systems, in which fluctuations control themselves. The dynamic response to a fluctuation is always such that it reduces the fluctuation.

2) Drifting systems, in which the population variable has no influence on the dynamic properties which fluctuate randomly without any control.

3) Instabilities, which are characterized by a selfamplification of fluctuations of a certain nature.

The stability criteria of Lyapunov (Zubov, 1964) can be applied to an analysis of these prototypes and their various combinations. The principal behavior can also be simulated by games which are discussed.

The first game corresponds to Ehrenfest's urn model (1907), representing stable distributions among a given number of states, the probability distributions being Gaussian. The second game, resembling drift behavior, is a certain version of "coin tossing" or "head or tail."

The third game—as a prototype of a whole series of non-deterministic "life-games" (Eigen and Winkler, 1973)—describes Darwinian behavior. It involves "instabilities" as a consequence of certain random fluctuations. These fluctuations amplify and hence appear at the macroscopic level. It is such a kind of behavior which proves to be suitable for the generation of information expressed in the "semantics" of functional organization. The "gain" of information occurs via discontinuous changes of the probability distribution. It originates from "random noise" which is evaluated by the dynamic properties of the system involving inherent selfreproduction as well as steady consumption of free energy in order to compensate for the steady entropy production. The decisive factors are the rate parameters for formation and decomposition, which must refer to—at least—two different reaction forces (affinities) in order to maintain the flow of free energy (metabolism). The rate term of formation has to be of a special nonlinear form in order to allow for nucleation of information, maintenance and further evolution. The accuracy of reproduction can be quantitatively related to the free energies of the interactions involved in the molecular recognition process. The thermal nature of this process provides the source of "random noise" from which certain fluctuations can be selected, causing a breakdown of the former steady state. The evolutionary process then represents a series of instabilities guided by certain optimization principles. Darwin's (1872) principle of natural selection appears as a derivable optimization principle bound to certain physical requirements. "Natural selection" therefore cannot be considered an irreducible phenomenon confined to the biosphere alone. In fact, it is based on more fundamental principles of non-equilibrium thermodynamics, in particular the stability criteria derived by Glansdorff and Prigogine (1971). Various molecular models have been tested on the basis of the dynamic theory of selection. It can be shown that the "formation of viable structures" (probably forerunners of unicellular organisms) is connected with special conditions of the selection mechanism, which cannot be fulfilled by the nucleic acids or the proteins alone. The selection mechanism, which is based on the reproduction process, itself has to be of non-linear nature. Both nucleic acids and proteins can reproduce themselves in a "quasi-linear" reaction mechanism. However, at the same time, there arise systems which contain too little information, as in the case of nucleic acids, or too much, as in the case of proteins.

"Too little" signifies that the sequences competing with each other are not in a position to collect enough "reproducible" information for the coding of selection-favorable functions. On the other hand, too much information means that the probability for a self-promoting mutation becomes too small or that the system is unable to free itself from the network of "parasitic" couplings. But it is also possible that a system containing nucleic acids as well as proteins could utilize the functional advantage of both types of substances for a stable selection. The advantageous characteristics are:

1) The inherent capability of self-instruction of the nucleic acids, with the help of which it is possible to reproduce not only each fully developed information state but also any further change.

2) The enormous functional capacity of the proteins (recognition, catalysis and regulation), which is indispensable for the coupling and correlation of single reactions in the synthesis of organized functional units.

The hierarchy of reaction cycles resulting from such coupling between nucleic acids and proteins and exhibiting non-linear growth kinetics already displays the essential characteristics of a "living" system and is "open" for further evolution up to the living cell. The origin of such a self-reproducing "hypercycle" (Eigen and Schuster, in preparation) depends on the development of a specific code system with unique assignments. Various assumptions can be tested by evolution experiments as first carried out by Spiegelman (1970) with Qβ-phages. The theory developed in the present paper provides the basis for carrying out reproducible measurements and their quantitative evaluation.

The results can be summarized thus:

1) Detailed analysis of the reproduction mechanism of nucleic acids and proteins does not offer any support for the hypothesis that life phenomena involve any forces or interactions not otherwise known in physics (Watson, 1970). The selection behavior which is characteristic for the evolution of living systems appears already at this stage as an inherent property of certain matter under special reaction conditions.

2) Every system obtained by mutation and selection is indeterminate with regard to its individual structure; nevertheless the resulting process of evolution is determinate, i.e., it is regulated by physical laws. The occurrence of a mutation with selective advantage corresponds to an instability and as such can be explained with the aid of Glansdorff and Prigorgine's principle for stationary irreversible thermodynamic processes. The optimization process of evolution is therefore inevitable but, with regard to the selection of the individual route, undetermined.

3) Finally, the origin of life depends on a series of characteristics which collectively can be clearly defined physically. The requirements for the formation of these characteristics have probably been fulfilled step by step so that the "origin of life," like the evolution of the species, cannot be represented as a single act of creation.

4) The very peculiar interplay of chance and law—as opposed to absolute chance

and complete determinacy (Gardner, 1971)—is the characteristic feature of the "life game," appearing at various levels of selforganization: in the interplay of macromolecules to form the cell; in the interplay of cells to form the organism; and in the interplay of cell circuits in the neural network to form our thoughts.

References

DARWIN, C., 1872. *The Origin of Species.* London (reprinted 1959, London: Collier MacMillan Ltd.).

EHRENFEST, P. and EHRENFEST, T., 1907. *Phys. Z.* **8**, 311.

EIGEN, M. 1971. *Naturwiss.* **58**, 465.

EIGEN, M. and WINKLER, R., 1973. Mannheimer Forum.

EIGEN, M. and SCHUSTER, P. *J. Mol. Evol.* (in preparation).

GARDNER, M., Febr. 1971. *Sci Amer.*, p. 112.

GLANSDORFF, P. and PRIGOGINE, I., 1971. *Thermodynamic Theory of Structure, Stability and Fluctuations.* New York: Wiley-Interscience.

RENYI, A., 1961. *Fourth Berkeley Symposium on Mathematical Statistics and Probability* (J. NEYMAN, ed.). Berkeley: University of California Press, p. 547.

SHANNON, C. E. and WEAVER, W., 1949. *The Mathematical Theory of Communication.* Urbana: University of Illinois Press.

SPIEGELMAN, S., 1970. In *The Neurosciences, 2nd Study Program.* New York: The Rockefeller Press.

SCHRÖDINGER, E., 1944. *What Is Life?* Cambridge University Press.

WATSON, J. D., 1970. *The Molecular Biology of the Gene.* New York: W. A. Benjamin Inc. (Recommended for introduction to facts in molecular biology.)

ZUBOV, V. I., 1964. *Methods of A. M. Lyapunov and Their Applications.* Groningen: Noordhoff Ltd.

The extracellular evolution of structure in replicating RNA molecules

S. SPIEGELMAN, D. R. MILLS AND F. R. KRAMER

Institute of Cancer Research
Department of Human Genetics and Development
College of Physicians & Surgeons, Columbia University, New York, New York 10032, U.S.A.

Introduction

It is plausible to assume that nucleic acids were the first chemical entities to emerge prebiotically that enjoyed and exploited the potential for evolution. However, it is far from evident what forces drove the replicating nucleic acids to evolve towards greater complexity, a necessary prelude to the invention of the cell. The present paper concerns this central issue and seeks an answer to the following question: What evolutionary potentials are possessed by replicating nucleic acid molecules *before* their sequences code for information translatable into proteins?

Heritable variations that do not express themselves as *phenotypic* differences *recognizable* by the environment cannot result in natural selection. The key question may therefore be posed in the following terms: How many *distinguishable* phenotypes can a non-translated but replicating nucleic acid molecule exhibit to its environment? If this number is very limited, then so is its potential for evolution. If it is large, conditions for evolving towards greater complexity can be readily discerned.

An opportunity to explore these and allied questions experimentally has been provided by the isolation of a purified enzyme capable of mediating extensive and continuous replication of RNA. Before proceeding to a description of the experiments thus made possible, it is instructive to begin with a more general discussion of the evolution of replicating entities, their essential properties and relation to "living" and "nonliving" matter.

The gene and the concept of "self-duplication"

We begin with a discussion of the minimal conditions required for evolution and the central contributions made to this problem by geneticists at the beginning of the present century. A more profound analytical discussion of many of these issues will be found in Manfred Eigen's contribution to this symposium. Here we focus only on those aspects relevant to an understanding of the rationale underlying the experiments we will describe.

A necessary condition for evolution is the existence of an entity capable of "self-duplication" or "self-replication." We use these terms because of their historical prevalence. They are semantically unfortunate, because they have generated a plethora of unnecessary difficulties. To some, these terms have implied that such entities must be able to function by themselves, unaided by other devices, in the production of copies. This line of reasoning led to holistic arguments in which the cell was the smallest unit capable of self-duplication. Such arguments are in accord with pregenetic thinking, which held that the cell and its capacities to produce copies of itself were consequences of a concatenation of interlocking reactions that served to maintain the phenotypic characteristics from one cell generation to the next.

With the recognition of mutations and their implications, this view became increasingly difficult to maintain. To support it, one would have to argue that a single modification in the complex of interacting reactions so modified the whole as to maintain and transmit this unique event, a viewpoint tenable logically, but implausible practically. Further, if this view were correct, unraveling the details of the duplicating mechanism would constitute an inordinately difficult task. In contrast, the geneticists offered a more acceptable alternative. They proposed the existence of cellular entities called *genes*, which alone possessed the capacity of self-duplication. All cellular properties, including self-duplication, were the consequence of the autosynthetic and heterosynthetic instructive capabilities of the genetic material. The implications of the geneticists were therefore both more optimistic and more understandable; they were saying that, if we ever identify the chemical nature of the gene, and thereby gain an understanding of its mode of replication, we will have understood the essence of replication. This prediction was fully confirmed with the announcement of the Watson–Crick model of DNA.

An operational definition of "self-duplication"

To be useful to biologists, one must extend the concept of self-duplication beyond that of copy generation. Any copying mechanism that is *absolutely* faithful has no evolutionary future and therefore serves no purpose as a biological model. Selection cannot operate on identical copies and as a consequence the potential for evolution demands variation. The existence of variants that permit selection demands a source of variability, which ultimately must come from inaccuracies in the copies. These inaccuracies can, in principle, occur either in the copying process or by modification of some of the copies. Whatever their origin, the changed copies must *retain* the capacity of replication if their occurrence is to have evolutionary consequences.

A usable operational definition of a self-duplicating entity has a similar requirement. It is important to recognize at the outset that self-duplication always involves two components. One is the entity being copied and the other is the copying machine. Essentially, we want to know which one of these two components is the *instructive* agent in the replicating process. It is conceivable that the copying machine is designed to turn out copies only of the particular entity being considered, the duplicated object serving merely as a stimulus for a preprogrammed synthetic activity. Were such the case, the entity being copied would not be the instructive agent in the duplicating process, and any mistakes made in the copying process would not be transmitted to future generations.

By a self-duplicating entity, we mean a self-instructive agent that carries the information in its structure for its own synthesis and can transmit this information to a passive synthesizing machine, which fabricates the kinds of copies dictated. A requirement for an operational definition of a self-instructive replicating agent is the availability of at least two distinguishable variants. Each is then given to the synthesizing machine.

If the product produced is always the same and independent of the particular entity used to initiate the synthesis, we do not have a self-duplicating object. On the other hand, if the copies generated by the synthesizing machine always correspond to the *initiating* variant, we have satisfied the operational definition of an object capable of self-instructive replication. We see, then, that evolutionary potential and a useful operational definition of the class of self-duplicating objects both require a source of variants.

To summarize, we define a self-duplicating—or, better, a "self-instructive replicating" (SIR)—object as one that, in the presence of a suitable synthesizing machine, possesses the following two properties: 1) at least one copy is required to get another, and 2) variants are occasionally produced and can function as instructive agents to initiate copies of themselves.

It should be noted that nothing is said about whether or not the synthesizing machine is also copied. Both situations satisfy our definition. In most biological systems, the copying machines are also copied, but other devices (e.g., for protein synthesis) are required for *their* fabrication, and all the information needed to synthesize these machines and their ancillary devices must be contained in the chemical *structure* of the self-instructive replicating object. These and other issues can be seen more clearly in terms of comparatively simple objects we consider below.

"Living" and "nonliving," "self-instructive" replicating objects

It is important to emphasize that our operational definition of self-instructive replication is not meant to provide an operational device to distinguish "living" from "nonliving" matter. In practice, it is not clear that an operational distinction is possible; we will return to this question later.

To make the central issue clear, it is instructive to consider some examples of objects that satisfy the two conditions specified and that are not generically derivable from what we would all accept as "living" material.

A somewhat trivial case is provided by Xerox copies in a universe of Xerox machines. One needs one copy to get another. Further, if a particular copy is modified and then fed back into the Xerox machine, the modifications will be reproduced in future copies. Of course, the permissible changes here are limited to those that can be made on paper and that can, in turn, be Xeroxed. All copying devices, however, have similar restrictions of greater or lesser complexity. This system clearly has an evolutionary potential, but of a relatively limited nature.

The Xerox example gains in interest to us, and greatly increases its evolutionary range, when we recognize that the material being copied can, in fact, contain information. In particular, this information could constitute the set of directions for making Xerox machines. We now have a potentially more interesting relation between the Xerox machine and the copies it is generating. If variant replicas are produced and these used to generate copies employed by engineers to build new Xerox machines, the consequence will be an evolution of the Xerox machines themselves. This sort of extension brings us close in principle to another class of devices, the "self-duplicating automata" conceived of by von Neumann (1966).

The von Neumann machines represent logical extensions of devices commonly found in the automatic tool-making industry. In principle, it is possible to devise a machine that can make copies of itself. Such a machine can be placed upon a pile of its component parts. The machine then selects the one required at the particular stage in the manufacture. So that the operations leading to the building of the new machine may contain no inconsistencies, the set of instructions can be put serially on a tape. In such a way, one avoids the possibility that the machine will attempt to do something physically impossible. Thus, if the machine must put a bolt in a hole, it is important that the instruction for drilling the hole precedes that for the insertion of the bolt.

After the set of instructions leading to the fabrication of the new machine is carried out, one will find a replica of the original machine sitting beside the old one. This is a copying device, but it does not have the essential property of transmitting mistakes, which we require of a self-instructive replicating object. Suppose, however, that subsequent to the instructions for the fabrication of the new machine, we give another set of instructions that say, "Make a copy of me, the tape, and then transfer it to the new machine." Once this last set of instructions is completed, the newly manufactured machine contains the set of instructions required to fabricate the next one. We now have a self-instructive replicating device in which everything, including the copying mechanism, is copied.

Imagine that a biologist were to come upon a group of machines such as these busily engaged in making new ones; if he were a geneticist, he would look for variants in the machines (e.g., an abnormal positioning of a hole on a particular plate). Once it was identified, he would isolate it and see if the next generation made by the variant machine would, in fact, possess the modification. If the new machine produced by the variant did not exhibit the variation, the geneticist would label it a "phenotypic" change and would conclude that the mistake was not made in *copying* the set of instructions, but rather in *following* the instructions. If, on the other hand, the variation *was* transmitted to the next generation, he would identify the variation as genetic, i.e., the mistake was made in copying the set of instructions transmitted to the new machine.

It should be noted that such machines could be instructed to undergo a process akin to that of recombination. Two machines could approach each other and align their tapes so that the two ends coincided. The tapes could then be compared and, if differences were found, the tapes could be cut at some point between the differences and rejoined so that each now possessed a portion of the other's tape.

It seems clear that, if a geneticist carried through the kinds of operations on these machines that he does with living matter, he would identify the tapes as the genetic material. The instructional tapes would satisfy his definition of self-duplicating entities, and errors in tape copying would be identified as genetic mutations.

We now come to the question of the relation of von Neumann machines, and other similar devices, to living material. It is of more than passing interest to note that any mechanism involving the transfer of information from one generation to the next can possess the basic feature of a self-instructive replicating entity. In fact, this feature has generated one of the most important difficulties facing man. When the first man scratched a piece of information on the wall of a cave, he invented a system that has its own genetics. Furthermore, the resulting evolution is *directed* by its creator and, as such, is far faster than the selection of random mutations, which has thus far characterized the biological evolution of man. As an example, one might cite the speed with which the memory capacity of computing machines has evolved in the past decade and a half. To achieve the same quantitative improvement by biological evolution would have required hundreds of millions of years. This dramatic difference in the rate of man's evolution compared to the evolution of the devices and artifacts he invents poses a central dilemma for man and the chances of his survival.

Although we are willing to place both the von Neumann self-duplicating automata and living material in the class of self-instructive replicating agents, there is an important difference of origin. We entertain as plausible the hypothesis that the archetype of living objects originated from nonreplicating subunits by polymerization into chains capable of making copies of themselves by a mechanism that initially employed only their own chemical reactivities. We see today, at least dimly, how the same sort of reactions that generate new copies of DNA enzymatically could have taken place nonenzymatically by chemical reactions, which we understand reasonably well. Thus, we are willing to grant the *spontaneous* origin of *living*, self-instructive replicating entities. It is difficult to see how the same sort of thing could occur with the tapes of the von Neumann automata. In other words, the basic difference is that, in the inanimate case, we are virtually forced to invoke an act of creation by some intelligent agent. The origin of living material demands no such postulate.

Precellular evolution

We now come to the central issue of our discussion, namely, the evolutionary events that precede the appearance of cells. As we have already noted, we assume that the first self-duplicating objects were genes, i.e., DNA or RNA replicating by means of the hydrogen bond recognition device. These are the only cellular macromolecules we know that contain in their chemical structures the inherent potential of carrying through a copying process. A given polynucleotide strand can, in principle, serve as a catalytic surface to attract complementary activated nucleotides that would be automatically aligned in a complementary array on the surface of the strand, ready for subsequent polymerization. That this can happen without the aid of protein catalysis is indicated by the recent experiments of Orgel and his collaborators (Sulston et al., 1969).

Cells, as we know them, can be looked upon as inventions of nucleic acids to provide themselves with a local environment optimally suited to provide the materials and conditions required for nucleic acid replication. Similarly, the evolution of the multicellular animals and plants can be interpreted as devices evolved to permit DNA to exploit all terrestrial space, including the land, the seas, and the air. Until the last few years, one might well have wondered why DNA invented man. It is now evident that man was invented to provide DNA with the opportunity to explore extraterrestrial possibilities for replication.

We have already mentioned that one of the most difficult evolutionary phases to understand is the period of precellular evolution when the environment was looking not at gene products, but at the genetic material itself in making decisions on which variant was more fit for survival.

An experimental approach to such problems would be available if a nucleic acid replicating mechanism could be isolated so that the replication of genetic material could be studied in the test tube. If this could be achieved, in terms of extensive and continuous synthesis, one could begin to vary the environmental parameters in an attempt to uncover the kinds of selective forces that can be imposed directly on replicating genetic material.

In the biological universe, there are two types of organisms; one uses DNA and another RNA as genetic material. The RNA class was the first, and to this day remains the only one, to yield the sort of extensive and unlimited *in vitro* synthesis required for the extracellular exploration of evolutionary problems. The probable reason is that the DNA duplicating mechanism has, through evolution, become so inextricably intermingled with repair and editing devices that the isolation of the enzymatic components required for *indefinite* extracellular replication has thus far defied all efforts.

The event that stimulated our group to institute an effort to study the RNA replicating system was the discovery by Loeb and Zinder (1961) of the RNA bacteriophage, f2. This was rapidly followed by other related phages such as MS2 (Strauss and Sinsheimer, 1963), R17 (Paranchych and Graham, 1962). The armamentarium accumulated during several decades of T-phage technology thus became available to those interested in the RNA viruses. Some of the difficulties and disadvantages inherent in the use of plant and animal systems could now be by-passed, and a number of laboratories, including our own, quickly took advantage of the opportunities provided.

In what follows I describe briefly the key experiments that led to the successful isolation of the RNA replicating system, and summarize our efforts to study the extracellular evolution that it made possible.

The problems of communication between an RNA virus and its host cell

All organisms that use RNA as their genomes are mandatory intracellular parasites. Therefore, they must carry out a major portion of their life cycle in cells that use DNA as genetic material and RNA as genetic messages. On entry, the viral RNA is faced with a problem of inserting itself into the flow pattern of cellular information in order to communicate its own instructions to the synthesizing machinery. A possibility one might entertain centers on whether an RNA virus employs the host DNA-to-RNA-to-protein pathway of information flow. This could take place either because the DNA of the host already contains a sequence homologous to the viral RNA (i.e., the escaped genetic message hypothesis) or because such DNA sequences are generated subsequent to infection by a reversal of transcription (i.e., of the RNA synthesizing reaction that is dependent on DNA). Both hypotheses predict that the DNA of infected cells should contain sequences complementary to viral RNA.

It is clear that a decision on the existence or nonexistence of homology between viral RNA and the infected host DNA is a necessary prelude to further experiments designed to delineate the molecular details of the life history of an RNA genome.

In an attempt to settle this issue, we (Doi and Spiegelman, 1962) employed the specific DNA-RNA hybridization test (Hall and Spiegelman, 1961) combined with the subsequently developed (Yankofsky and Spiegelman, 1962a) use of ribonuclease to eliminate nonspecific pairing. The sensitivity required had already been attained in the course of experiments that identified the DNA complements of ribosomal RNA (Yankofsky and Spiegelman, 1962a, b) and tRNA (Giacomoni and Spiegelman, 1962; Goodman and Rich, 1962). Under conditions in which the expected hybrid complexes were observed between 23*S* rRNA and DNA, none was detected between the RNA of the bacteriophage MS2 and the DNA derived from cells infected with this virus. The viral RNA used in these experiments was labeled with ^{32}P at a specific activity which would have permitted hybrid detection even if the DNA had contained only one-tenth of an RNA equivalent per genome. This insured a meaningful interpretation of the negative outcome.

It must be recognized that, in a logically rigorous sense, all the evidence provided is negative and, as such, cannot be used to eliminate a proposed mechanism in a global biological sense. Nevertheless, the absence of any evidence of DNA involvement was generally accepted to imply that RNA bacteriophages had evolved a DNA-independent mechanism of generating RNA copies from RNA. The next obvious step was to identify and then to isolate the new type of RNA-dependent RNA polymerase (replicase) predicted by this line of reasoning.

The search for RNA replicase and template specificity

The search for a replicase unique to cells infected with an RNA virus is complicated by the presence of a variety of enzymes that can incorporate the ribonucleotides either terminally or subterminally into preexistent RNA chains. Further, there are others (e.g., transcriptase, RNA phosphorylase, polyadenylate synthetase) that can mediate extensive synthesis of polynucleotide chains. The complications introduced by these and other enzymes can be obviated by suitable adjustments of the assay conditions to minimize their activity. It was clear at the outset, however, that claims for a new type of RNA polymerase would ultimately have to be supported by evidence for RNA dependence and a demonstration that the enzyme possesses some unique characteristic that differentiated it from the known RNA polymerases.

In addition to these enzymological difficulties, we recognized another potential complication inherent in the fact that an RNA virus must always operate in a heterogenetic environment replete with strange RNA molecules. The point at issue may perhaps best be described in rather naive and admittedly anthropomorphic terms. Consider an RNA virus approaching a cell some 10^6 times its size and into which the virus is going to inject its only strand of genetic information. This strand codes for the new kind of enzyme required for the replication of the viral genome. Even if the protein-coated ribosomal RNA molecules are ignored, the cell cytoplasm still contains thousands of RNA molecules of various sorts. *If the new RNA replicase were indifferent and copied any RNA molecule it happened to meet, what chance would the original strand injected have of multiplying?*

Admittedly, there are several ways out of this dilemma. One could somehow sequester the new enzyme and the viral RNA so that replication could occur undisturbed by the mass of the cellular RNA components. Because it had important experimental consequences, however, we entertained the then unusual hypothesis that the virus is ingenious enough to design a replicase that would recognize only the genome of its origin and ignore all other RNA molecules.

Initially, we could not know which solution had been adopted by the virus to solve the dilemma we posed, or even if the dilemma were real. The possibility that it did exist, however, and that template selectivity by the replicase might be the chosen solution demanded that its implications should not be ignored; *for if true, to disregard them would guarantee that the attempt to isolate the relevant enzyme would inevitably end in failure.* In particular, this meant that, in the search for replicase, we could not afford the luxury of employing any RNA conveniently available in the usual biochemical laboratory. It required the use of the specific viral RNA as the challenging template at all stages of enzyme purification. Finally, this line of reasoning could be pushed to its ultimate pessimistic conclusion. It might indeed be that fragments do not possess the proper recognition structures; in that case a further demand would have to be imposed, that the viral RNA employed in the assays during enzyme purification must be monitored for its size to insure intactness. This meant that assays in the presence of contaminating nucleases might be completely meaningless.

The MS2 replicase

Despite all these potential obstacles, many of which were actually realized, our first success was achieved in 1963 with *E. coli* infected with the MS2 phage (Haruna et al., 1963). A procedure involving negative protamine fractionation combined with column chromatography yielded what resembled a relevant enzyme activity. Most important of all, the preparation exhibited a virtually complete dependence on added RNA for activity, permitting a test of the expectation of specific template requirements. The results of these tests are summarized in Table 1. It will be seen that the responses of the MS2 replicase to the various kinds of nucleic acids revealed a striking preference for its own RNA. No significant response was observed with the tRNA or the ribosomal RNA of the host cell. It would appear that our intuitive guess was confirmed. Producing a polymerase that ignores cellular RNA components guarantees that replication is focused on the single strand of the incoming viral RNA, the ultimate origin of progeny.

The announcement of the specific template requirement of the RNA replicase was greeted with what may best be described as "well-controlled enthusiasm." It should

Table 1

TEMPLATE SPECIFICITY OF PURIFIED RNA-DEPENDENT POLYMERASE

(Haruna et al., 1963)

The standard reaction of 0.25 ml contained the following in μM: Tris–HCl pH 7.5, 21; $MgCl_2$, 1.4; $MnCl_2$, 1.0; KCl, 3.75; mercaptoethanol, 0.65; spermine, 2.5; phosphoenolpyruvate (PEA), 1.0; $(NH_4)_2LO_4$, 70; CTP, ATP, GTP, and UTP, 0.5 each. In addition, it contained pyruvate kinase (PEA kinase), 5 μg; DNase, 2.5 μg and, where indicated, 10 μg of the polynucleotide being tested as template. Enzyme was assayed at levels between 50 and 300 μg of protein per sample. DNase was always omitted in assays for DNA-dependent polymerase activity. Incubations were carried out at 35°C for 10 min and terminated by placing the reaction mixture in an ice bath and by the addition of 0.15 ml of neutralized saturated pyrophosphate, 0.15 ml of neutralized orthophosphate, and 0.1 ml of 80% trichloracetic acid (TCA). The precipitate was washed onto a millipore filter and washed five times with 10 ml of cold 10% TCA containing 0.9% of Na pyrophosphate. The millipore membrane was then dried and counted in a liquid scintillation counter. The pyrophosphate is included in the wash to depress the zero time backgrounds to acceptable levels (40–70 cpm per sample containing input counts of 1×10^6 cpm).

Template all at 10 γ/0.25 ml	NT incorporated in mμM/min/mg protein
0	0.08
MS2 RNA	8.5
sRNA	0.09
Ribosomal RNA	0.06
Ribosomal RNA + MS2 RNA	8.0
TMV RNA	0.3
TYMV RNA	2.2
CT–DNA*	0.11

*DNase omitted from assay mixture

be noted that, thus far, template specificity had not been observed in any of the known nucleic acid polymerases, including the Kornberg enzyme, the RNA phosphorylase, and the RNA polymerase that is dependent on DNA. It should be noted, however, that all these cellular enzymes were evolved in an essentially closed genetic system. These enzymes were rarely faced (except in the case of virus infection) with the problem of deciding whether a particular nucleic acid was genetically related to it or not. In contrast, the RNA-dependent RNA polymerases induced by RNA viruses *always* had to function in a heterogenetic environment and therefore evolved under conditions in which they were continually required to ask the question, "Do you belong to me?"

Confirmation of specific template requirements with the Qβ replicase

Template specificity was not accepted in 1963, and further evidence was clearly needed to support so novel a concept. Our line of reasoning predicted that replicases

Table 2

RESPONSE OF Qβ REPLICASE TO DIFFERENT TEMPLATES

(Haruna and Spiegelman, 1965a)

In all cases, assay for DNA-dependent activity is carried out at 10 μg of DNA per 0.25 ml of reaction mixture. Input levels of template Qβ RNA were 1 μg per 0.25 ml. Control reactions containing no template yielded an average of 30 cpm. Numbers represent counts per minute (cpm).

Template	Incorporation
Qβ	4929
TYMV	146
MS2	35
Ribosomal RNA	45
sRNA	15
Bulk RNA from infected cells	146
Satellite virus	61
DNA (10 μg)	36

induced by other viruses would possess a similar preference for their homologous templates. It seemed desirable, therefore, to check this with another virus, preferably one unrelated to the MS2 group. The Qβ phage of Watanabe (1964) was chosen because of its immunological and other chemical differences (Overby et al., 1966a, b). In addition, it had the advantage of an excellent difference in the molar ratios of adenine (A) and uracil (U). As a result, one could readily distinguish the viral (plus) strand from its complement (minus) strand, a possibility not available with any of the RNA phages discovered previously. Finally, an unexpected windfall was the remarkable stability of the Qβ replicase, which made it much easier to handle enzymologically than the replicases of the MS2 group.

The isolation and purification of the Qβ replicase followed, with slight modifications, the procedures worked out earlier for the MS2 replicase. The general properties of the Qβ replicase were similar to those observed with the MS2 replicase, including requirements for all four riboside triphosphates and magnesium. The ability of various RNA molecules to stimulate the Qβ replicase to synthetic activity at saturation concentrations of homologous RNA is recorded in Table 2. The response of the Qβ replicase is in accord with what we have previously noted for the MS2 enzyme, the preference being clearly for its own template. The heterologous viral RNAs, MS2, and STNV are completely inactive, as are the ribosomal and transfer RNA species of the host cell. It is important to note that, as in the case of MS2 replicase, the absence of response to DNA shows that our purification procedure eliminates detectable evidence of active transcriptase from our enzyme preparations.

The nature of the product synthesized

By 1965, the purification of Qβ replicase had been brought to a stage at which it was largely freed of contaminating nucleases. The enzyme was virtually completely dependent on added RNA and could mediate almost unlimited syntheses of RNA over long periods of time. Further, the synthetic material was indistinguishable physically and chemically from the RNA obtained from virus. Naturally, we wanted more subtle information, and turned to our enzyme to provide it.

The ability of replicase to distinguish one RNA molecule from another can be used to provide information pertinent to the question of the similarity between the original template and the synthesized product. Two sorts of readily performed experiments can show whether the product is recognized by the enzyme as a template. One approach is to examine the kinetics of RNA synthesis at template concentrations that start below those required to saturate the enzyme. If the product can serve as a template, a period of autocatalytic increase of RNA should be observed. Exponential kinetics should continue as long as there are enzyme molecules unoccupied by template. When the product saturates the enzyme, the synthesis should become linear.

A second type of experiment is a direct test of the ability of the synthesized product to function as an initiating template. Here a synthesis of sufficient extent is carried out to insure that the initial input of RNA becomes a quantitatively minor component of the end product. The synthetic RNA can then be purified and examined for its template functioning capacities, a property readily examined by a saturation curve. If the response of the enzyme to variations in concentration of product is the same as that observed with authentic viral RNA, one would have to conclude that the product generated in the reaction is as effective a template for the replicase as is RNA obtained from the mature virus particle. We performed both types of experiments (Haruna and Spiegelman, 1965a). Autocatalytic synthesis was observed until the amount of RNA produced reached the saturation levels, whereupon the kinetics become linear. It was further shown (Haruna and Spiegelman, 1965b) that the response of the enzyme to increasing levels of synthetic RNA is identical with that observed with authentic viral RNA isolated from the Qβ particles.

The data obtained supported the assertion that the reaction produces a polynucleotide of the same molecular weight (1×10^6) as viral RNA, and one the replicase cannot distinguish from its homologous genome. Evidently the enzyme in the test tube is, at the very least, faithfully copying the recognition sequences employed by the replicase to distinguish one RNA molecule from another.

Evidence for the synthesis of an infectious viral RNA

The next question concerns the extent of the similarity between product and template. Have identical duplicates been produced? The most decisive test would determine whether the product contains all the information required to program the synthesis of complete virus particles in a suitable test system. The success recorded above encouraged us to attempt the next phase of the investigation, i.e., to subject the synthesized RNA to this more rigorous challenge (Spiegelman et al., 1965).

The ability to perform these experiments depended on the fact that viral nucleic acid can be introduced into protoplasts of bacteria, i.e., bacterial cells from which the cell wall has been removed. Once inside the cells, the RNA is translated and replicated to form complete virus particles which can be assayed by the usual plaque assay on intact cells.

As a first approach, the appearance of newly synthesized RNA was compared with the number of infectious RNA strands during the course of an extensive synthesis. We took aliquots from a reaction mixture to determine the amount of radioactive RNA that was synthesized, and phenol purification of the product was performed for the infectivity assay. As illustrated in Fig. 1, increase in RNA is paralleled by a rise in the number of infectious units.

This kind of experiment offers plausible evidence for the infectivity of newly synthesized and radioactively labeled RNA. It is not, however, conclusive because there is still a possibility that the agreement is fortuitous. It must be recalled that the infective efficiency of RNA strands is low being of the order of 1 in 10^6. One could therefore argue, however implausibly, that the enzyme is activating the input RNA to higher levels of infective efficiency while it is synthesizing new, non-infectious RNA. The concordance of the rather complex combination of exponential and linear kinetics of the two processes would then be just an unlucky coincidence.

To answer issues raised by such arguments, one must design experiments to eliminate the possibility that the input RNA is involved in the infectivity tests. There are several possible approaches to this problem, including the use of heavy isotopes and the separation of newly formed strands by density difference. One can, however, perform a much simpler experiment by taking advantage of the fact that we are dealing with a presumed self-propagating entity.

Consider a series of tubes, each containing 0.25 ml of the standard reaction mixture, but with no added template. The first tube is inoculated with 0.2 μg of Qβ RNA and incubated for a period adequate for the synthesis of ten times as much radioactive RNA. A one-tenth aliquot is transferred to the second tube, which is in turn permitted to synthesize ten times the input RNA, a portion of which is transferred to a third tube, and so on. If each successive synthesis produces RNA that can serve to initiate the next one, the experiment can be continued indefinitely and, in particular, until the point is reached at which the initial RNA of tube one has been diluted to an insignificant level. In fact enough transfers can be made to insure that the last tube contains less than one strand of the input template. *If in all the tubes, including the last one, the number of infectious units corresponds to the amount of radioactive RNA found, convincing evidence is offered that the newly synthesized RNA is indeed infectious.*

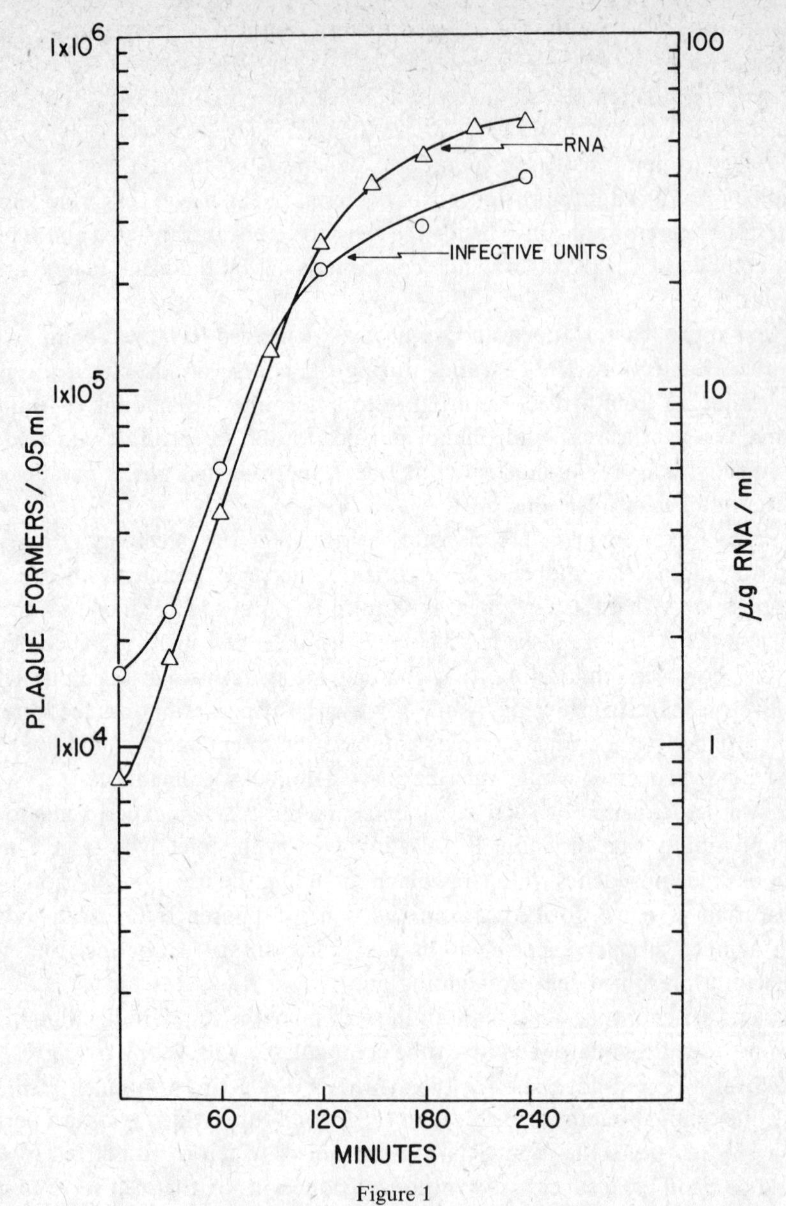

Figure 1
Kinetics of RNA synthesis and formation of infectious units.
An 8-ml reaction mixture was set up and samples were taken as follows: 1 ml at 0 time and 30 min; 0.5 ml at 60 min; 0.3 ml at 90 min; and 0.2 ml at all subsequent times. 20 λ were removed for assay of incorporated radioactivity. The RNA was purified from the remainder, radioactivity being determined on the final product to monitor recovery (Spiegelman et al., 1965).

A complete account of such a serial transfer experiment can be found in Spiegelman et al. (1965), and Fig. 2 describes the outcome. We knew the molecular weight and the amount of RNA put into the initial tube, so we could readily calculate how many tubes should be used to insure that the last tube contained less than one strand of the initial input. In this experiment, aside from the controls, 15 transfers were made, each resulting in a dilution of one to six. By the eighth tube there was less than one infectious unit ascribable to the initiating RNA, and the fifteenth tube contained less than one strand of the initial input. Nevertheless, every tube, including the last,

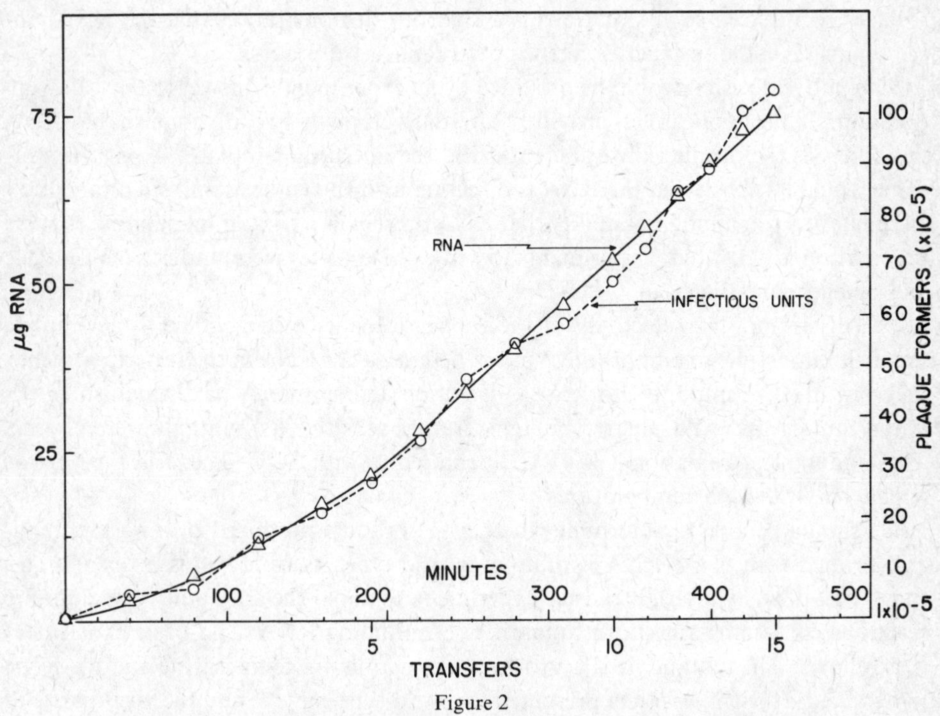

Figure 2

RNA synthesis and formation of infectious units in a serial transfer experiment. Sixteen reaction mixtures of 0.25 ml were set up, each containing 40 γ of protein and the other components specified for the "standard" assay. 0.2 γ of template RNA was added to tubes 0 and 1; RNA was extracted from the former immediately, and the latter was allowed to incubate for 40 min. The 50 λ of tube 1 was transferred to tube 3, and so on, each step after the first involving a 1 to 6 dilution of the input material. Every tube was transferred from an ice bath to the 35°C water bath a few minutes before use to permit temperature equilibration. After the transfer from a given tube, 20 λ were removed to determine the amount of ^{32}P-RNA synthesized and the product was purified from the remainder. Control tubes incubated for 60 min without the addition of the 0.2 γ of RNA showed no detectable RNA synthesis, nor any increase in the number of infectious units. All recorded numbers are normalized to 0.25 ml. The ordinates represent cumulative increases of infectious units and radioactive RNA in each transfer; the abscissae record elapsed time and the transfer number. Further details are to be found in Spiegelman et al. (1965).

showed an increment in infectious units corresponding to the amount of radioactive RNA found. In other words, the newly synthesized RNA was as infectious as the material obtained originally from the virus particle.

A rigorous proof that the added RNA is the self-duplicating entity

The central issue I now consider is one which I have already noted in my introductory remarks. It stems from the fact that two informed components are present in the reaction mixture—replicase and RNA template. None of the experiments thus far described *proved* that the RNA synthesized in this system is a self-duplicating entity. What is required in such situations is a rigorous demonstration that the RNA, not the replicase, is the instructive agent in the replicative process.

A definitive decision would be provided by an experimental answer to the following question: If the replicase is provided alternatively with two distinguishable RNA molecules, is the product always identical to the initiating template? A positive outcome would establish that the RNA is directing its own synthesis and simultaneously eliminate any remaining possibility that RNA present as a contaminant of the enzyme preparation is activated. Experiments to settle these issues were undertaken by Pace and Spiegelman (1966a, b).

The discriminating selectivity of the replicase for its own genome as a template makes it impossible to employ any kind of heterologous RNA in the test experiments, and we turned to mutants. For ease in isolation and simplicity in distinguishing between mutant and wild phenotype, temperature-sensitive (ts) mutants were chosen. The ts mutant grows poorly at 41°C as compared with 34°C. The wild type grows equally well at both temperatures.

We should be able to determine whether the product produced by a normal replicase primed with ts-Qβ RNA is mutant or wild type. As in previous investigations, this is best done by a serial transfer experiment, to avoid the ambiguity of examining reactions containing significant quantities of initiating RNA. The basic experiment is as follows: The mutant virus is grown at the permissive temperature and the RNA is extracted. This RNA is then presented to a wild-type enzyme and the virus particles induced by the newly synthesized RNA in bacterial protoplasts are examined. If they retain the temperature-sensitive phenotype, it is clear that the RNA is the instructive agent. If, on the other hand, the RNA produced is wild type, it is the enzyme that is geared to synthesize RNA, and the initiating template simply serves as a stimulator of synthesis.

Accordingly, we prepared seven standard mixtures, each containing 60 μg of Qβ replicase, isolated from cells infected with *normal* virus. To the first reaction mixture we added 0.2 μg of ts RNA and allowed synthesis to proceed at 35°C. After a suitable interval, we used one-tenth of this reaction mixture to initiate a second reaction, which, in turn, was diluted into a third, and so on for seven transfers. A control series

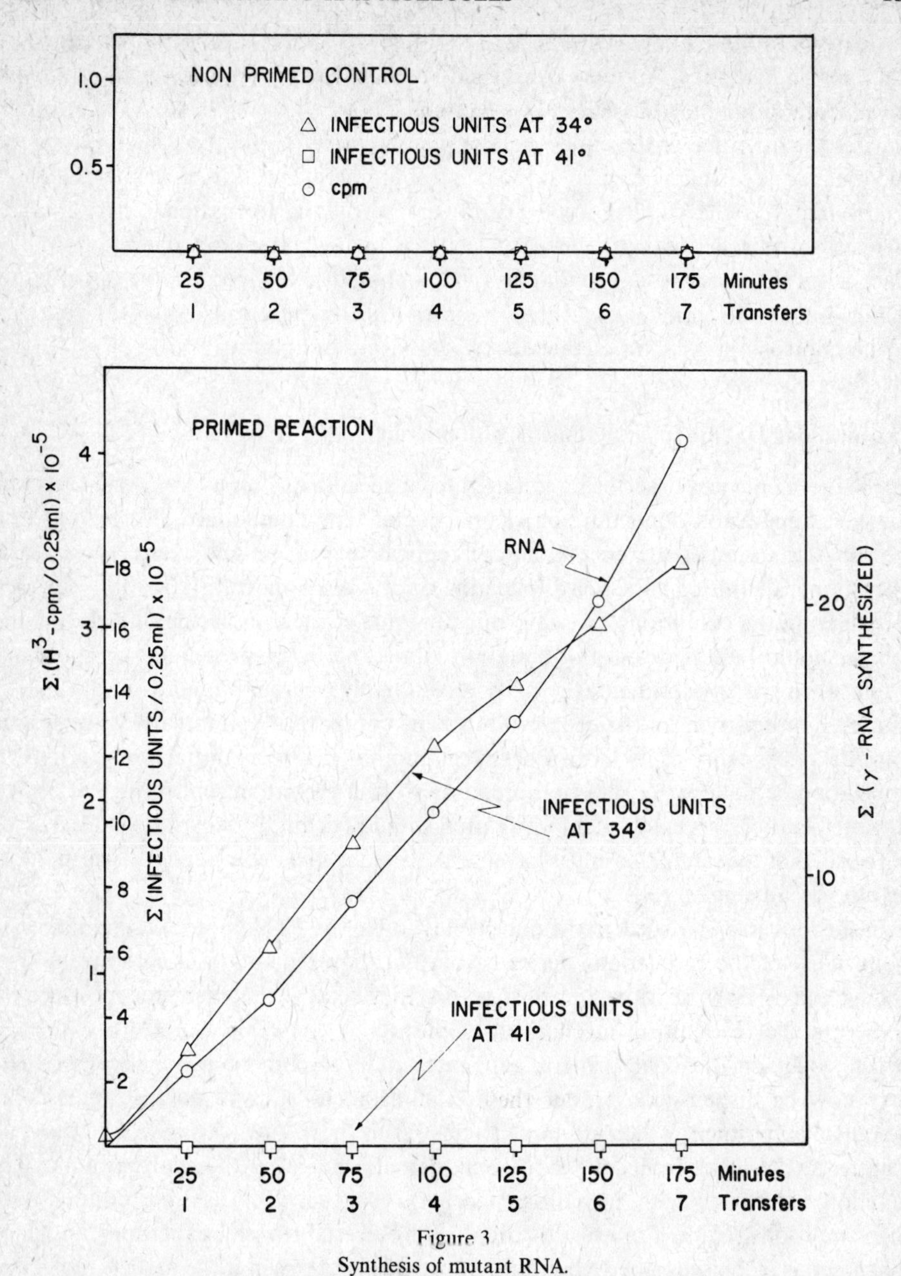

Figure 3

Synthesis of mutant RNA.

Each 0.25-ml reaction contained 60 μg of Qβ replicase purified through CsCl and sucrose centrifugation. The first reaction was initiated by addition of 0.2 μg of tsRNA. Each reaction was carried out at 35°C for 25 min, whereupon 0.02 ml was withdrawn for counting and 0.025 ml was used to prime the next reaction. All samples were stored frozen at $-$ 70°C until infectivity assays were carried out at 41°C and 34°C. A control series was carried out in which no initiating RNA was added (Pace and Spiegelman, 1966b).

was carried out in a manner identical to that just described, except that we added no RNA to the first tube. Aliquots from each reaction were examined for radioactivity in the material and also assayed for infectious RNA at 34° and 41°C. Figure 3 summarizes the outcome of the experiment in a cumulative plot of the RNA synthesized and the plaque formers detected at the two test temperatures. It is clear that the RNA synthesized has the ts phenotype. Although the plaque formation tested at 34°C increases in parallel with the new RNA that is synthesized, no such increase takes place when the tests are carried out at 41°C. It should be noticed (on the upper panel of Fig. 3) that no significant synthesis of either RNA or infectious units was observed in the control series of tubes that lacked initiating templates.

Extracellular Darwinian experiments with the replicating RNA molecules

The experiments just described demonstrated a specific response of one and the same enzyme preparation depending on the particular template added. This proved that the RNA is the instructive agent in the replicative process and hence satisfies the operational definition of a self-duplicating entity. An opportunity is thus provided for studying the evolution of a self-replicating nucleic acid molecule outside a living cell. It should be noted that this situation mimics an early, precellular evolutionary event, when the environmental selection presumably operated directly on the genetic material, rather than on the gene product. The comparative simplicity of the system and the accessibility of its known chemical components to manipulation permit the imposition of a variety of selection pressures during growth of replicating molecules. As noted earlier, this experimental situation should permit us to explore at least some of the rules of the game that must have been played in the evolution of living material before the advent of cells.

In the universe provided to them in the test tube, the RNA molecules are liberated from many of the restrictions derived from the requirements of the complete viral life cycle. The only restraint imposed is that they retain whatever sequences are involved in the recognition mechanism employed by the replicase. Thus, sequences which code for the coat protein, replicase, and attachment protein components may now be dispensable. Under these circumstances, it is of no little interest to design an experiment which attempts an answer to the following question: What will happen to the RNA molecules if the only demand made on them is the biblical injunction, "*Multiply*," with the biological proviso that they do so as rapidly as possible? The conditions required are readily obtained by a serial transfer experiment in which the intervals of synthesis between transfers are adjusted to select the earliest molecules completed. If the sequences coding for the three viral proteins are unnecessary for the replicative act, these sequences become so much excess genetic baggage. Because the longer a chain is, the longer it takes to complete, molecules would gain an advantage by discarding any unneeded genetic information to achieve a smaller size and therefore a more rapid completion.

Isolation of fast-growing variant (*V*-1). Mills, Peterson and Spiegelman (1967) performed a series of seventy-five transfers (Fig. 4), during which each reaction mixture was diluted 12.5 times in the next tube. The incubation intervals were reduced periodically as faster and faster growth was achieved. As can be seen from the insert in Fig. 4,

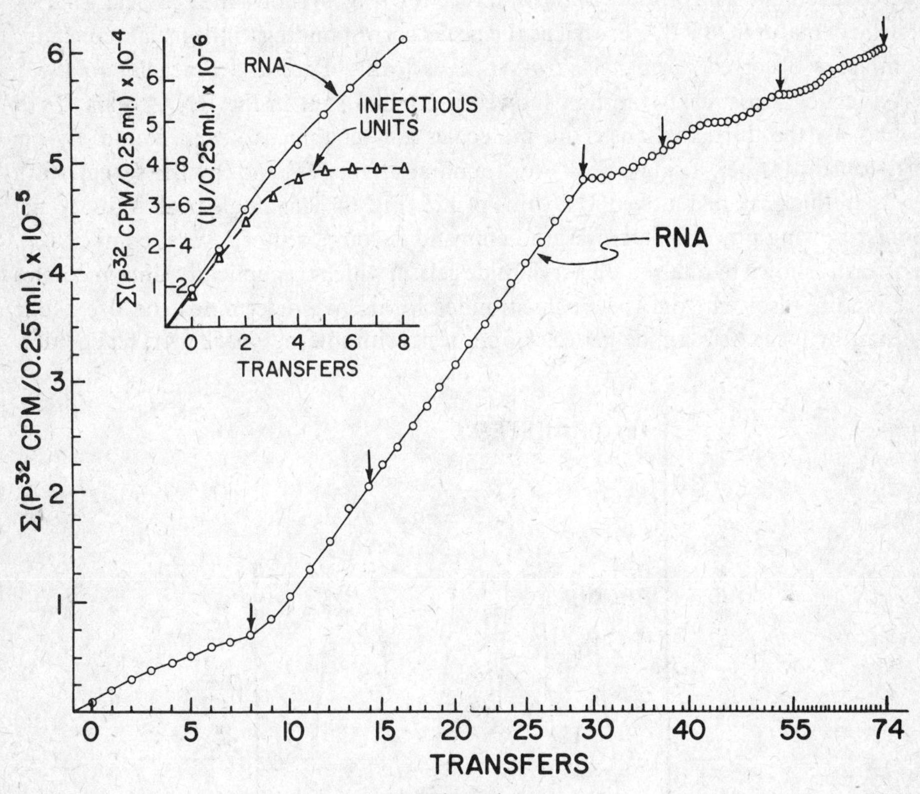

Figure 4

Serial transfer experiment.

Each 0.25-ml standard reaction mixture contained 40 μg of Qβ replicase and ^{32}P-UTP. The first reaction (0 transfer) was initiated by the addition of 0.2 μg ts-1 (temperature-sensitive RNA) and incubated at 35°C for 20 min, whereupon 0.02 ml was drawn for counting and 0.02 ml was used to prime the second reaction (first transfer), and so on. After the first 13 reactions, the incubation periods were reduced to 15 min (transfers 14–29). Transfers 30–38 were incubated for 10 min. Transfers 39–52 were incubated for 7 min, and transfers 53–74 were incubated for 5 min. The arrows above certain transfers (0, 8, 14, 29, 37, 53, and 73) indicated where 0.001–0.1 ml of product was removed and used to prime reactions for sedimentation analysis on sucrose. The inset examines both infectious and total RNA. The results show that biologically competent RNA ceases to appear after the 4th transfer (Mills et al., 1967).

although RNA continued to be synthesized, the formation of infectious RNA ceased after the fourth transfer. A dramatic increase in the rate of RNA synthesis occurred after the eighth transfer.

It should be noted that this sort of evolutionary experiment has its own built-in paleontology, for each tube can be frozen and its contents expanded whenever one decides to examine what happened at that particular period in the evolutionary process. We illustrate this by analyzing the products of several of the transfers in a sucrose gradient. The product of the first reaction (Fig. 5) shows the 28*S* peak characteristic of mature Qβ RNA, as well as the peaks corresponding to the usual replicative complexes observed during *in vitro* synthesis (Mills, Pace and Spiegelman, 1966). Products of subsequent transfers showed a gradual shift in the RNA to smaller *S* values. By the thirtieth transfer no molecules greater than 16*S* were found. By the fifty-fourth transfer, a single peak was found at about 15*S*, and, by the seventy-fifth transfer, this peak had moved to a value of 12*S* (Fig. 6). This single peak is due to the poor resolving power of sucrose gradients and is composed of several components, as is easily shown by analysis on acrylamide gels, in which the replicative intermediates are readily resolved from the single-stranded forms. We determined the size of the variant by using acrylamide gel electrophoresis with internal markers. The position

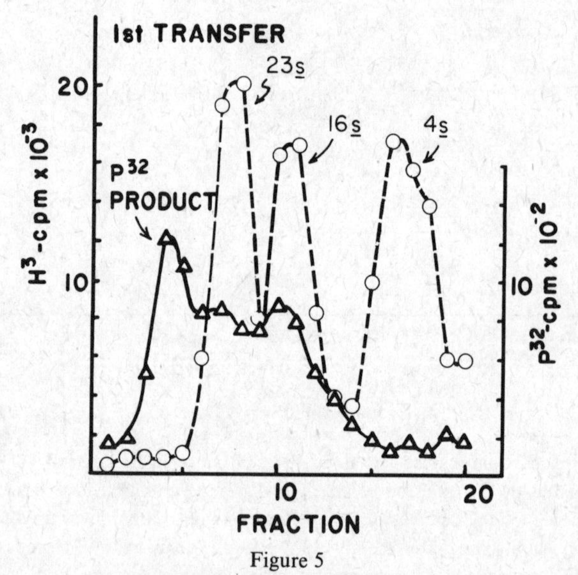

Figure 5

Sedimentation analysis of first transfer reaction.

0.002 ml of the 0 reaction was used to initiate a reaction for a first transfer reaction product. After completion, this reaction was adjusted to 0.2% SDS, an aliquot was withdrawn, diluted to 0.2 ml in TE (tris–EDTA) buffer (0.01 M tris, pH 7.4, 0.003 M EDTA), and then layered onto a 5-ml linear sucrose (2–20% in 0.1 M tris, pH 7.4, and 0.003 M EDTA). ^{3}H-labeled bulk RNA of *E. coli* was included as internal size markers (Mills et al., 1967).

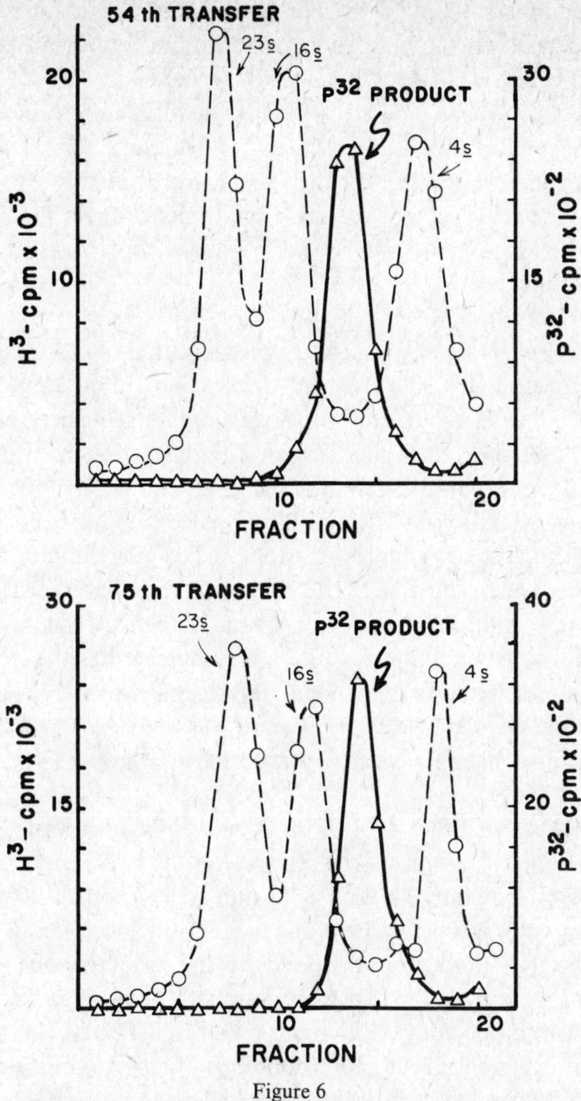

Figure 6
Sedimentation analysis of the 54th and 75th transfers.
Details as in Fig. 10 (Mills et al., 1967).

of the variant found in the seventy-fourth transfer (V-1) indicates that it has a molecular weight of 1.7×10^5 daltons, corresponding to 550 nucleotides. This means that, in response to the selection pressure for fast growth, the replicating molecule discarded 83% of its genetic information.

The principal purpose of the experiment was to illustrate the potentialities of the replicase system for examining the extracellular evolution of a self-replicating nucleic acid molecule. The seven samples we examined indicate that progress to a smaller size occurs in a series of steps.

The availability of a molecule that has eliminated large and unnecessary segments provides an object with obvious experimental advantages for the analysis of many aspects of the replicative process. Further, these abbreviated RNA molecules have a high affinity for the replicase but are no longer able to direct the synthesis of virus particles. This opens up a pathway towards a highly specific device for interfering with viral replication.

Cloning of self-replicating molecules. In principle, the Qβ replicase system should be capable of generating clones descended from individual strands. The resulting clones would provide the sort of uniformity required for sequence and genetic studies. Levisohn and Spiegelman (1968) succeeded in demonstrating that this can be realized.

As a first step, we isolated a new mutant, characterized by an ability to initiate synthesis at a low level of input. This was clearly necessary for our attempt at cloning. The selection of this type of mutant was accomplished by modifying the earlier, serial-selection procedures described in Fig. 2. Where the time intervals between transfers were decreased, the dilution remained constant. In the experiments to be described, the time was kept constant and the dilution was increased as the transfers were continued. During a serial transfer experiment, the incubation interval was held constant at 15 min, and the desired selection pressure was achieved by recurrent sharp increases in the dilution experienced by successive transfers from 1.25×10^{-1} to 2.5×10^{-10}.

The variant RNA (V-2) that evolved by the seventeenth transfer was selected for further studies. Figure 7 compares the response of the two variants, V-1 and V-2, to low-level inputs of template RNA in a 15-min incubation at 30°C. V-2 is clearly superior in growth at levels below 100 $\mu\mu\mu$g of RNA. In fact, V-2 can initiate synthesis with 0.29 $\mu\mu\mu$g of RNA, which corresponds to the weight of one strand. A kinetic analysis reveals that, during the exponential growth phase, V-2 has a doubling time of 0.403 min as compared with 0.456 min for V-1. In a 15-min period of logarithmic growth, V-2 can experience four more doublings (i.e., a 16-fold increase) than V-1. V-1 and V-2 have a similar electrophoretic mobility on polyacrylamide gels and the same base composition. They do, however, differ in sequences, as determined by oligonucleotide fingerprint patterns.

We obtained clones of RNA molecules *in vitro* by using an approach that depends on a straightforward comparison of the observed frequency distribution in a series of replicate syntheses with that expected from the Poisson distribution. If one strand is sufficient to initiate synthesis, the proportion of tubes showing no synthesis should correspond to e^{-m}, m being the average number of strands per tube. Further, if the onset and syntheses are adequately synchronized, one should be able to identify tubes

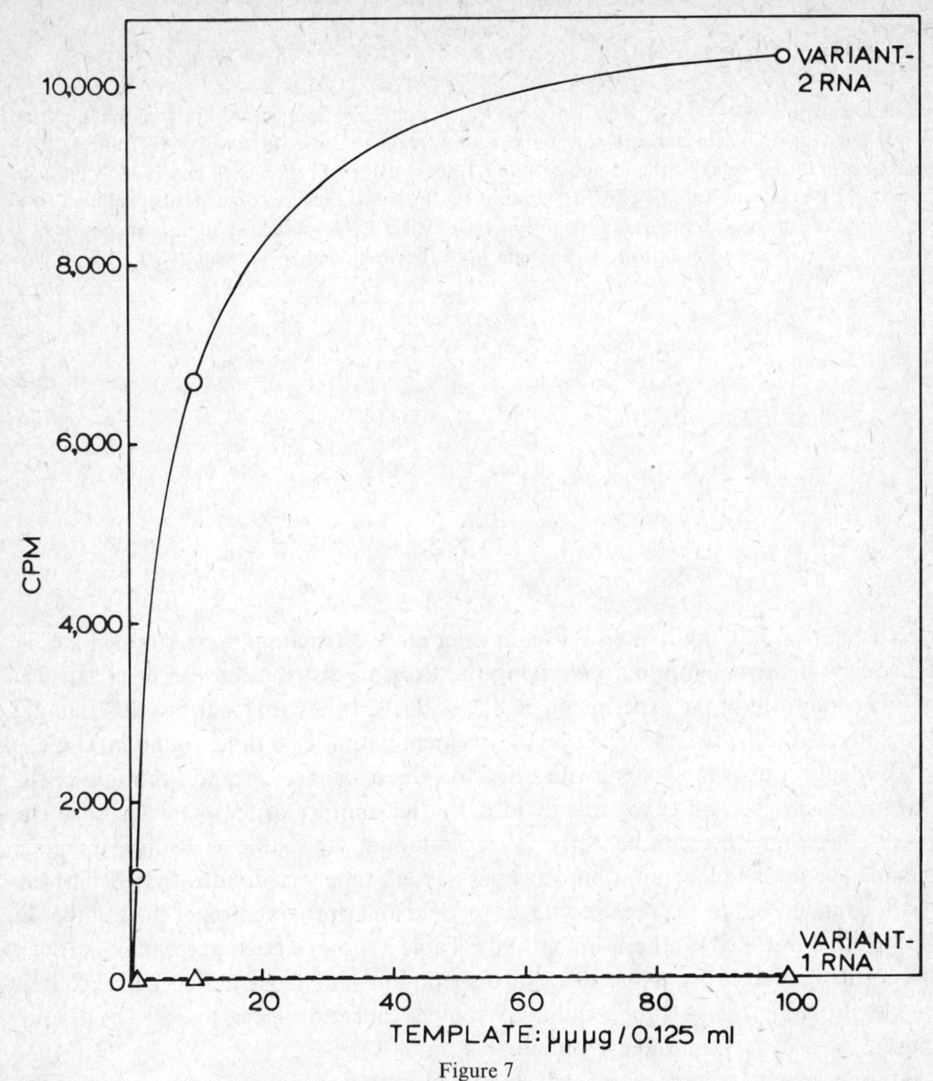

Figure 7

Comparison between the synthesis of variant-1 (V-1) RNA and V-2 RNA when used as templates at very low concentrations.

The indicated concentrations of V-1 RNA and V-2 RNA were added as templates to 0.125 ml of standard reaction mixture. After a 15-min incubation at 38°C, the ^{32}P–UTP incorporated into acid-insoluble material was determined (Levisohn and Spiegelman, 1968).

that received initially one, two, or three strands. These tubes should appear with frequencies corresponding to me^{-m}, $(m^2/2!)e^{-m}$ and $(m^3/3!)e^{-m}$, respectively.

A reaction mixture containing 0.29 $\mu\mu\mu$g of V-2 RNA per 0.1 ml was distributed at 0°C in 0.1 ml portions into each of 82 tubes. The tubes were placed simultaneously

Table 3

DISTRIBUTION OF TEMPLATE STRANDS AT AN AVERAGE INPUT OF ONE TEMPLATE STRAND PER TUBE

A reaction mixture in which 13.2 μg Variant-2 RNA were synthesized was diluted to a final concentration of 0.29 $\mu\mu\mu$g/ml RNA into standard reaction mixture containing 7.8×10^5 cpm ^{32}P-GTP per 0.25 ml. Quantities of 0.1 ml were distributed into each of 82 tubes. After incubation of 30 min at 38°C, the acid-insoluble ^{32}P-GTP was determined. The average is for the actual number of counts after subtraction of the average of nine control tubes (22 cpm) to which no variant RNA was added. In this experiment 3240 cpm is equivalent to 1 μg of variant RNA (Levisohn and Spiegelman, 1968).

Strands per tube	$P(r)$	Among 82 tubes	
		expected	found
0	0.368	30.2	30
1	0.368	30.2	29
2	0.184	15.1	19
3	0.0613	5.4	3
4	0.0153	1.3	1

into a bath of 38°C and, after a 30-min interval, the reactions were stopped simultaneously. If the assumption underlying the Poisson distribution has been satisfied by the conditions of the experiment, 36.8% of the 82 tubes, that is to say 30.2, should show no synthesis. This is in excellent agreement with the 30 tubes found in Table 3.

To identify tubes inoculated with one, two, three, or more strands, the sum of the counts observed in all tubes was divided by the number of template strands. The result, 309 counts per minute, is the average amount ascribable to a single template strand. The actual incorporation observed in each tube was divided by 309 and the result rounded out to the nearest integer to yield an approximation of the number of strands that initiated synthesis in that tube. Table 3 shows a good agreement between the results expected from the Poisson distribution and those actually found. It is highly probable that a tube exhibiting an incorporation close to 309 counts per minute was, in fact, initiated by a single strand.

An interesting outcome of the last experiment is that 0.29 $\mu\mu\mu$g indeed satisfied the Poisson expectation for an average of one strand of RNA. This permits an independent estimation of molecular weight as 174,000 daltons for V-2 RNA, in good agreement with the value deduced from gel electrophoresis.

The ability to clone variant RNA molecule *in vitro* was useful for sequence analysis of V-2 (Bishop, Mills and Spiegelman, 1968) and provided a necessary prerequisite for a genetic analysis.

Diverse variants isolated under different selective conditions

The experiments thus far described were concerned with the isolation of mutants possessing increased growth rates under standard conditions. We now turn our attention to the central question: Can *qualitatively distinguishable* phenotypes be exhibited by a nucleic acid molecule under conditions in which its information is replicated, but never translated?

Selection of "nutritional mutants". One general approach (Levisohn and Spiegelman, 1969) for obtaining a variety of mutants is to run the syntheses under less-than-optimal conditions with respect to a component or parameter of the reaction. If a variant arises that can cope with the imposed suboptimal condition, continued transfers should lead to its selection over the wild type. This can be done with variations in the level of the ribonucleaside triphosphates. It can be seen in Fig. 8 that the rate of synthesis of V-2 begins to decrease sharply as the level of cytidine triphosphate (CTP) drops below 20 nmoles per 0.125 ml. At a CTP concentration of 2 nmoles, the rate of synthesis of V-2 is only 25 per cent of normal. At 1 nmole of CTP,

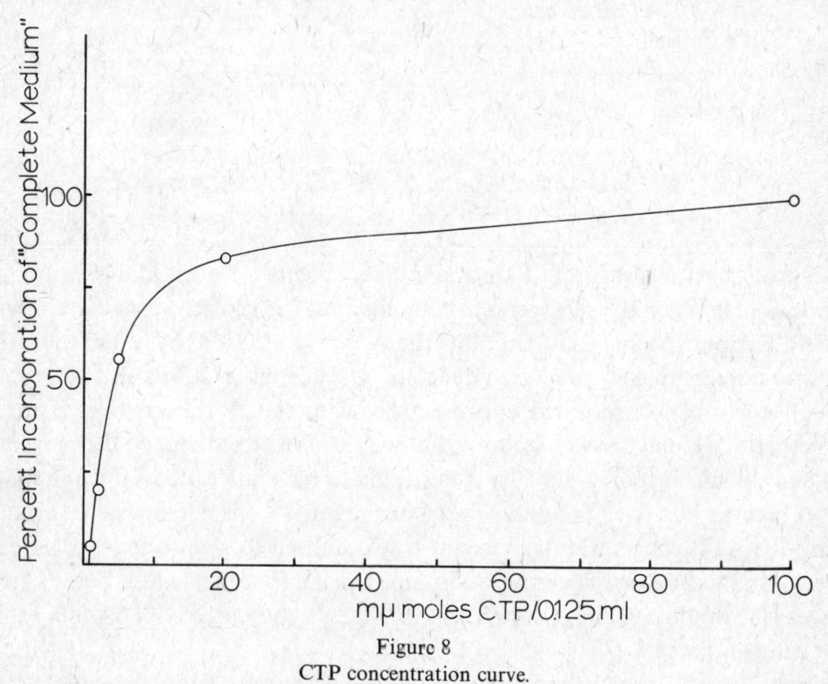

Figure 8
CTP concentration curve.

Reaction mixtures of 0.125 ml containing the indicated concentrations of CTP and otherwise identical to "standard reaction mixture" were incubated for 30 min at 38°C with 20 μg Qβ replicase and 0.01 μg Variant-2 RNA. Subsequently, the acid-insoluble radioactivity was determined. In the "complete medium" (containing 100 mμM CTP) 1.6 μg of V-2 RNA were synthesized (Levisohn and Spiegelman, 1969).

the rate decreases to 5% of normal. With such information available, a search was made for variants that could replicate better than V-2 on limiting levels of CTP. A serial transfer experiment at 2 nmoles of CTP per reaction was initiated with Qβ RNA, culminating after 10 transfers in the appearance of V-4. A second series of transfers at one nmole of CTP per reaction was then started with V-4 and after 40 transfers led to the isolation of V-6.

Variants V-4 and V-6 replicate 28% and 56% better, respectively, than does V-2 at low levels of CTP. This increased capacity might be explained on the basis of smaller sizes or modification of base composition toward a lower cytosine content. However, on examination neither V-2 nor V-6 showed any significant difference in either size or base composition (Table 4). The identification of the changes will require more subtle examination by sequence determinations.

Table 4
BASE RATIOS OF VARIANTS
The base composition of purified strands of Variant-2 and Variant-6. The numbers represent mole per cent (Levisohn and Spiegelman, 1969).

Base	V-2	V-6
C	24.8	24.8
A	23.2	23.5
B	26.6	26.7
U	25.4	25.1

Selection of a variant resistant to an inhibitory analogue. Tubercidin is an analogue of adenosine in which the nitrogen atom in position 7 is replaced by a carbon atom. Tubercidin triphosphate (TuTP) inhibits the synthesis of Qβ RNA *in vitro.* However, TuTP cannot completely replace adenosine triphosphate (ATP) in the reaction, although it can be incorporated into the growing chains. It was of some interest to see whether one could derive a mutant that would show resistance to the presence of this agent. In such experiments it was desirable to have the ratio of the analogue to ATP as high as possible. To attain this more readily, we isolated a new variant by limiting the ATP concentration. Variant 6 was chosen to start a series of transfers in a reaction mixture that contained 1.5 nmoles of ATP, which led to the isolation of variant 8, the doubling time of which was 2.8 min as compared with 8.4 min for V-6, the starting variant.

The replication rate of V-8 in a reaction mixture that contained 5 nmoles of ATP was inhibited fourfold with the addition of 30 nmoles of TuTP. A serial transfer was initiated with V-8 in the inhibitory medium and led to the isolation of V-9. The doubling time of V-9 in the presence of TuTP was 2.0 min as compared with 4.1

min for V-8. In the absence of TuTP, both variants were synthesized with 1.0-min doubling time. It is clear that V-9 exhibits a specifically increased resistance to the inhibitory effect of TuTP.

Selection of variants resistant to ethidium bromide. The variants we have thus far described solved the problems posed them by changes that were, for the most part, phenotypically expressed by interaction with the replicase. It was of obvious interest to see whether a class of mutants could be derived that overcomes an imposed difficulty more directly without involving the enzyme. An obvious approach is to use agents that exert this inhibitory influence by interaction with the RNA molecules. A plausible candidate is ethidium bromide (2,7-diamino-50-ethyl-phenyl phenanthridium bromide). This compound is known to inhibit the DNA-dependent RNA polymerase of *E. coli* and is believed to do so by intercalating in double-stranded base-paired regions.

Transfers were initiated with V-2 and continued in the presence of increasing levels of ethidium bromide (Saffhill et al., 1970). Resistant variants appeared early during the serial selection and the degree of resistance increased as the level of the ethidium bromide was increased. Figure 9 compares the resistance to ethidium bromide of the starting V-2 with variants isolated in the presence of 20 μg and 40 μg of the inhibitory agent. Both the V (20 μg) and V (40 μg) variants can replicate at levels of ethidium bromide that completely inhibit the wild-type V-2 from which they were derived. That the resistance of the variants is accompanied by a decreased ability to bind ethidium bromide is shown in Fig. 10, which compares the fluorescence intensity with ethidium bromide concentration in the presence of V-2 and V(40 μg). The fact

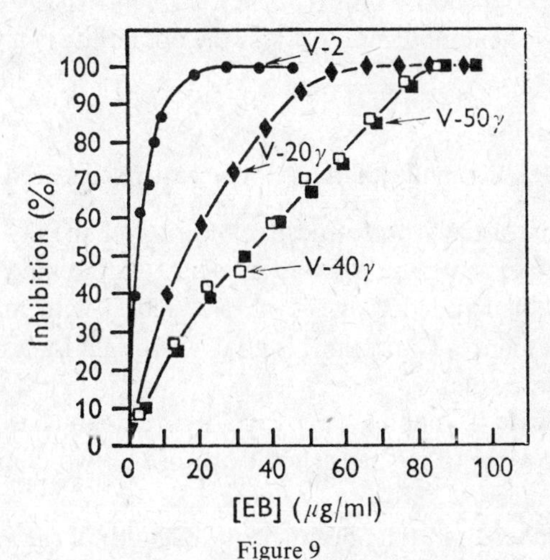

Figure 9
Inhibition of replication of V-2, V-20 γ, V-40 γ and V-50 γ by ethidium bromide. (Saffhill et al., 1970).

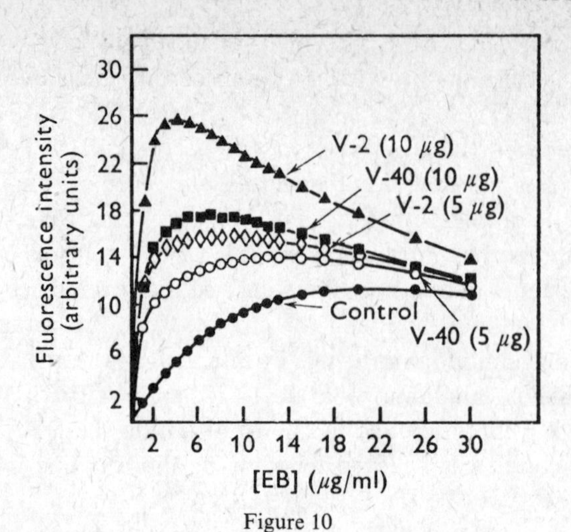

Figure 10
Fluorescence of ethidium bromide in the presence of V-2 and V-40 γ in 0.1 M tris buffer (pH 7.4). The control has no added RNA (Saffhill et al., 1970).

that V(40 μg) enhances the fluorescence of ethidium bromide less than V-2 indicates that V(40 μg) binds with ethidium bromide less than V-2.

It is evident from even the limited set of successful examples, and the ease with which they are attained, that the number of identifiably different mutant molecules possessing prespecified phenotypes is virtually unlimited. The only restriction would appear to stem from the ingenuity exercisable by the experimenter in devising the appropriate selective conditions. This diversity of phenotypes would become understandable if the nucleic acid molecules possess the possibility of a variety of secondary and tertiary structures.

The search for a replicating molecule that can be easily sequenced

It became evident that a truly profound exploitation of the insight inherent in this *in vitro* system, and one that would bring us closer to the kind of chemical understanding we desired, demanded that we start selection with an RNA molecule of completely known sequence. To attain this goal, we sought for a replicating molecule with the following properties:

a) It should replicate a manner recognizably analogous to that of Qβ RNA;

b) It should be possible to isolate at least one of the two complementary strands for independent sequencing, and

c) It should be smaller than the 550-nucleotide length of the variants we had thus far studied so that its absolute sequence, and those of its mutants, could be obtained without inordinate time and effort.

We have recently reported (Kacian et al., 1972) the isolation of a replicating RNA molecule (MDV-1) that contains 218 nucleotides and possesses the other features justifying the effort required to obtain its absolute sequence. We will now describe the studies that led to a knowledge of its complete sequence. In doing so we will briefly note only the key experiments with particular emphasis on new sequencing technology devised in the course of our studies (Mills, Kramer and Spiegelman, 1973).

The strategy of sequencing RNA molecules synthesizable in vitro

If the molecule being sequenced can be synthesized extracellularly, the logic of the sequencing procedure can be markedly modified to ease the labor and, to a large extent, avoid the difficulties inherent in dealing with uniformly labeled material. The determination of the composition, multiplicity, and sequence of a given oligonucleotide is readily accomplished by running four separate reactions, each containing one of the ribosidetriphosphates labeled with ^{32}P in the α-position and a fifth reaction containing all four riboside triphosphates similarly labeled to the same specific activity. The latter reaction serves to locate all the oligonucleotides produced and at the same time yields their composition. The four individually labeled reactions provide nearest neighbor information (Josse, Kaiser and Kornberg, 1961) from which the sequences of many of the oligonucleotides can be directly deduced (Bishop, Mills and Spiegelman, 1968).

The use of nearest neighbor information

Here advantage is taken of the fact that in a cleavage of 3′–5′ phosphodiester bonds, on the 5′ side of the phosphorous, whether by enzyme or alkali, the phosphorous of the resulting 2′ or 3′ nucleotides derives from the 5′ α-phosphorous of their nearest 3′ neighbors. If this type of analysis is carried out on a particular oligonucleotide derived from each of the four individually labeled syntheses, the data obtained almost always lead to the deduction of its unique sequence.

To exemplify this methodology, consider the ribonuclease A fragment pppGpGpGpGpApCp/C. This is of course the 5′ terminus of the original molecule since the terminal G carries the triphosphate group, a fact established by the presence in subsequent alkaline or T_1 hydrolysates of the tetraphosphate pppGp. The composition of the fragment was determined from a uniformly labeled synthesis by separating and counting the ^{32}P-labeled nucleotides produced by alkaline hydrolysis. The results indicated that the fragment contained four residues of G and one each of A and C. The sequence of the six residues was then established from the individually labeled syntheses as follows. Alkaline hydrolysis of the terminal fragment derived from the synthesis containing [α-^{32}P]GTP revealed that only G residues were labeled with ^{32}P. Consequently, no G occurred on the 3′ side of either A or C and all four G residues must therefore adjoin each other at the 5′ end of the fragment. The reaction containing [α-^{32}P]ATP yielded a fragment that contained one G residue with ^{32}P, establishing that the A residue follows immediately on the 3′ side of the four G residues. Finally, the A was labeled with ^{32}P only in the synthesis containing [α-^{32}P]CTP, placing the C after the A, thus completing the sequence of the fragment. Confirmation of this can be obtained by digestion of this fragment with RNase T_1, which should, and does, lead to the production of ApCp/C.

It should be further noted that nearest neighbor information helps in ordering the oligonucleotides. For example, the 3′ terminal Cp of the 5′ terminal fragment was labeled with ^{32}P only in the synthesis containing [α-^{32}P]CTP. This implies that the next oligonucleotide on the 3′ side of this fragment must contain a C on its 5′ end. This type of information usefully limits the number of ways sequences can be assembled.

Use of complementary strands for sequence information

Another advantage derivable from an *in vitro* synthesis is the availability of the plus and minus strands produced in the synthetic reaction. We have shown that MDV-1 (Kacian et al., 1972), like Qβ (Spiegelman et al, 1968; Weissmann, Feix and Slor, 1968) and V-2 (Mills, Bishop and Spiegelman, 1968) replicates via a complementary antiparallel intermediate. Therefore, every purine-rich segment of the plus strand must be represented by an antiparallel pyrimidine-rich segment in the minus strand. By comparing oligonucleotides obtained from the plus strand with those from its complement, one can find overlapping complementary matches of sufficient length to extend and confirm known sequences.

The use of synchronized in vitro reactions

In their elegant sequence studies of Qβ RNA, Weissmann and his colleagues (Billeter et al., 1969) introduced the use of another benefit inherent in an *in vitro* synthesis. Knowing that the first six residues at the 5′ end of Qβ consist only of G and A, they initiated the synthesis at 37°C with a reaction mixture lacking CTP and UTP. This allows all the molecules to begin and then halt after the insertion of the first six residues. The temperature was then lowered to 20°C to discourage new initiations and to slow down subsequent polymerization. The missing CTP and UTP were then added resulting in a synchronized onset of synthesis from the same point in all the molecules. Samples were taken at 5-second intervals and the time required for a particular oligonucleotide to appear in the growing molecule was used to locate its position in the sequence. This procedure permitted the ordering of the first 300 nucleotides, although larger fragments obtained by partial nuclease digestion were still required to sort out the sequence.

The method of "fragment length mapping" of oligonucleotides

Ingenious as it is, the use of timed samples of synchronized polymerizations depends for success on two assumptions that are not always, or indefinitely, satisfied. One demands that the replicase molecules traverse the polynucleotide template at constant speed during the sampling period, and the second requires that the reinitiated polymerizations remain in synchrony. The latter demand can hardly be maintained indefinitely and a distance of about 300 nucleotides appears to be the limit. Beyond this, the synchrony becomes sufficiently blurred to interfere seriously with the resolving power of the method. The requirement of constant speed is likely to be satisfied only along polynucleotide stretches lacking significant secondary structures in the form of base-paired helixes. This was a complication that was probably aggravated in our instance, since a major portion of our molecule is involved in such duplex-containing structures.

In any event, we quickly found that timed samples of synchronized reactions contained partially synthesized fragments of a rather broad size class that could not be arranged in any internally consistent linear array. We therefore devised a method of mapping by fragment length, a procedure that does not depend on constant rates of polymerization, avoids the ambiguities generated by loss of synchrony, and provides a higher degree of resolution. The method is exemplified in detail below and we need here only note its simple logic. In any nucleic acid synthesis mediated by a nuclease-free polymerase, every fragment will contain the 5′ terminus. To locate a specific oligonucleotide, one need only determine the minimal fragment length required to insure its presence. Separation of partially synthesized RNA fragments into narrow size classes can be achieved by electrophoresis through polyacrylamide gels.

We now describe each of the sequencing steps dictated by the particular strategy adopted. Emphasis will be focused on the plus and minus strand oligonucleotide catalogues, the construction of extended sequence blocks by the use of complementary oligonucleotides, the logic used in deciding which blocks can be joined, and finally on the "fragment length mapping" of the extended sequence blocks with the aid of high resolution gel electrophoresis. We conclude with the complete sequences of the two complementary strands and a discussion of their probable secondary structures.

Synthesis of the plus and minus strands of MDV-1 RNA

Figure 11 shows a 4.8% polyacrylamide gel profile of a representative self-annealed product from a reaction initiated with MDV-1 (+) RNA. We have previously shown (Kacian et al., 1972) that peak I contains mature single strands, peak II consists of double-stranded duplexes of antiparallel complements, and peak III comprises multistranded complexes with both complementary strands present in various

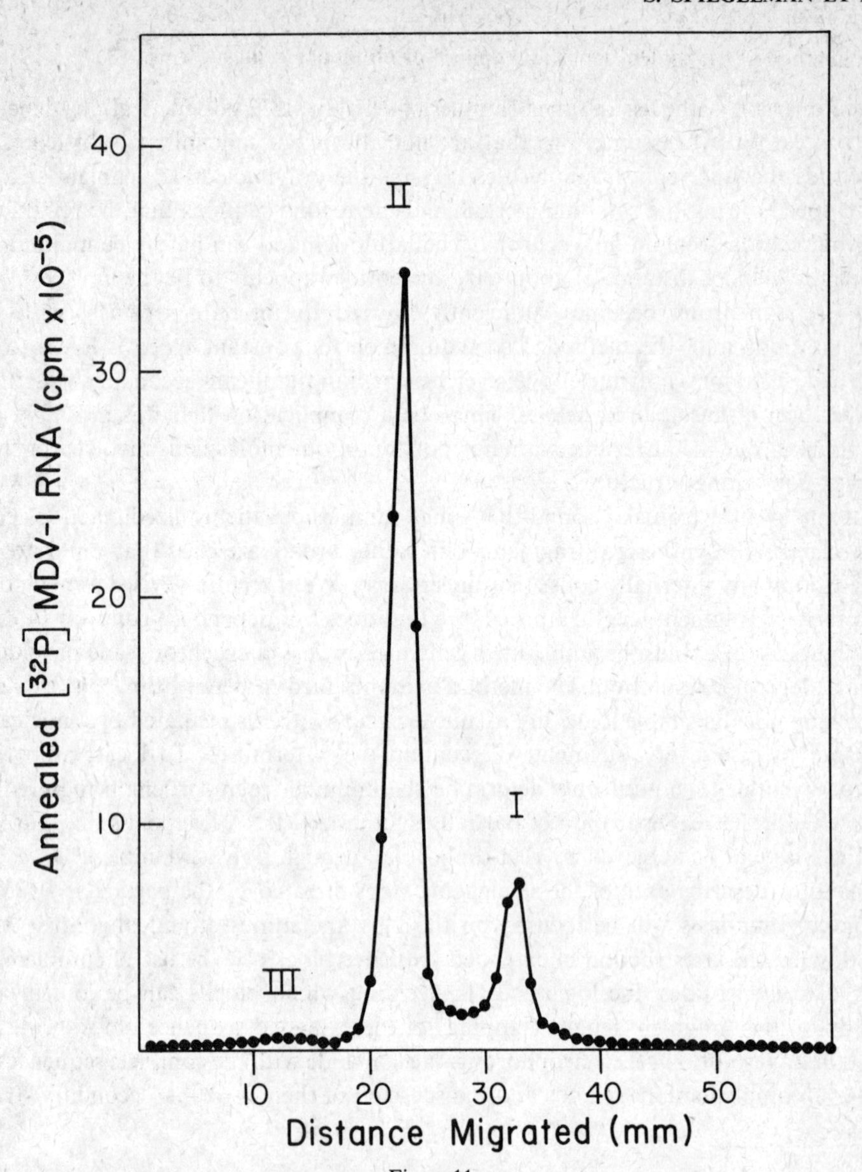

Figure 11

Polyacrylamide gel electrophoresis of self-annealed MDV-1 RNA.

RNA product was synthesized and isolated as detailed previously (Kacian et al., 1972). Prior to gel electrophoresis, the RNA was self-annealed for 60 min at 65°C in 400 mM NaCl, 3 mM EDTA. RNA species were then analyzed on 4.8% polyacrylamide gels (Bishop, Claybrook and Spiegelman, 1967) by electrophoresis at 10 mA/gel for 120 min. The RNAs from peaks I and II were eluted and precipitated with ethanol (Kacian et al., 1972). Peak I contained only MDV-1 (+) RNA, the strand synthesized in excess. Peak II contained duplex structures formed by the annealing of the complementary MDV-1 (−) and MDV-1 (+) strands (Mills, Kramer and Spiegelman, 1973).

proportions. Since the material was self-annealed prior to electrophoretic separation, peak I contains the single strand made in excess and which we have designated as the plus strand, in accordance with the situation that obtains in the *in vitro* synthesis of Qβ RNA. All three peaks are easily eluted from the gels in high yield for further characterization.

RNase A and T_1 oligonucleotide catalogues of MDV-1 (+) RNA

Four separate syntheses were run, each containing one of the four ribosidetriphosphates labeled with ^{32}P in the α-position, and the RNA in peak I was isolated. Aliquots were subjected individually to either ribonuclease A or T_1 digestion. The oligonucleotides were then separated by the Sanger, Brownlee and Barrell (1965) two-dimensional electrophoresis procedure. In those few cases where oligonucleotides were not adequately resolved (i.e., the very large oligonucleotides derived from RNase A digestion), the digests were first separated according to size (isopleths) by urea–DEAE cellulose chromatography (Tomlinson and Tener, 1963), and then each isopleth was resolved into its component oligonucleotides by two-dimensional electrophoresis.

When the chain length of a particular oligonucleotide was in doubt, it was exposed to alkaline phosphatase to remove charges due to terminal phosphates followed by rechromatography on urea–DEAE cellulose with known chain length markers (Bishop, Mills and Spiegelman, 1968).

Every oligonucleotide from the four syntheses was eluted and subjected to alkaline hydrolysis. The resulting 2′ and 3′ nucleotides were separated electrophoretically and the distribution of ^{32}P was determined, in order to yield the nearest neighbor analysis of each residue.

As a further aid, each oligonucleotide was digested with the nuclease not used to produce it. Thus, T_1 oligonucleotides were digested with RNase A and *vice versa.* Electrophoretic separation of these digests yielded predictable sequence elements of each oligonucleotide. They served to confirm, and in many cases helped to establish, the sequence of each oligonucleotide. In certain instances partial digestion with spleen phosphodiesterase (Sanger, Brownlee and Barrell, 1965) was also used to obtain sequence elements of oligonucleotides.

The 3′ terminal oligonucleotide was identified by alkaline hydrolysis of a product synthesized with [3H]cytidine triphosphate. The 3′ oligonucleotide terminus yielded [^{3}H]cytosine, whereas the other fragments gave only nucleotides on hydrolysis. The 5′ terminus was identified as the only RNase A fragment that was labeled by [γ-^{32}P]GTP and by the fact that it yielded pppGp on alkaline hydrolysis. Once the sequence of each oligonucleotide was ascertained, its molar frequency in the molecule was determined by comparing its ^{32}P content with that of the entire digest.

Table 5 summarizes the complete T_1 and RNase A oligonucleotide catalogues

Table 5

MDV-1 RNA PLUS STRAND CATALOGUES

The RNA from peak I was analyzed by digestion with RNase T_1 and RNase A. The oligonucleotide fragments obtained from these digestions are listed below. The actual fragments are contained within the verticle bars. The residue outside each bar indicates the identity of the nearest neighbor nucleotide. Note that the nearest neighbor on the 5′ side of RNase A fragments can be either C or U. The number of times each fragment occurs in the RNA is indicated to the left of its sequence. The occurrence of many small fragments is due to the high G–C content (70 %) of the RNA (Mills et al., 1973).

<table>
<tr><th colspan="3">RNase T₁ digest</th><th colspan="3">RNase A digest</th></tr>
<tr><td>5 G|G|C</td><td>2 G|ACG|C</td><td>2 G|CUG|C</td><td>11 C/U|U|C</td><td>1 C/U|AC|C</td><td>3 C/U|GAC|C</td></tr>
<tr><td>4 G|G|A</td><td>2 G|ACG|A</td><td>1 G|CUAG|C</td><td>1 C/U|U|A</td><td>4 C/U|AC|G</td><td>3 C/U|GAC|G</td></tr>
<tr><td>13 G|G|G</td><td>1 G|AAG|A</td><td>1 G|UACG|G</td><td>2 C/U|U|G</td><td>2 C/U|GC|C</td><td>1 C/U|AGC|C</td></tr>
<tr><td>4 G|G|U</td><td>2 G|AAG|G</td><td>½ G|CCUCG|A</td><td>6 C/U|U|U</td><td>3 C/U|GC|A</td><td>2 C/U|AGC|G</td></tr>
<tr><td>5 G|CG|C</td><td>1 G|CCCG|C</td><td>½ G|CCUCG|U</td><td>13 C/U|C|C</td><td>7 C/U|GC|G</td><td>1 C/U|GGC|C</td></tr>
<tr><td>3 G|CG|A</td><td>1 G|ACCG|U</td><td>1 G|UCACG|G</td><td>3 C/U|C|A</td><td>5 C/U|GC|U</td><td>1 C/U|GGC|G</td></tr>
<tr><td>1 G|CG|G</td><td>1 G|CACG|A</td><td>1 G|UUCG|A</td><td>13 C/U|C|G</td><td>1 C/U|GU|A</td><td>1 C/U|GGGC|A</td></tr>
<tr><td>2 G|CG|U</td><td>1 G|AACCG|C</td><td>1 G|CUUCG|C</td><td>5 C/U|C|U</td><td>3½ C/U|GU|G</td><td>1 C/U|GGGC|U</td></tr>
<tr><td>1½ G|AG|A</td><td>1 G|CCACG|C</td><td>1 G|UUUCG|G</td><td>1 C/U|C-OH</td><td>1 C/U|GU|U</td><td>1 C/U|GAGU|C</td></tr>
<tr><td>3 G|AG|G</td><td>2 G|UG|C</td><td>1 G|CUUUCG|C</td><td>1 pppGGGGAC|C</td><td></td><td>1 C/U|AGGU|G</td></tr>
<tr><td>2 G|AG|U</td><td>3½ G|UG|A</td><td>1 pppG|G</td><td>½ C/U|GAAGAGGC|G</td><td></td><td>2 C/U|GAGGU|G</td></tr>
<tr><td>2 G|CAG|C</td><td>1 G|UG|G</td><td>1 G|UUCCCC-OH</td><td>½ C/U|GAGAAGAGGC|G</td><td></td><td>1 C/U|GGGAGU|U</td></tr>
<tr><td>1 G|ACCCCCG|A</td><td></td><td>1 G|CCCUUCG|C</td><td>1 C/U|GAAGGGGGU|U</td><td></td><td>1 C/U|GAGAAC|C</td></tr>
<tr><td>1 G|ACCCCCCG|G</td><td></td><td>1 G|ACCUUCG|U</td><td>1 C/U|GGAAGGGGGAC|G</td><td></td><td></td></tr>
<tr><td>1 G|CACCUCG|U</td><td></td><td>1 G|CUCUCCAG|G</td><td></td><td></td><td></td></tr>
</table>

determined in this manner. All oligonucleotides in this and subsequent tables and figures are written in the 5′ to 3′ direction. The residues outside the vertical bar on the 5′ side are deduced from the known properties of the enzyme used to obtain the fragment. The residues on the 3′ side of the vertical bar are derived from nearest neighbor information. The multiplicity of each oligonucleotide is indicated to the left of its sequence. Of interest is the appearance of a few fragments with nonintegral values, a puzzling feature that was ultimately resolved by the discovery that our variant contained a mixture of two mutants that differed at residue 104 of the plus strand.

RNase A and T_1 oligonucleotide catalogues of MDV-1 (−) RNA

Since the complementary MDV-1 (−) RNA is not obtainable in adequate yield, the oligonucleotide catalogues of this species were obtained by a rather different device. The double-helical RNA (peak II) of Fig. 11, containing both plus and minus strands, was eluted from the gel, concentrated and melted thermally. Aliquots of the denatured equimolar mixture of plus and minus strands [MDV-1(+/−)RNA] were then digested individually with RNase A and RNase T_1 as previously described for MDV-1 (+) RNA. The resulting digest contained a mixture of fragments derived from plus and minus strands.

As an example, Fig. 12 compares the plus strand and duplex RNase T_1 fingerprint patterns of reactions labeled with [α-^{32}P]CTP. Some of the sequences found only in the minus strand are circled. Others are clearly common to both. To sort out the set of oligonucleotides complementary to those found for the plus strand, four duplex preparations were synthesized each labeled with a different [α-^{32}P]riboside-triphosphate. The RNase A and T_1 digests were then used to construct the complete oligonucleotide catalogues for the combined equimolar mixture of plus and minus

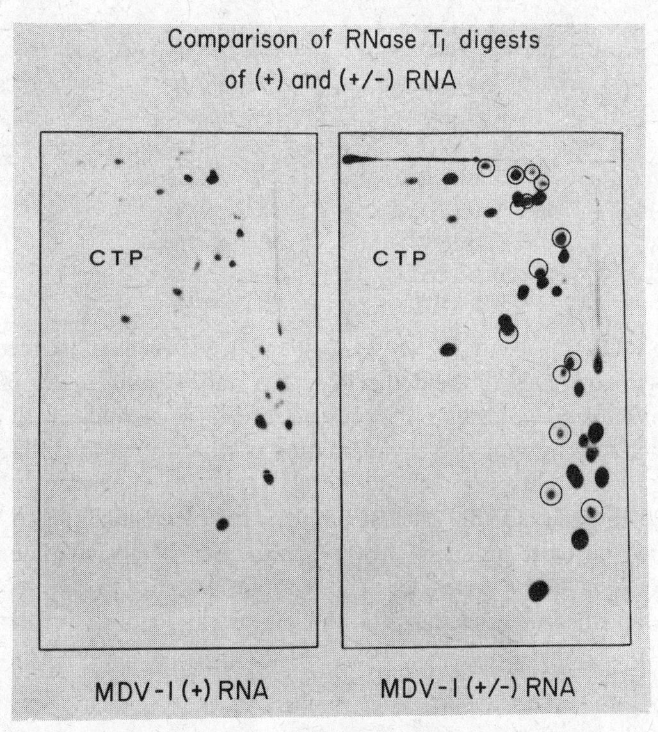

Figure 12

Comparison of RNase T_1 digests of MDV-1 (+) and MDV-1 (+/−) RNA.

RNA from peak II was melted to separate the strands of the duplex into an equimolar mixture of plus and minus RNA, digested with ribonuclease, and analyzed by two-dimensional electrophoresis (Sanger, Brownlee and Barrell, 1965). The resulting fingerprint patterns were more complicated than their plus strand counterparts, as they contained the fragments of both strands. The minus strand oligonucleotides were identified by a comparison of the duplex and single-stranded fingerprints. The RNase T_1 fingerprints shown above were obtained from [α-^{32}P]CTP-labeled RNA. Some of the fragments found only in the minus strand are circled. Many other minus strand oligonucleotides, however, are identical to those found in the plus strand (Mills, Kramer and Spiegelman, 1973).

Table 6

MDV-1 RNA MINUS STRAND CATALOGUES

These catalogues were prepared by deleting oligonucleotides found in each plus strand catalogue from the corresponding catalogue of fragments obtained from MDV-1 (+/−) RNA (Mills et al., 1973).

<table>
<tr><th colspan="6">RNase T_1 digest</th><th colspan="6">RNase A digest</th></tr>
<tr><td>4</td><td>G|G|C</td><td>2</td><td>G|CAG|C</td><td>2</td><td>G|CUG|C</td><td>10½</td><td>C/U|U|C</td><td>3</td><td>C/U|AC|C</td><td>1</td><td>C/U|GAC|U</td></tr>
<tr><td>2</td><td>G|G|A</td><td>1</td><td>G|AAG|C</td><td>2</td><td>G|UCG|C</td><td>1</td><td>C/U|U|A</td><td>4½</td><td>C/U|AC|G</td><td>1</td><td>C/U|AGC|C</td></tr>
<tr><td>11</td><td>G|G|G</td><td>2</td><td>G|AAG|G</td><td>1</td><td>G|UCG|A</td><td>3</td><td>C/U|U|G</td><td>3</td><td>C/U|GC|C</td><td>2</td><td>C/U|AGC|G</td></tr>
<tr><td>6</td><td>G|G|U</td><td>1</td><td>G|CCCG|C</td><td>1</td><td>G|CUAG|C</td><td>3</td><td>C/U|U|U</td><td>3</td><td>C/U|GC|A</td><td>1</td><td>C/U|GGC|G</td></tr>
<tr><td>5</td><td>G|CG|C</td><td>1</td><td>G|CCCG|U</td><td>1</td><td>G|UACG|A</td><td>17</td><td>C/U|C|C</td><td>7</td><td>C/U|GC|G</td><td>1</td><td>C/U|GGU|C</td></tr>
<tr><td>3</td><td>G|CG|A</td><td>1</td><td>G|CACG|A</td><td>1</td><td>G|UCACG|G</td><td>4½</td><td>C/U|C|A</td><td>2</td><td>C/U|GC|U</td><td>1</td><td>C/U|GGU|U</td></tr>
<tr><td>2</td><td>G|CG|G</td><td>1</td><td>G|AAAG|C</td><td>1</td><td>G|ACUCG|U</td><td>13½</td><td>C/U|C|G</td><td>4</td><td>C/U|GU|C</td><td>1</td><td>C/U|GGGC|U</td></tr>
<tr><td>3</td><td>G|CG|U</td><td>1</td><td>G|CCACG|C</td><td>1</td><td>G|CACCUCG|U</td><td>8½</td><td>C/U|C|U</td><td>1</td><td>C/U|GU|A</td><td>1</td><td>C/U|GAAC|U</td></tr>
<tr><td>1</td><td>G|AG|C</td><td>1</td><td>G|AAACG|C</td><td>1</td><td>G|UCACCUG|G</td><td>1</td><td>C/U|C-OH</td><td>3</td><td>C/U|GU|G</td><td>1</td><td>C/U|GAGGC|G</td></tr>
<tr><td>1</td><td>G|AG|A</td><td>2</td><td>G|UG|C</td><td>1</td><td>G|UUCUCG|U</td><td>1</td><td>pppGGGGAAC|C</td><td></td><td></td><td>1</td><td>C/U|GAAGC|A</td></tr>
<tr><td>2</td><td>G|AG|G</td><td>1</td><td>G|UG|A</td><td>1</td><td>pppG|G</td><td>1</td><td>C/U|GAGGU|G</td><td></td><td></td><td>1</td><td>C/U|GAAAC|G</td></tr>
<tr><td>1</td><td>G|CCG|A</td><td>1</td><td>G|UG|G</td><td>1</td><td>G|UCCCC-OH</td><td>1</td><td>C/U|GGGGGU|C</td><td></td><td></td><td>1</td><td>C/U|GAAAGC|G</td></tr>
<tr><td>1</td><td>G|AACUCCCG|U</td><td></td><td></td><td>½</td><td>G|CCUCUUCUCG|A</td><td>1</td><td>C/U|GAAGGU|C</td><td></td><td></td><td>1</td><td>C/U|GAAGGGC|C</td></tr>
<tr><td>1</td><td>G|UCACCUCG|C</td><td></td><td></td><td>½</td><td>G|CCUCUUCACG|A</td><td>1</td><td>C/U|GGGGGGU|C</td><td></td><td></td><td>1</td><td>C/U|GGAGAGC|G</td></tr>
<tr><td>1</td><td>G|AACCCCCCUUCG|G</td><td></td><td></td><td>1</td><td>G|UCCCCCCUUCCG|G</td><td></td><td></td><td></td><td></td><td></td><td></td></tr>
</table>

strands. The MDV-1(+) catalogues of Table 5 were then subtracted from the MDV-1 (+/−) catalogues to yield the RNase A and T_1 catalogues of the minus strand listed in Table 6. Note again the presence of oligonucleotides with nonintegral molar frequencies, a consequence of the mixture of the two variants differing at one residue.

It should be emphasized that the plus and minus catalogues of Tables 5 and 6 were established virtually independently of each other. Almost without exception, the elements of each strand were identified without reference to the elements of the other. The complementarity of the plus and minus catalogues is evidence of their correctness.

The construction of extended sequence blocks

The T_1 and RNase A fragments of Tables 5 and 6 can be extended into larger sequence blocks by the use of the following devices:

(a) If a T_1 oligonucleotide uniquely overlaps an RNase A oligonucleotide, the one may be used to extend the other;

(b) Nearest residue neighbors on the 5′ end of oligonucleotides are deductible from the enzyme used, and nearest neighbors on the 3′ end are determined by identifying the [α-^{32}P]ribosidetriphosphate that labels the 3′ terminal phosphate. This nearest neighbor information logically limits the choice of the fragments to be placed on either side of a given oligonucleotide;

(c) Since MDV-1 (+) and MDV-1 (−) are antiparallel complements, one can use sequences unique in one strand to extend sequences in the other.

For example, the plus strand catalogue of Table 1 contains the unique RNase A oligonucleotide Py/GAGAAC/C, where Py denotes a pyrimidine. Searching the minus strand catalogue for unique antiparallel complements, one finds G/UUCUCG/U among the T_1 fragments. This immediately allows us to extend the plus RNase A fragment to include ACGAGAAC/C. This last sequence can be further lengthened on the 3′ side by recognizing that a T_1 cleavage of this segment should yield a fragment beginning with G/AACC . . ./, and there is only one such T_1 fragment in the plus catalogue, namely, G/AACCG/C. Since this must be on the 3′ side of ACGAGAAC/C, the sequence becomes ACGAGAACCGC.

Using these devices, all of the oligonucleotides of the plus and minus catalogues were assembled into the extended sequence blocks recorded in Fig. 13. It should be noted that for convenient referencing the blocks are not arranged in the order of their identification in the course of our analysis, but rather, according to their ultimate linear arrangement in the molecule.

The conventions used in representing these blocks may be briefly noted. Both minus (lower two) and plus (upper two) strands are represented, the latter being inverted to indicate the antiparallel structure of the two complements and to permit both to be read in the 5′ to 3′ direction. Plus and minus strands are represented twice so that the sequence information from the two nuclease catalogues could be explicitly indicated. The first and the fourth rows correspond to the RNase A fragments, and the second and third rows correspond to the T_1 oligonucleotides, the commas indicating points of enzymatic cleavage. Only the residues enclosed within the outlined figures belong to the final sequence. The outside residues on either side of each block are deduced from identity, complementarity, enzyme specificity, and nearest neighbor information. These external residues are brought in by the neighboring block. They, and the end shape of a block, help in choosing its nearest neighbor. The actual fragments that will "replace" the external residues of a block must be found within the outlined figure of the adjoining block.

Logical restrictions in connecting sequence blocks

In sequencing one strand of RNA, nearest neighbor information narrows the choice of which oligonucleotide can be placed next to another. In joining sequence blocks, nearest neighbor data play a similar but far more selective role because each sequence block contains the two complementary strands and the demands made by the nearest neighbor of both strands must be satisfied. Two blocks can join together only if the external residues of the plus and minus strands of one block are precisely replaced by

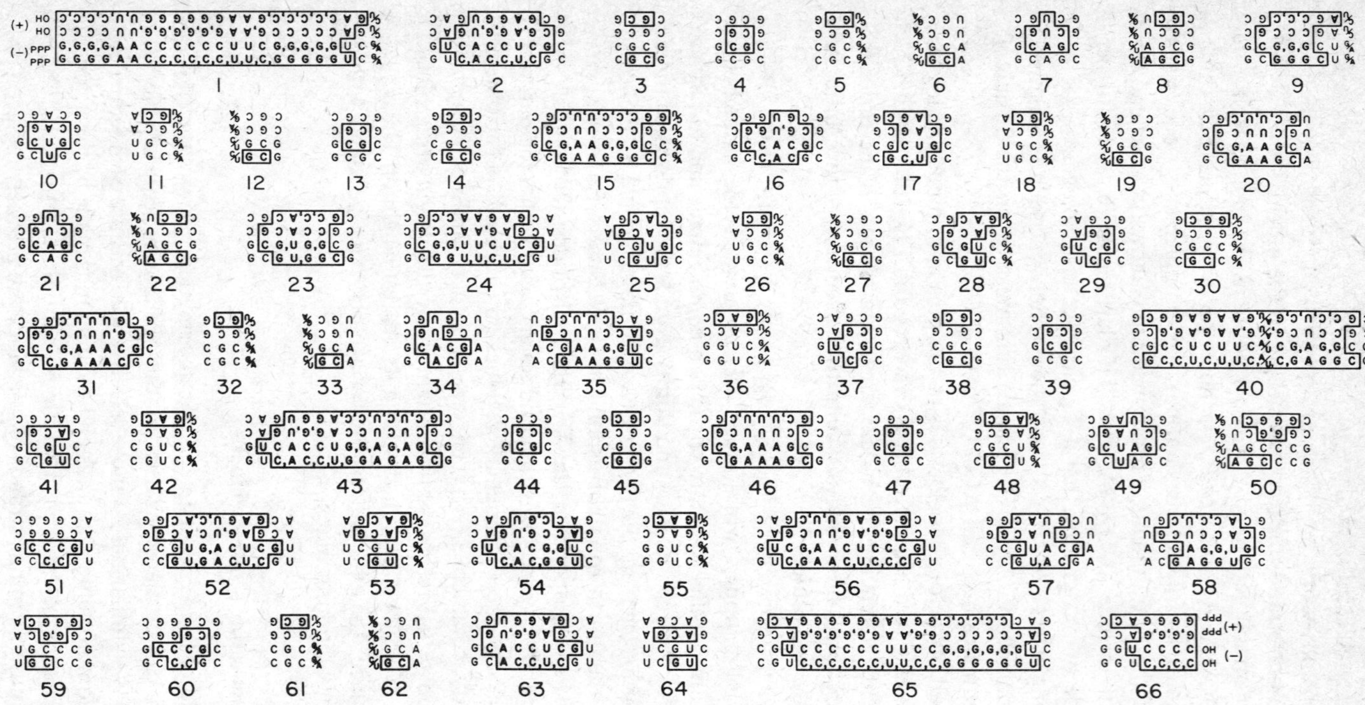

Figure 13

The extended sequence blocks.

Information obtained from enzymatic digestion (Tables 5 and 6) was combined to yield 66 sections of duplex MDV-1 RNA. Each strand is represented twice to accommodate fragments from the two enzymatic digests. The actual fragments are contained within the outline of each block, and the points of enzymatic cleavage are indicated by commas. In assembling the blocks, the smaller fragments of one catalogue were "used up" by the larger fragments of another catalogue. In some blocks, however, the overlapping of larger fragments (by complementarity of identity) extended the length of the block. The size of each block was limited by the proviso that only unique sequences could be used to extend them. If two or more different oligonucleotides could be added in the same location, the choice could not be made. The advantage of assembling catalogue data into block form is that each block cannot be combined with every other block. The identity of each block's nearest neighbor is logically restricted to only a few other blocks. For instance, the right sides of blocks 24, 52, and 63 are identical. These blocks can only be linked with the left sides of blocks 25, 53, and 64. The problem at this point was to determine which blocks on the left paired with which blocks on the right. In all, there were 15 sets of choices (Mills, Kramer and Spiegelman, 1973).

the identical internal residues of the plus and minus strands of the neighboring block.

For example, consider in Fig. 13 the problem of identifying the righthand neighbor of block 34. Examination of all the sequence blocks reveals that only two blocks, 35 and 58, fulfill the requirements. Thus, the problem has been reduced to a choice between two alternatives. Once it is known which block pairs with block 34, the remaining block can be joined with the right side of block 57, as it is the only other block with a right side identical to block 34. Having narrowed the choice this far, it is only necessary to determine which pairs are close together in the molecule. To make this and similar choices, we developed the technique of "fragment length mapping."

"Fragment length mapping" of the extended sequence blocks

We now describe the device for ordering sequence blocks that is independent of the time of their appearance or the rate at which the corresponding segment is synthesized. Since our enzyme preparation is effectively free of nuclease, every partially completed molecule found in our reaction mixture will contain the 5′ terminus. If a collection of fragments synthesized haphazardly with respect to time or speed can be separated according to size, the following question can be answered: What is the minimal fragment length required to guarantee that it contain a specified oligonucleotide? If made sufficiently precise, it is evident that the answer locates the oligonucleotide in question in terms of distance from the 5′ end of the molecule. Separating RNA fragments according to size is readily achieved by electrophoresis through polyacrylamide gels. Further, the resolution capacities can be easily adjusted by varying gel strength and degree of crosslinking to suit the RNA size class.

To insure a complete collection of fragment sizes, syntheses were carried out for periods varying from 20 to 200 seconds. The products were then pooled and purified. To make certain that fragments of only the minus strand were synthesized, two precautions were taken. The reactions were carried out with a vast excess of the plus strand to minimize initiation on any completed minus strands. Further, after the 37°C initiation step, the temperature was dropped to 20°C to discourage any new starts. Two-dimensional fingerprint patterns of the pooled products revealed only minus strand oligonucleotides.

The mixture of products was electrophoresed through a 7.2% polyacrylamide gel and yielded the size classes shown in Fig. 14A. Each of the 22 gel slices was assigned a letter. The smallest (A) contained fragments of approximately 30 nucleotides in length and the largest (V) consisted of nearly complete chains.

An interesting feature of the size distribution is the discontinuity evident in the accumulation of particular size classes. These irregularities can be explained by one or more of the following: a) Aborted syntheses might occur with higher probability at certain sites in the template; b) Particular regions might be more difficult for the enzyme to traverse, thus slowing it down; and c) The electrophoretic mobility of

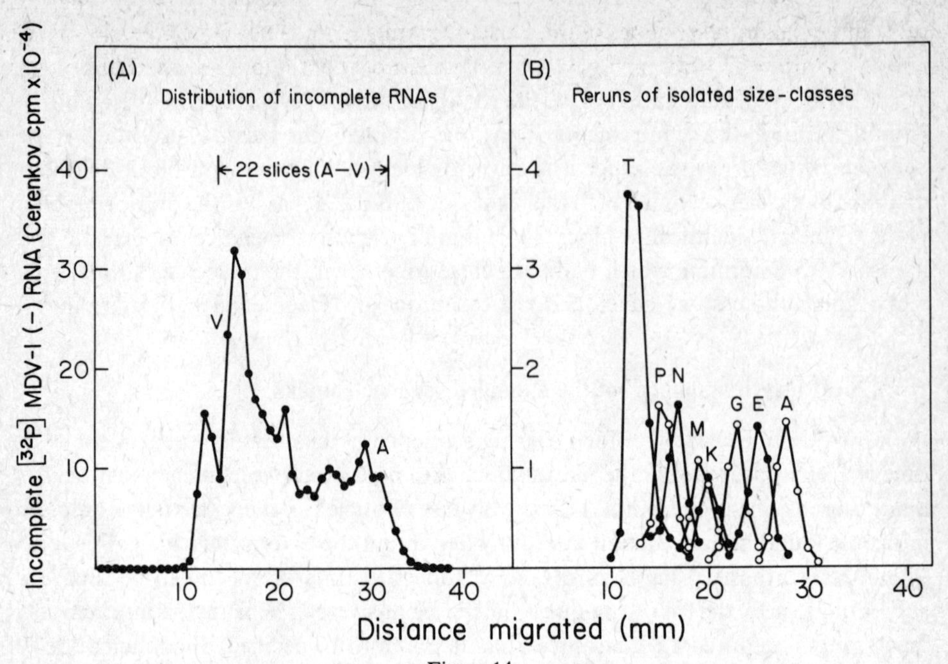

Figure 14

Isolation of incomplete MDV-1 (−) RNAs by gel electrophoresis.

A. A reaction, templated with an excess of MDV-1 (+) RNA, was run at 38°C in the absence of CTP and UTP, to assure that chain initiation would occur without chain elongation beyond the initial GGGGAA sequence. The temperature was dropped to 20°C and then CTP and UTP were added. The reaction was sampled at intervals as the chains elongated. Samples were pooled, resulting in a mixture of incomplete minus strands, all containing the 5′ end. The RNA was then purified, melted at 100°C, and separated according to length by electrophoresis through a 7.2% polyacrylamide gel. The RNA in each of 22 gel slices (labeled A through V) was isolated and rerun on 22 separate 7.2% gels.

B. The selected regels plotted above show the narrowness of the RNA size classes isolated from each slice of the initial gel. However, only RNA isolated from the peak slice of each regel was used to obtain the fragment length mapping data. The resulting 22 RNA preparations were each exceptionally homogeneous in length (Mills, Kramer and Spiegelman, 1973).

molecules in different states of completion may not be a smooth function of length. All three explanations invoke secondary structure. In any event, the observed bias toward certain size classes explains the difficulties we noted earlier in ordering timed samples from synchronized syntheses. However, the bias toward certain size classes did not interfere with the use of fragment size to locate sequence blocks.

The products contained in each of the 22 gel slices were eluted separately, precipitated with ethanol, and each was rerun on another gel. A number of these are exemplified in Fig. 14B. The RNAs in the peak slices of each of these 22 gels were eluted and purified.

In this manner, quite homogeneous size classes of MDV-1 (−) fragments, all containing the 5′ terminus, were collected. Each of these 22 size classes was divided into two aliquots for digestion, one with RNase T_1 and the other with RNase A. The molar frequency of each oligonucleotide present in the resulting two-dimensional fingerprint patterns was determined for every size class.

We illustrate how this information was used for ordering sequence blocks with several examples. Consider the RNase A oligonucleotide GAAAGC, which occurs only once in the minus strand (Table 6) and is a unique marker of sequence block 46 close to the 141st nucleotide of the minus strand. It is evident from Fig. 15 that GAAAGC rises from a molar frequency of zero to virtually one between the size classes of N and P. It is clear from the normalization to residue number shown in the lower portion of Fig. 15 that this corresponds to an average chain length of 141. We can thus locate block 46 close to the 141st nucleotide of the minus strand.

A more complicated situation is illustrated with the oligonucleotide AGC, which occurs three times in the minus strand. The AGC sequences in blocks 8 and 22 have a 3′ neighbor G, whereas the AGC in block 50 has a C neighbor. Figure 16 shows

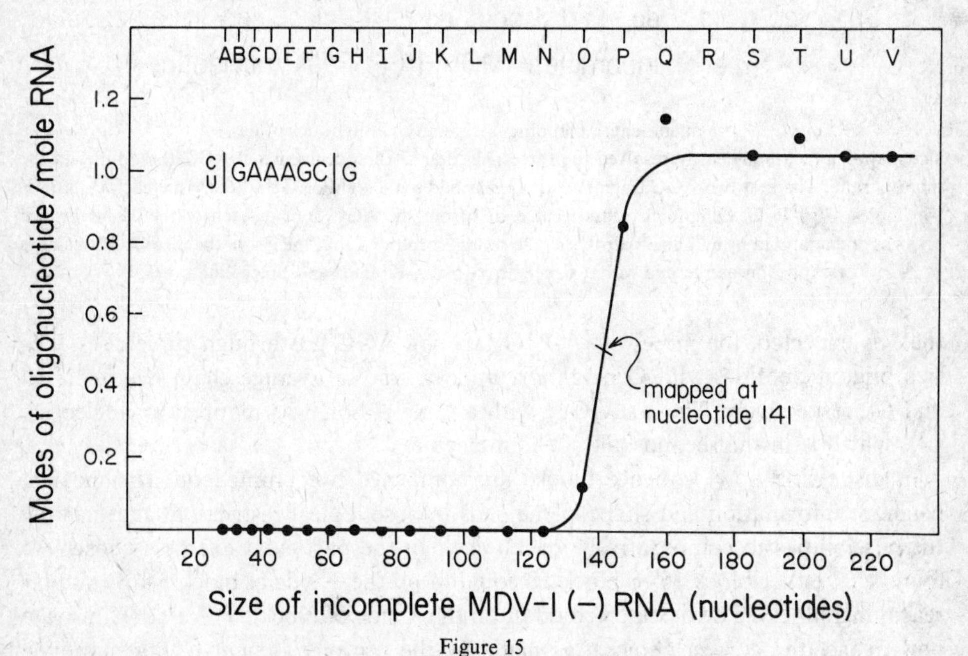

Figure 15

Fragment length mapping of an oligonucleotide.

Each of the 22 size classes of incomplete MDV-1 (−) RNA was digested with nuclease and analyzed by two-dimensional electrophoresis. The sequence GAAAGC is found as a unique spot in the RNase A fingerprint. This spot is not present in digests of incomplete RNAs smaller than size class N, but is fully present in digests of RNAs larger than size class P. By measuring the moles of GAAAGC present in each fingerprint, it was found that a minimum fragment length of 141 nucleotides was required to assure the presence of this oligonucleotide in the molecule (Mills, Kramer and Spiegelman, 1973).

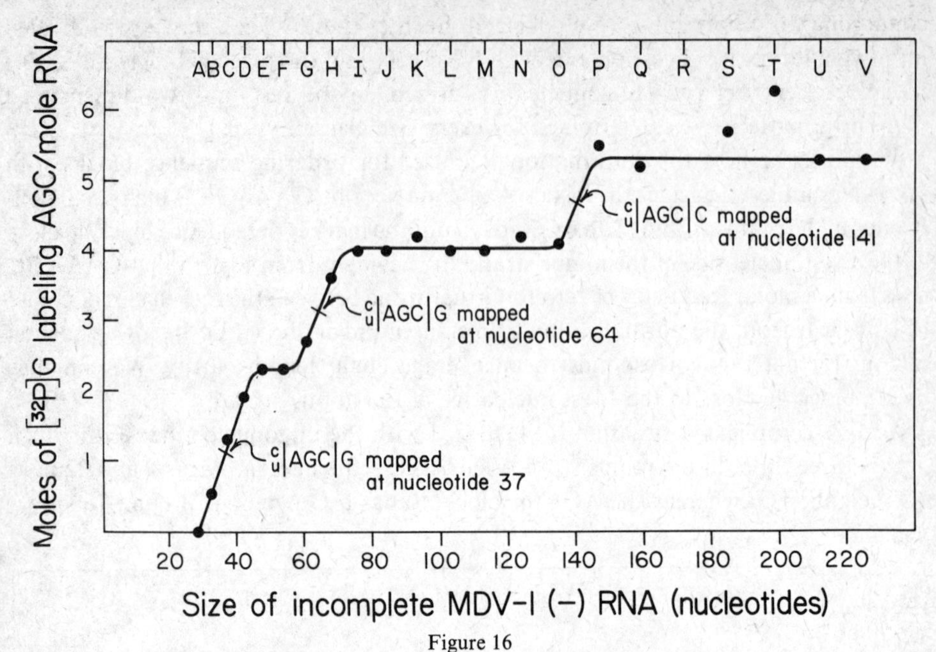

Figure 16

Fragment length mapping of three AGC oligonucleotides.

The mapping technique easily resolved the three rises due to the occurrence of AGC three times in the minus strand. The first two AGC fragments to appear had a 3′ neighbor G and therefore each contained two moles of [^{32}P]G. Oligonucleotides were eluted from the AGC spot in each of the 22 fingerprint patterns and analyzed by alkaline hydrolysis. The results confirmed that the last of the three AGC residues to appear was the one with a C neighbor (Mills, Kramer and Spiegelman, 1973).

that, as expected, the moles of [^{32}P]G labeling AGC go through three rises. The two oligonucleotides with G neighbors appear first, at average chain lengths of 37 and 64, respectively. The last AGC, with a C neighbor, was mapped at nucleotide 141, which is in the second half of the molecule.

In discussing how sequence blocks are connected, we pointed out that nearest neighbor information and shape of the ends impose logical restrictions that narrow the possibilities to comparatively few choices. In the particular example chosen we found that either block 35 or block 58 could fit on the 3′ side of block 34. By similar reasoning, block 25 or block 58 could fit on the 3′ side of block 57. We now illustrate how in fact the correct choice was made by the use of fragment length mapping.

In what follows, the oligonucleotides and the partially synthesized fragments being analyzed belong to the minus strand. Each of the four blocks we are concerned with contains a unique and identifying oligonucleotide. Thus, 34 contains the RNase T_1 fragment CACG, 35 includes the RNase A piece GAAGGU, 37 is identified by the RNase T_1 segment UACG, and finally 58 contains GAGGU, an RNase A oligonucleotide. Since each of these occurs only once in the minus strand catalogue, they

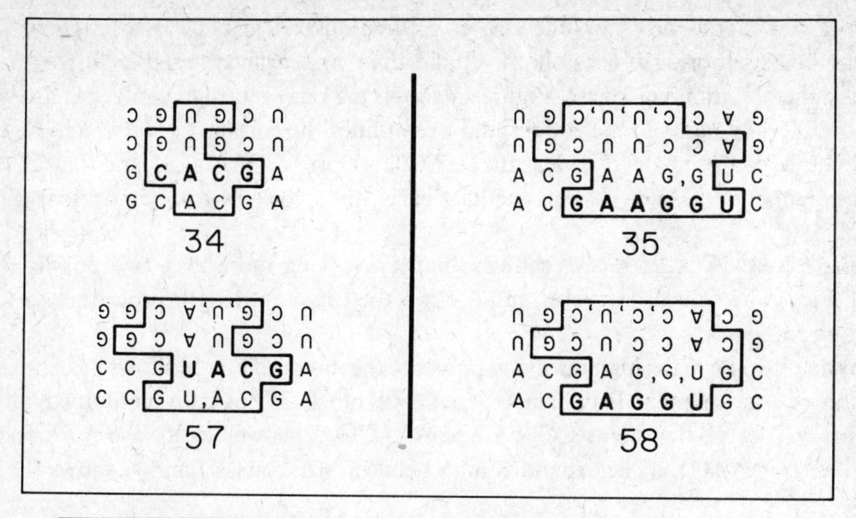

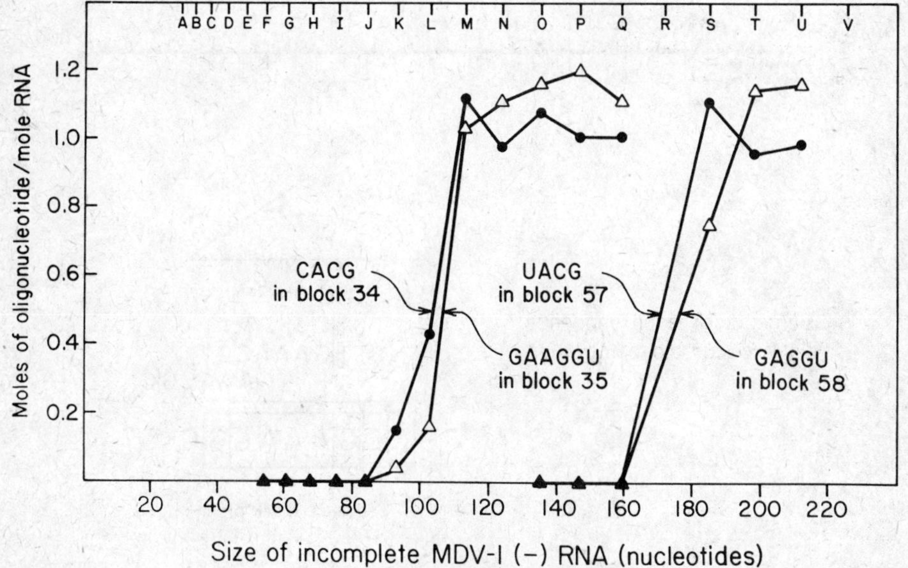

Figure 17

Solving a logical choice set.

The data obtained from fragment length mapping would not have been of sufficient resolution to have unambiguously ordered all 66 sequence blocks in Fig. 13. Fortunately, every sequence block cannot logically be linked with every other block. The permissible block pairings could be divided into 15 different sets of logical choices. The mapping data were of sufficient resolution to link the blocks in each of these choice sets. One such set is illustrated above. Note that each block contains a unique oligonucleotide (indicated in bold type), whose position in the minus strand was determined by fragment length mapping. Block 34 could logically pair with either block 35 or block 58. However, its marker oligonucleotide CACG was mapped close to GAAGGU, the marker of block 35. In addition the markers of blocks 57 and 58 map close to each other, thus solving this logical choice set (Mills, Kramer and Spiegelman, 1973).

will each undergo precisely one rise as a function of fragment length. Those that occur in neighboring blocks should make their appearance together in fragments of about equal minimal length. Figure 17 shows the relevant fragment length analyses for the identifying oligonucleotides and exemplifies the certainty with which the correct choice can be made. It is clear that CACG is close to GAAGGU and that UACG is a neighbor of GAGGU. The results clearly link block 34 to 35 and block 57 to block 58.

There were 14 other sets of choices, many involving more than two possibilities, that had to be solved. However, in all cases the fragment length procedure yielded unique solutions.

As an example of the high resolving power of the method, Fig. 18 shows an expanded region of the fragment length map. Sequence block 31 logically contained minus strand RNase T_1 fragments CCG/A and AAACG/C, as well as RNase A fragments C/G and GAAAC/G. The expanded map between size classes J and K shows the oc-

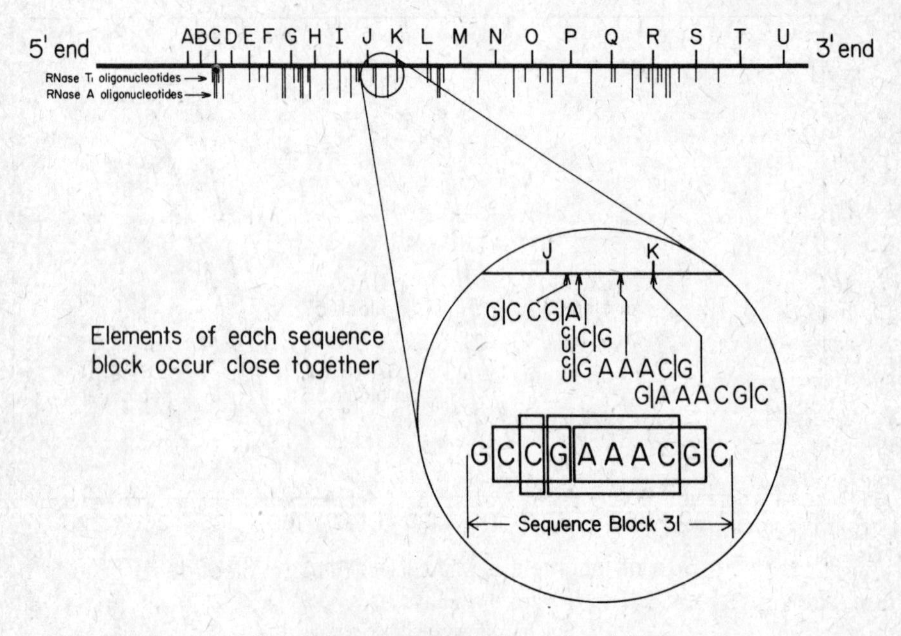

Figure 18

Confirmation of the identity and location of each block.

The oligonucleotides of each sequence block mapped close together. In no case were the predicted components of a block found in distant map locations. This confirms the sequence of each block, since the oligonucleotides were assembled into blocks without any knowledge of their map locations. For example, the oligonucleotides of block 31, shown above, were all mapped between size classes J and K. Furthermore, the close mapping of many oligonucleotides from the same block increased the certainty with which a block could be located in the sequence (Mills, Kramer and Spiegelman, 1973).

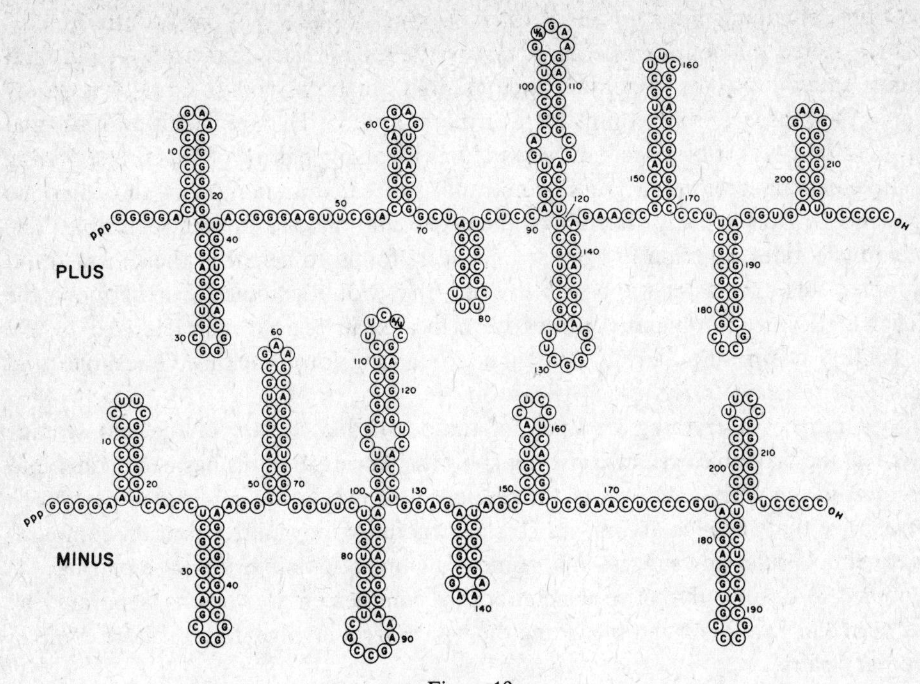

Figure 19
The complete sequences of the complementary strands of MDV-1 RNA
(Mills, Kramer and Spiegelman, 1973).

currence of each of the predicted elements. It must be emphasized that in this type of analysis the attempt is not to order the individual oligonucleotides, but rather the sequence blocks they identify. In no case, however, did the resulting map locations of the oligonucleotides suggest that the already established internal sequences of the blocks were not correct. Once the order of the sequence blocks had been established, the final sequences are determined and Fig. 19 shows the complete nucleotide sequences of both strands of MDV-1 RNA.

The existence and implications of secondary structure

In the course of our studies a variety of features appeared pointing to the conclusion that the plus and minus strands were highly structured. Both complements had a high G–C content (70%) and a self-complementary nucleotide composition, with guanosine and adenosine each being equal in molar frequency to their respective complements. We also observed that the plus strand was very resistant (70%) to RNase A under salt conditions that encouraged base pairing. Further, during our sequence studies we noted that some T_1 oligonucleotides were found in lower than expected yield at their correct position in the two-dimensional fingerprint, the missing portions remaining in the nuclease-resistant "core" at the top of the pattern. Finally, we discovered that a number of large oligonucleotides occurred in both the plus and minus catalogues. Any oligonucleotide common to both strands must of necessity imply the presence of its complement in both strands. This immediately generated the possibility of forming secondary structures involving pairing of the corresponding complementary regions in the same strand. Indeed, examination of the complete sequences made it likely that all of these expectations are in fact realizable. The oligonucleotides common to both strands were found to neighbor their intrastrand complements, increasing the possibility that they would complex. Furthermore, the deficient RNase T_1 oligonucleotides were located in regions characterized by the possibility of forming hairpin structures containing loops lacking G residues and therefore resistant to RNase T_1 digestion.

In summary, everything we knew of the detailed chemistry of the two strands dictated the probable existence of extensive base pairing and this has been taken into account in the representations of the sequences shown in Fig. 19. It is important to emphasize that the plus and minus strand structures shown are based on comparatively simple rules and suggest only some of the many possibilities. Base pairing was assumed to occur only between neighboring complementary regions, separated by between 3 to 7 residues, and providing the helical stem involves four or more complementary pairs.

Consequences of complementarity for the evolution of structure

As we pointed out in our opening paragraph, it is plausible to argue that the ability of complementary bases to pair via hydrogen bonds made possible the emergence of nucleic acids as self-instructive replicating entities. The use of base pairing demands that replication occur via complementary intermediates and this in turn introduces restraints on the evolutionary pathway. Thus, if a given variant is superior because of a particular sequence in its plus strand, optimization of the selective advantage would encourage the inclusion of the same sequence in the minus strand. This can be assured *only* if the plus strand contains the complement of the advantageous sequence. Note, however, that as soon as this device is employed both strands will necessarily contain intrastrand complements and a new possibility emerges, namely, the formation of helixes between the intrastrand complementary regions. The recognition and use of such structures literally adds new dimensions to the types of selective interactions that can occur. The replicating molecules would no longer be limited to exploiting differences in the linear properties of their sequences but could in addition explore the usefulness of their two- and three-dimensional consequences.

In the course of evolution, advantageous structures would be conserved. A structure retained in one strand would be expected to occur in the other. Thus, two complementary strands can have very different primary sequences (genotypes) and yet possess similar secondary and tertiary structures (phenotypes). As a result of this evolutionary process, self-replicating molecules would become highly structured and would possess many regions of intrastrand complementarity.

It will not escape the reader's notice that the double-stranded DNA we know today could not have participated in this kind of evolutionary game. A replicating molecule, such as DNA, probably arose later when environmental selection operated not on the gene but on the gene product. Under these circumstances secondary structures of the genetic material would become irrelevant.

The structures described in Fig. 19 illustrate many of the consequences of intrastrand complementarity. The helical portion of each hairpin in the plus strand has an identical counterpart in a complementary stem in the minus strand. For example, the helix between nucleotides 148 and 170 of the plus strand is identical to the helix in the minus strand between nucleotides 49 and 71. Further, it will be noted that the 5′ and 3′ ends of each strand are extensively complementary. One can therefore construct a distant intrastrand match involving 16 of the 20 nucleotides of each end. The fact that this and other structures can be formed with both strands is related to the ability of the replicase to accept either strand as a suitable template. It is of interest to recall that an end-to-end duplex of this nature was proposed (Haruna and Spiegelman, 1965b) to explain the observation that fragments of Qβ RNA could not be duplicated by Qβ replicase.

A striking characteristic of MDV-1 is its extensive secondary structure, a feature to be expected of a highly evolved replicating molecule. Indeed, if MDV-1 had been brought back in a moon sample, its sequence would have revealed it as a molecule with an evolutionary past. As such, MDV-1 would have been identified not as a random polymer but either as a self-instructive replicating entity or a derivative of one.

Summary

The primary purpose of the present paper was to describe an extracellular experimental approach that would permit one to explore the forces that could drive replicating nucleic acid molecules to evolve towards greater complexity. We demonstrated that the Qβ-replicase system possessed the necessary requisites and showed that mutant RNA molecules with prespecified phenotypes are readily isolated by exerting the appropriate selective pressures. The ease with which such mutants could be obtained suggested that secondary and tertiary structures were being recognized by the environmental selective agents.

To fully exploit the chemical information inherent in this system, it was essential that such evolutionary experiments be carried out with a replicating molecule of known sequence. This task has been accomplished and the complete sequence of MDV-1, a 218-nucleotide variant, was obtained.

A number of implications for the rules of precellular evolution emerged from the sequence and its structure, which we may briefly note.

1. The primary sequence contains a surprising number of neighboring intrastrand complements, a feature generating the potentiality for extensive secondary and tertiary structures containing loops and antiparallel stems.

2. The existence of these structures would permit these molecules to go beyond their primary sequences and exploit the selective advantages of their two- and three-dimensional consequences.

3. Environmental selection of structure and the consequent distinction between genotype and phenotype could thus occur before the primary sequence was used for translation.

4. To maximize selective advantages, antiparallel complementary copying would require similar advantageous structures in plus and minus strands. a situation that would inevitably lead to stem and loop formation and, hence, to greater structural complexity.

5. The use of these kinds of evolutionary forces necessarily had a determining effect on the structure of the genetic language and its syntax even before translation to proteins evolved. Thus, complementary copying requires at least two bases. The use of structure as a selection agent demands at least *four*, the ultimate basic number of the genetic alphabet. Further, even restraints on certain sequence runs could occur at this pretranslational stage to stabilize certain stems and loops.

6. Finally, so long as this sort of structural selection was employed, the replicating mechanism had to remain one that emphasizes the existence of open, predominantly single-stranded replicating structures. This process is probably going on in the present-day biological universe during the intracellular replication of RNA viruses and the translation of genetic messages where secondary and tertiary structures could still exert a selective role.

7. It is only with the appearance of the first cell that the structural "phenotype" of the cellular gene would become irrelevant as a selection element. One could then afford to store translatable genetic information in the perfectly paired double helix we know today, a structure that can no longer personally participate as an environmentally recognized entity in the game of evolution.

Acknowledgments

This work was supported by National Science Foundation grant GB-17251X2, and National Institutes of Health, National Cancer Institute research grant CA-02332.

References

BILLETER, M. A., DAHLBERG, J. E., GOODMAN, H. M., HINDLEY, J. and WEISSMANN, C., 1969. "Sequence of the first 175 nucleotides from the 5 terminus of Qβ RNA synthesized *in vitro*." *Nature* **224**: 1083–1086.

BISHOP, D. H. L., CLAYBROOK, J. R. and SPIEGELMAN S., 1967. "Electrophoretic separation of viral nucleic acids on polyacrylamide gels." *J. Mol. Biol.* **26**: 373–387.

BISHOP, D. H. L., MILLS, D. R., and SPIEGELMAN, S., 1968. "The sequence at the 5′ terminus of a self-replicating variant of viral Qβ-RNA." *Biochem.* **7**: 3744–3753.

DOI, R. H. and SPIEGELMAN S., 1962. "Homology test between the nucleic acid of an RNA virus and the DNA in the host cell." *Science* **138**: 1270–1272.

GIACOMONI, D. and SPIEGELMAN, S., 1962. "Origin and biologic individuality of the genetic dictionary." *Science* **138**: 1328–1331.

GOODMAN, H. M. and RICH, A., 1962. "Formation of a DNA-soluble RNA hybrid and its relation to the origin, evolution, and degeneracy of soluble RNA." *Proc. Nat. Acad. Sci. USA* **48**: 2101–2109.

HALL, B. D. and SPIEGELMAN, S., 1961. "Sequence complementarity of T2-DNA and T2-specific RNA." *Proc. Nat. Acad. Sci. USA* **47**: 137–146.

HARUNA, I. and SPIEGELMAN, S., 1965a. "Specific template requirements of RNA replicase." *Proc. Nat. Acad. Sci. USA* **54**: 579–587.

HARUNA, I. and SPIEGELMAN, S., 1965b. "Recognition of size and sequence by an RNA replicase." *Proc. Nat. Acad. Sci. USA* **54**: 1189–1193.

HARUNA, I., NOZU, K., OHTAKA, Y. and SPIEGELMAN, S., 1963. "An RNA 'replicase' induced by and selective for a viral RNA: Isolation and properties." *Proc. Nat. Acad. Sci. USA* **50**: 905–911.

JOSE, J., KAISER, A. D. and KORNBERG, A., 1961. "Frequencies of nearest neighbor base sequences in deoxyribonucleic acid." *J. Biol. Chem.* **236**: 864–875.

KACIAN, D. L., MILLS, D. R., KRAMER, F. R. and SPIEGELMAN, S., 1972. "A replicating RNA molecule suitable for a detailed analysis of extracellular evolution and replication." *Proc. Nat. Acad. Sci. USA* **69**: 3038–3042.

LEVISOHN, R. and SPIEGELMAN, S., 1968. "The cloning of a self-replicating RNA molecule." *Proc. Nat. Acad. Sci. USA* **60**: 866–872.

LEVISOHN, R. and SPIEGELMAN, S., 1969. "Further extracellular Darwinian experiments with replicating RNA molecules: diverse variants isolated under different selective conditions." *Proc. Nat. Acad. Sci. USA* **63**: 805–811.

LOEB, T. and ZINDER, N. D., 1961. "A bacteriophage containing RNA." *Proc. Nat. Acad. Sci. USA* **47**: 282–289.

MILLS, D. R., BISHOP, D. H. L. and SPIEGELMAN, S., 1968. "The mechanism and direction of RNA synthesis templated by free minus strands of a 'little' variant of Qβ RNA." *Proc. Nat. Acad. USA* **60**: 713–720.

MILLS, D. R., KRAMER, F. R. and SPIEGELMAN, S., 1973. "Nucleotide sequence of a replicating RNA molecule with evolutionary potential." *Science* **180**: 916–928.

MILLS, D. R., PACE, N. R. and SPIEGELMAN, S., 1966. "The *in vitro* synthesis of a noninfectious complex containing biologically active viral RNA." *Proc. Nat. Acad. Sci. USA* **56**: 1778–1785.

MILLS, D. R., PETERSON, R. L. and SPIEGELMAN, S., 1967. "An extracellular Darwinian experiment with a self-duplicating nucleic acid molecule." *Proc. Nat. Acad. Sci. USA* **58**: 217–224.

OVERBY, L. R., BARLOW, G. H., DOI, R. H., JACOB, M. and SPIEGELMAN, S., 1966a. "Comparison of two serologically distinct ribonucleic acid bacteriophages. I. Properties of the viral particles." *J. Bact.* **91**: 442–448.

OVERBY, L. R., BARLOW, G. H., DOI, R. H., JACOB, M. and SPIEGELMAN, S., 1966b. "Comparison of two serologically distinct ribonucleic acid bacteriophages. II. Properties of the nucleic acids and coat proteins." *J. Bact.* **92**: 739–745.

PACE, N. R. and SPIEGELMAN, S., 1966a. "The synthesis of infectious RNA with a replicase purified according to its size and density." *Proc. Nat. Acad. Sci. USA* **55**:1608–1615.

PACE, N. R. and SPIEGELMAN, S., 1966b. "*In vitro* synthesis of an infectious mutant RNA with a normal RNA replicase." *Science* **153**:64–67.

PARANCHYCH, W. and GRAHAM, A. F., 1962. "Isolation and properties of an RNA-containing bacteriophage." *J. Cell. Comp. Physiol.* **60**:199–208.

SAFFHILL, R., SCHNEIDER-BERNLOEHR, H., ORGEL, L. E. and SPIEGELMAN, S., 1970. "*In vitro* selection of bacteriophage Qβ ribonucleic acid variants resistant to ethidium bromide." *J. Mol. Biol.* **51**:531–539.

SANGER, F., BROWNLEE, G. G. and BARRELL, B. G., 1965. "A two-dimensional fractionation procedure for radioactive nucleotides." *J. Mol. Biol.* **13**:373–398.

SPIEGELMAN, S., 1971. "An approach to the experimental analysis of precellular evolution." *Quart. Rev. of Biophysics* **4**:213–253.

SPIEGELMAN, S., HARUNA, I., HOLLAND, I. B., BEAUDREAU, G. and MILLS, D., 1965. "The synthesis of a self-propagating and infectious nucleic acid with a purified enzyme." *Proc. Nat. Acad. Sci. USA* **54**: 919–927.

SPIEGELMAN, S., PACE, N. R., MILLS, D. R., LEVISOHN, R., EIKHOM, T. S., TAYLOR, M. M., PETERSON, R. L. and BISHOP, D. H. L., 1968. "The mechanism of RNA replication." *Cold Spring Harbor Symposia on Quant. Biol.* **33**:101–124.

STRAUSS, J. H. and SINSHEIMER, R. L., 1963. "Purification and properties of bacteriophage MS2 and of its ribonucleic acid." *J. Mol. Biol.* **7**:43–53.

SULSTON, J., LOHRMANN, R., ORGEL, L. E., SCHNEIDER-BERNLOEHR, H., WEIMANN, B. J. and MILES, H. T., 1969. "Non-enzymic oligonucleotide synthesis on a polycytidylate template." *J. Mol. Biol.* **40**:227–234.

TOMLINSON, R. V. and TENER, E. M., 1963. "The effect of urea, formamide and glycols on the secondary binding forces in the ion exchange chromatography of polynucleotides on DEAE cellulose." *Biochemistry* **2**:697–702.

VON NEUMANN, J., 1966. *Theory of Self-reproducing Automata.* Urbana: University of Illinois Press.

WATANABE, I., 1964. "Persistent infection with an RNA bacteriophage." *Nihon Rinsho* **22**:1187–1195.

WEISSMANN, C., FEIX, G. and SLOR, H., 1968. "*In vitro* synthesis of phage RNA: The nature of the intermediates." *Cold Spring Harbor Symp. Quant. Biol.* **33**:83–100.

YANKOFSKY, S. A. and SPIEGELMAN, S., 1962a. "The identification of the ribosomal RNA cistron by sequence complementarity. I. Specificity of complex formation." *Proc. Nat. Acad. Sci. USA* **48**:1069–1078.

YANKOFSKY, S. A. and SPIEGELMAN, S., 1962b. "The identification of the ribosomal RNA cistron by sequence complementarity. II. Saturation of and competitive interaction at the RNA cistron." *Proc. Nat. Acad. Sci. USA* **48**:1466–1472.

Antigen design and immunological recognition*

MICHAEL SELA

Department of Chemical Immunology, The Weizmann Institute of Science, Rehovot, Israel

The immune response is specific, and it is based on the process of selection rather than instruction. While there is still some uncertainty concerning the nature of development of the capacity to respond immunologically by antibody formation in mature animals, there is no doubt that the information, as is the case for the biosynthesis of other proteins, is encoded in DNA. There are several levels of control of gene expression, resulting in an increase or decrease in the formation of antibodies or of a cell-mediated immune response. The specific genetic control of immune response is presently an active field of research that will hopefully yield a better understanding of the regulation of immune response.

The immune response is triggered by the reaction of an immunogen with a specific cell receptor. The idea that this triggering is mediated by allosteric changes within the receptor molecule is an appealing one, but the possibility that the triggering occurs via a multivalent attachment of the immunogen to several receptors should not be excluded, especially in view of the fluidity of proteins within a membrane.

Changes in the antibody molecule

Little experimental evidence has been brought forward until this day concerning conformative changes within the active site of an antibody as a result of its reaction with a hapten. In a collaborative study with Prof. Kratky and Prof. Pilz, Mr. Licht and I have shown recently that, as measured by small angle X-ray scattering, the anti-poly-D-alanyl antibody molecule becomes more compact upon reacting with tetra-D-alanine (Pilz, Kratky, Licht and Sela, 1973).

* This paper was prepared for publication while the author was a Fogarty Scholar-in-Residence at the Fogarty International Center, National Institutes of Health, U.S. Public Health Service, Bethesda, Maryland.

Small angle X-ray scattering is a useful technique for the investigation of the structure of biopolymers in solution. Pilz et al. (1970) have used it to investigate a homogeneous human immunoglobulin IgGl, a myeloma protein of known primary structure. Their results suggested that the Fab and Fc regions of the molecule are relatively compact, but that the whole molecule has an extended structure in solution, fitting a T-shaped model in which the fragments are linked by a flexible hinge. This T-shaped model was confirmed by David Davies and his colleagues (Sarma et al., 1971) in the first crystal X-ray study of an immunoglobulin G.

We have now enquired whether there are any changes upon reaction of an antibody with its hapten, and found that a volume contraction occurs upon the interaction of anti-poly-D-alanyl antibodies with the tetra-D-alanine hapten (Pilz et al., 1973). As seen in Fig. 1, the anti-poly-D-alanyl antibodies obtained by immunization with poly-D-alanyl diphtheria toxoid showed a decrease of the volume by 10% and a decrease of the radius of gyration by 7.7% when 90% of the binding sites were occupied by hapten. Since other data determined for the antibodies, such as molecular weight, radii of gyration of the cross-section, and form of the scattering curve, were unchanged upon interaction with hapten within the errors of measurement, it must be assumed that the volume contraction is due to a change in conformation upon interaction.

It seems that the changes in the antibody molecule resulting from its interaction with the hapten may be due either to a contraction involving the less densely packed

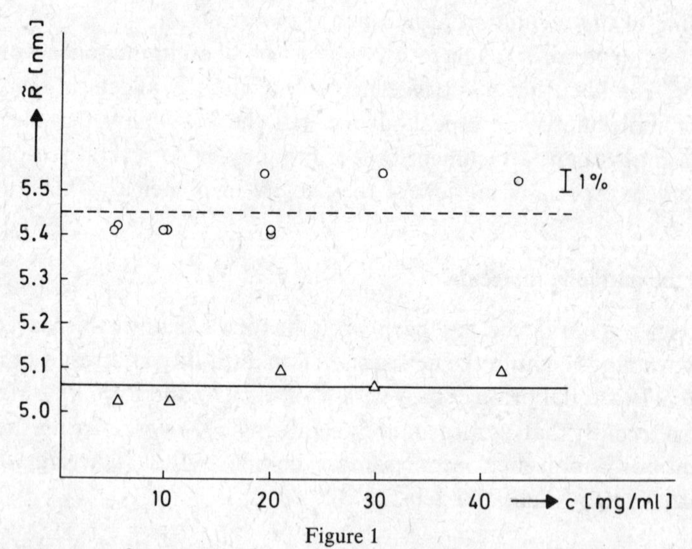

Figure 1

Apparent radii of gyration $\tilde{R}$ plotted versus the concentration c of anti-poly-D-alanyl diphtheria toxoid antibodies in solution: ○ values of the free antibodies; △ values of the antibodies having 90% of the active sites occupied by hapten. From Pilz et al. (1973).

regions of the molecule, or to a decrease of flexibility in the "hinge" region. An investigation by small angle X-ray scattering of the Fab and $(Fab')_2$ fragments of the antibody, in the presence and absence of hapten, may throw additional light on this problem.

Changes in the conformation of antigen

It is very difficult to show clearly conformational changes induced within the antibody combining site by an antigenic determinant, and the evidence accumulated until now is rather fragmentary, with the data described above, obtained by small X-ray scattering, being among the most suggestive. On the other hand, there is much more solid evidence on the transconformation within an antigen induced by antibodies.

In protein and polypeptide antigens it is possible to distinguish between "sequential" determinants and conformation-dependent determinants (Sela, Schechter, Schechter and Borek, 1967; Sela, 1969). I would like to call a sequential determinant one due to an amino acid sequence in a random coil form, and antibodies to such a determinant are expected to react with a peptide of identical, or similar, sequence. On the other hand, a conformational determinant results from the steric conformation of the antigenic macromolecule, and leads to antibodies which would not necessarily react with peptides derived from that area of the molecule. It seems that antibodies to native proteins are directed mostly against conformational rather than sequential determinants. By the way, similar argumentation makes it possible to distinguish between sequential and conformational determinants in other macromolecules, such as nucleic acids.

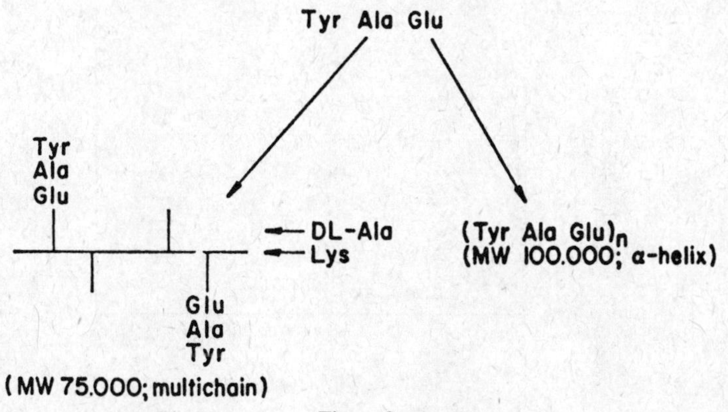

Figure 2
Synthetic branched polymer in which peptides of sequence Tyr-Ala-Glu are attached to the amino termini of polymeric side chains in multipoly-DL-alanyl–poly-L-lysine (left) and of a periodic polymer of the tripeptide Tyr-Ala-Glu (right).

In contrast to antibodies against sequential determinants, antibodies against conformation-dependent determinants usually do not react with small peptides derived from the same area of the molecule but devoid of the characteristic conformation. The same tripeptide, Tyr-Ala-Glu, may be either attached to a branched polymer of alanine (and then it behaves as a sequential determinant) or may be polymerized to yield a high molecular weight periodic polymer which exists in an α-helical form under physiologic conditions (Sela et al., 1967; Schechter et al., 1971a). The two polymers are denoted, respectively, TAG-A–L and $(TAG)_n$ (Fig. 2). There is no immunological cross-precipitation between the two polymeric systems. The peptides Tyr-Ala-Glu and $(Tyr\text{-}Ala\text{-}Glu)_2$ are efficient inhibitors of the TAG-A–L homologous system (Fig. 3), but not of the $(TAG)_n$ system. Even $(Tyr\text{-}Ala\text{-}Glu)_3$ is a poor inhibitor of the reaction of $(TAG)_n$ with anti-$(TAG)_n$ antibodies, whereas $(Tyr\text{-}Ala\text{-}Glu)_7$ and $(Tyr\text{-}Ala\text{-}Glu)_9$ are effective inhibitors of this system (Fig. 4). It has to be assumed that only those peptides of $(Tyr\text{-}Ala\text{-}Glu)_n$ structure that can fold themselves into an α-helical conformation will react with anti-$(TAG)_n$ antibodies. Indeed, the smaller peptides ($n = 1$ to 4) show a very weak interaction with these antibodies. On the other hand, peptides with $n = 7$ and $n = 9$ are quite efficient inhibitors (Fig. 4). This may mean that either these peptides possess, under physiological conditions, a structure similar to that of $(TAG)_n$ or that they are devoid of such structure, but capable of acquiring it upon interaction with the specific combining site on the antibody.

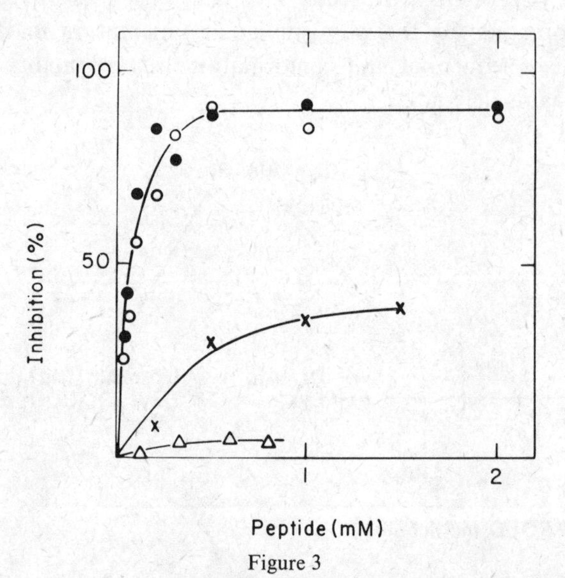

Figure 3

Inhibition of the precipitates obtained by reacting (TAG)-A–L (100 μg) with a purified anti-(TAG)-A–L antibody solution (1 ml, 0.9 OD/ml at 280 nm) by the peptides Tyr-Ala-Glu (●);$(Tyr\text{-}Ala\text{-}Glu)_2$ (○); t-Boc-Tyr-Ala-Glu (X), and $(DAla)_4$ (△). From Schechter et al. (1971a).

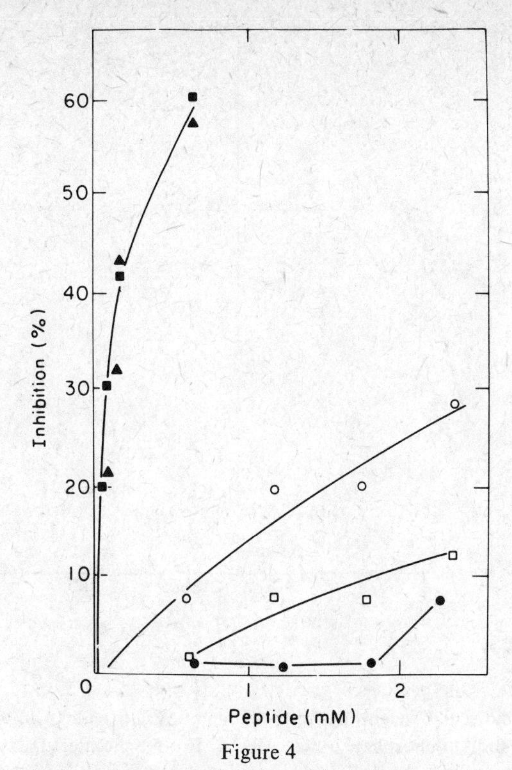

Figure 4

Inhibition of the homologous reaction in the $(TAG)_n$-anti-$(TAG)_n$ system, by the peptides Tyr-Ala-Glu (●); $(\text{Tyr-Ala-Glu})_2$ (□); $(\text{Tyr-Ala-Glu})_3$ (○); $(\text{Tyr-Ala-Glu})_7$ (▲), and $(\text{Tyr-Ala-Glu})_9$ (■). The last two peptides were obtained by polymerization techniques, and they represent average values. From Schechter et al. (1971a).

Circular dichroic studies (Fig. 5) showed clearly that the peptides with $n = 7$ and $n = 9$ possess very little α-helical conformation, if at all (Schechter, et al., 1971b), and led to the conclusion that a transconformation into the correct antigenic determinant occurs in these peptides only upon reaction with antibodies, most probably due to the decrease in entropy, accompanying the antibody-peptide reaction. We have examined this problem experimentally by checking whether the addition of Fab fragments derived from anti-$(TAG)_n$ antibodies to the peptide $(\text{Tyr-Ala-Glu})_{13}$—which has a very low helical content but is able to form precipitates with anti-$(TAG)_n$ antibodies—could transconform this oligopeptide into a structure more similar to the high molecular weight polymer (Schechter, Conway-Jacobs and Sela, 1971).

The circular dichroic measurements illustrated in Fig. 6 detected profound changes upon the interaction of $(\text{Tyr-Ala-Glu})_{13}$ with the Fab fragment of antibodies against the helical $(TAG)_n$. No such changes occurred upon reacting tetra-L-alanine with antibodies against a non-helical short poly-L-alanyl segment, nor were any changes

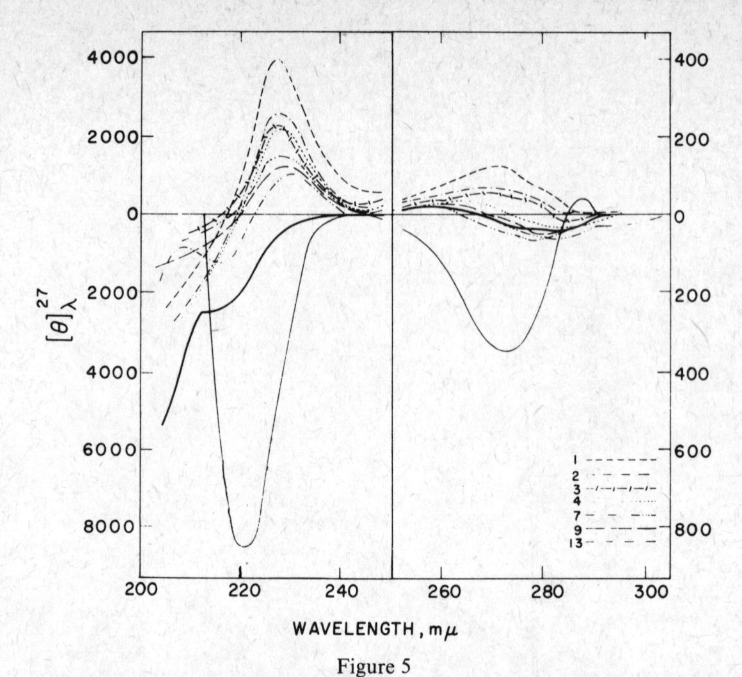

Figure 5

Circular dichroic spectra of the peptides (Tyr-Ala-Glu)$_n$, n = 1, 2, 3, 4, 7, 9, 13, the random copolymer (TAG)$_r$ and of the high molecular weight ordered polymer (TAG)$_n$ (n = 200), in 0.15 M sodium chloride–0.02 M sodium phosphate buffer, pH 7.4. From Schechter et al. (1971 b).

in ellipticity observed upon mixing (Tyr-Ala-Glu)$_{13}$ with the Fab fragment of normal IgG. The difference spectrum observed with a mixture of (Tyr-Ala-Glu)$_{13}$ and anti-(TAG)$_n$-Fab shows a large peak at 278 nm corresponding to the ellipticity band of (Tyr-Ala-Glu)$_{13}$ (Fig. 5). When Fab and (Tyr-Ala-Glu)$_{13}$ were mixed in a molar ratio of 1 to 6, a 3-fold increase in the ellipticity value of (Tyr-Ala-Glu)$_{13}$ at 278 nm was observed. At a ratio of 1 to 1.6 a 6-fold increase was observed.

The difference spectrum observed could be due to an effect of Fab on (Tyr-Ala--Glu)$_{13}$ or *vice versa.* It seems very unlikely, however, that if a tyrosine chromophore of Fab was affected by binding to (Tyr-Ala-Glu)$_{13}$ the change in the resultant circular dichroism spectrum would be so large, precisely at the ellipticity band of (Tyr-Ala--Glu)$_{13}$ and in the direction of more ordered (Tyr-Ala-Glu)$_{13}$. For this reason it was concluded that the ellipticity changes shown in Fig. 6 are mainly due to an increase in the helical content of (Tyr-Ala-Glu)$_{13}$.

Thus, antibodies against a helical structure may induce the helical tendency of a polypeptide which is largely in a random coil form when in free solution. This has been concluded also by Crumpton and Small (1967) for a peptide derived from myoglobin which, although in a random coil form, was capable of reacting with antibodies against a helical segment within this protein. In more precise language,

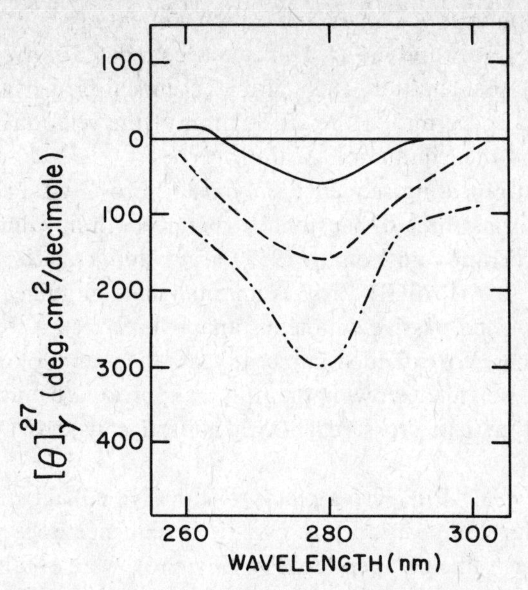

Figure 6

Circular dichroism spectra of (Tyr-Ala-Glu)$_{13}$ in the presence and absence of Fab fragments derived from anti-(TAG)$_n$ antibodies in 0.15 M sodium chloride—0.02 M sodium phosphate, pH 7.4. The ellipticity values of the Fab fragment were subtracted from those of the mixture. Solid line, (Tyr-Ala-Glu)$_{13}$ alone; dashed, (Tyr-Ala-Glu)$_{13}$ with Fab in a ratio of 6 to 1 (mole/mole); dash-dotted, (Tyr-Ala-Glu)$_{13}$ with Fab in a ratio of 1.6 to 1 (mole/mole). From Schechter et al. (1971c).

this may mean that the antibodies affect the equilibrium between the random coil and the helical forms, or – more precisely – between the random coil forms and those forms which possess at least a nucleation leading to the helical shape.

Collagen model

Another model, which we have studied immunologically in detail in recent years, is that of collagen (Fuchs, Maoz and Sela, 1974). Collagen is a structural scleroprotein which constitutes about 30% by weight of all proteins in the mammalian body, in diverse tissues. The immunological properties of collagen present an intriguing problem for several reasons. It is the most abundant protein, it has a unique amino acid sequence and conformation, and, besides being a structural protein, collagen may be involved in several biological functions such as autoimmune conditions, senescence, wound healing, etc.

In order to elucidate the nature of the collagen fold conformation, collagen models were synthesized in the last decade by polymerization of tri- and hexa-peptides (Engel, Kurz, Katchalski and Berger, 1966; Segal, 1969). The resulting ordered polypeptides have in common a sequential occurrence of glycine at every third position along the polypeptide chain and a high content of proline. The availability of these ordered polymers, and the data already accumulated concerning the relationships between their sequence and their physicochemical properties in solution and in the solid state, also permitted the elucidation of the relationship between their chemical nature and their immunological properties.

The first ordered repeating sequence polymer (L-Pro-Gly-L-Pro)$_n$ was shown to resemble collagen in its three dimensional structure both in solution and its X-ray diffraction pattern (Traub and Yonath 1965; Traub and Yonath, 1966). Borek et al. (1969) have shown that (Pro-Gly-Pro)$_n$ is immunogenic in guinea pigs and rabbits, and demonstrated, using passive cutaneous anaphylaxis test, a weak cross-reaction with several collagens. We extended in recent years the immunological characterization of this ordered polymer, showing that it is an appropriate immunological model for collagen and thus can cross react with natural collagens (Maoz, Fuchs and Sela, 1973a, 1973b).

The ordered (Pro-Gly-Pro)$_n$ cross-reacts with native collagens, whereas a copolymer with a similar composition but a random sequence does not. Specific antibodies to the collagen-like polypeptide (Pro-Gly-Pro)$_n$ were obtained, in high titers, by immunization with covalent conjugates of the polymer with carrier proteins. The antibodies to (Pro-Gly-Pro)$_n$ which were obtained were shown to be specific to the unique collagen-like conformation of the polymer. They cross-react with other collagen-like polyhexapeptides; they cross-react very poorly with the random copolymer (Pro^{66}, Gly^{34})$_n$ which is similar to the ordered polymer in amino acid composition, but is not collagen-like; they do not cross-react with the tripeptide Pro-Gly-Pro, which is the building unit of the ordered polymer; collagenase digestion of (Pro-Gly-Pro)$_n$ abolishes its binding to the antibodies. It was shown that the cross-reaction between (Pro-Gly-Pro)$_n$ and related polymers depends on the size and conformation of the polymers, both in the humoral and cellular levels of the immune response (Maoz, Fuchs and Sela, 1973a). Thus, e.g., the capacity of the ordered polymer to react with anti-"ordered polymer" antibodies decreased markedly as a result of digestion with collagenase (Fig. 7).

The collagen-like polypeptide (Pro-Gly-Pro)$_n$ elicits in guinea pigs an immune response which cross-reacts with collagens of several species by delayed and immediate skin reactions (Maoz, Fuchs and Sela, 1973a). The broad range cross-reactivity with natural collagens includes *inter alia* cross-reactions with collagen of the same animal species, namely, guinea pig skin collagen. No immunological cross-reactivity was detected between the random copolymer (Pro^{66}, Gly^{34})$_n$ and collagens. The cellular cross-reactions with collagens, manifested by delayed type sensitivity precedes the appearance of cross-reacting antibodies. The cross-reactivity with collagens increases with the molecular weight of the immunizing polymer (16).

The cross-reaction in guinea pigs between (Pro-Gly-Pro)$_n$ and collagen from the same species is of special interest. By immunization with the synthetic polymer, the

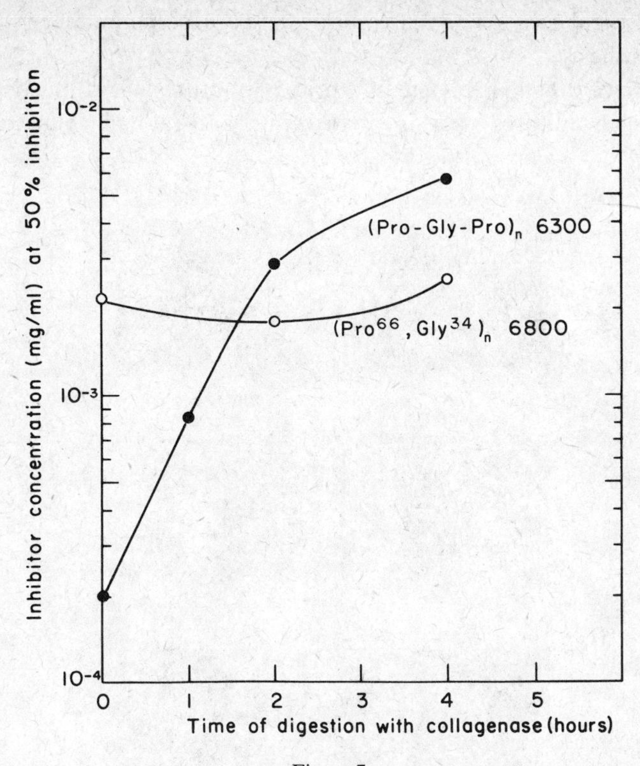

Figure 7

The effect of collagenase digestion of the ordered (Pro-Gly-Pro)$_n$ and the random (Pro^{66}, Gly^{34})$_n$ on their inhibitory capacity of the inactivation of (Pro-Gly-Pro)$_n$-RNase-T4 by rabbit anti-(Pro-Gly-Pro)$_n$-ovalbumin serum. Samples of (Pro-Gly-Pro)$_n$ (M. W. 6300) and of (Pro^{66}, Gly^{34})$_n$ (M. W. 6800) were incubated at 37°C with collagenase (2 % by weight of the polymer). Aliquots were removed from the digestion mixture at various times of incubation and diluted into Ca^{++} free medium for inhibition. From Maoz et al. (1973a).

tolerance to self collagen could be broken, and the animal became sensitive to its own collagen. Thus, the synthetic polymer offers a good system for inducing experimental autoimmune conditions which can be helpful in understanding the involvement of collagen in autoimmune diseases.

The immunological cross-reactivity between (Pro-Gly-Pro)$_n$ and natural collagens makes it possible to apply antibodies against the synthetic model to biological studies of collagen. We have shown that antibodies to (Pro-Gly-Pro)$_n$, as well as antibodies to rat tail collagen, have a similar specific cytotoxic activity, which is complement dependent, on primary cultures of rat muscle cells and on rat embryo fibroblasts (Maoz, Dym, Fuchs and Sela, 1973c). The reaction of antibodies to the synthetic polypeptide (Pro-Gly-Pro)$_n$ with fibroblasts reflects an immunological cross-reaction with mammalian collagen. These antibodies were shown to react with fibroblasts of several species and may thus be used as a common reagent for studying collagen biosynthesis.

We made use of the *in vitro* system of primary cultures of rat muscle cells (Yaffe, 1969). When such cultures were incubated with anti-rat tail collagen antiserum,

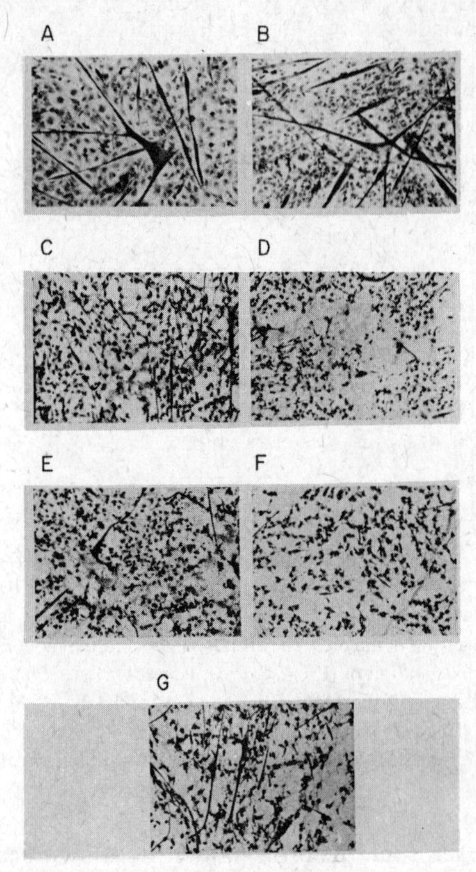

Figure 8

Rat muscle cells in culture treated as follows:

A. Normal muscle cells.
B. Muscle cells with complement alone.
C. Muscle cells exposed to a 1000-fold dilution of rabbit anti-acid soluble rat tail tendon collagen, 37 hours after complement addition.
D. Muscle cells exposed to a 100-fold dilution of rabbit anti-acid soluble rat tail tendon collagen, 37 hours after complement addition.
E. Muscle cells exposed to antibodies to $(\text{Pro-Gly-Pro})_n$ in a concentration of 4 μg/ml, 22 hours after complement addition.
F. Same as E, 37 hours after complement addition.
G. Muscle cells exposed to antibodies to $(\text{Pro-Gly-Pro})_n$, 2 μg/ml. 37 hours after complement addition. From Maoz et al. (1973c).

followed by the addition of complement, we observed significant destruction of the culture. The cells sequentially present irregularities in shape and granulation within the cell. Later on, general cell damage and complete destruction of the cultures could be observed. A similar toxic effect was observed by antibodies to the ordered polymer $(Pro\text{-}Gly\text{-}Pro)_n$ (Figs. 8 and 9). The extent of the cytotoxic effect was shown to increase with time as well as with concentration of the specific antibodies.

The interaction of antibodies to the synthetic periodic polypeptide $(Pro\text{-}Gly\text{-}Pro)_n$ with some cell membranes would suggest localization of similar structures in fibroblast membranes. There may be several advantages in using antibodies to the synthetic polymer rather than antibodies to collagen in such a system. With antibodies to the synthetic collagen-like polymer, one can be sure that the antibodies are specific just for collagen and not to other protein contaminants which may accompany preparations of natural collagens. In addition the barrier of species specificity is overcome and one may use the same reagent, namely, antibodies to the synthetic model, for fibroblasts from any source.

Synthetic collagen models were of interest also in studies on the relation between steric conformation and the need for cooperation between cells derived from thymus and from bone marrow in order to attain an efficient immune response. These studies will be discussed later.

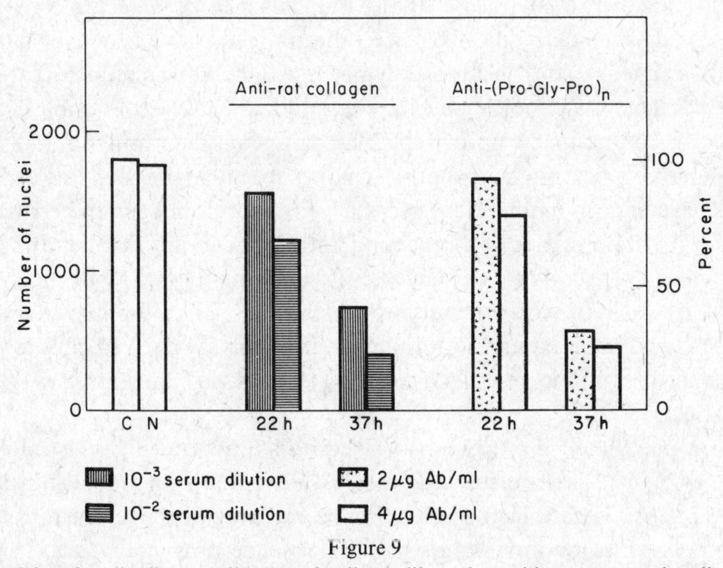

Figure 9

Cytotoxic activity of antibodies to collagen and collagen-like polypeptide on rat muscle cells.

C. Muscle cells with complement alone.

N. Muscle cells treated with normal rabbit serum (100-fold dilution), 22 hours after complement addition. Each value for the number of nuclei represents the total number of nuclei in 5 fields observed randomly under the microscope. From Maoz et al. (1973c).

Synthetic antigens provoking anti-lysozyme antibodies

Immunological studies of a unique region of hen egg-white lysozyme, denoted "loop", led to the synthesis of a peptide corresponding to this region. The peptide was attached to a branched amino acid copolymer, and the resulting conjugate led in several animal species to the formation of specific antibodies cross-reacting efficiently with lysozyme. Thus, we were able to obtain a totally synthetic molecule provoking antibodies reacting specifically with a unique region of a native protein, which is conformation-dependent (Arnon, Maron, Sela and Anfinsen, 1971). I have summarized this study in previous lectures, and shall not repeat it here (Sela, 1971; Sela, 1973).

Thymus-independence of antigens

The phenomenon of cell-to-cell interaction has occupied an important place in studies of the cellular nature of the immune response. Cooperation between thymus-derived and marrow-derived cellular components of the immune system of mice has been demonstrated for humoral antibody responses to several immunogens including heterologous erythrocytes, serum proteins, and synthetic polypeptides. Immunogens have been described that elicit humoral antibody responses in heavily irradiated, bone-marrow reconstituted mice without the presence of detectable numbers of thymocytes or thymus-derived cells. These immunogens have been called "thymus-independent antigens," and include polymerized flagellin, pneumococcal polysaccharide, *Escherichia coli* lipopolysaccharide, and polyvinylpyrrolidone. Since they consist of long chains made up of repeating antigenic determinants (polymerized flagellin contains repeating polypeptide chain subunits), the suggestion has been made that the structural requirement necessary for an immunogen to be independent of the thymus is the presence of a repeating antigenic subunit (Möller and Michael, 1971). This may be a necessary but insufficient requirement for thymus-independence, since several multichain synthetic polypeptide immunogens, all possessing repeating antigenic determinants, require both thymus- and marrow-derived cells for eliciting efficient humoral immune responses in mice (Mozes and Shearer, 1971; Shearer, Mozes and Sela, 1972).

It has been postulated that thymus-derived cells function to concentrate antigen (Mitchison, 1971). It, therefore, seemed possible that in addition to repeating antigenic determinants, slow metabolism might also be required for efficient antibody responses in mince to some immunogens in the absence of thymus cells. In order to investigate this hypothesis, use was made of the fact that immunogens composed of D amino acids are slowly and incompletely metabolized.

The multichain synthetic copolypeptides used in the study I am discussing here (Sela, Mozes and Shearer, 1972) can be prepared in such a manner that different parts of the molecule may be built from amino acids that differ in their optical configuration. The role of thymus in antibody responses to a series of four synthetic polypeptide immunogens of the general formula multi-copoly(Tyr, Glu)-poly(Pro)-poly(Lys) was investigated as a function of the optical activity of the amino acids composing their structure. Irradiated nonthymectomized and thymectomized SJL mice were injected with thymocytes, marrow cells, or a mixture of both. Each group of recipients was immunized with the following copolymer enantiomorphs: all L-amino acids; L-amino acids outside and D inside; D-amino acids outside and L inside; all D-amino acids. The antibody response to the immunogen composed of all L-amino acids was thymus-dependent, whereas the responses to the other three copolymers were all independent of the thymus. Similar cell transfers were performed in DBA/1 mice immunized with multi-copoly(L-Phe, L-Glu)-poly(D-Pro)-poly(D-Lys). This mouse strain produces specific antibodies against the (Phe, Glu) region and against the poly(D-prolyl) region. The immune response to the determinant with only L-amino acids on the outside was thymus-dependent, whereas the response to the inside immunopotent region with only D-amino acids was thymus-independent.

Since earlier studies have demonstrated that synthetic polypeptide antigens that contain D-amino acids are poorly metabolized, the thymus-independence of the antibody responses to these multichain synthetic polypeptides that possess repeating antigenic determinants was correlated with the metabolizability of the immunogens or their component determinants. It is likely that for the category of thymus-independent immunogens that possess repeating antigenic determinants, slow metabolism may be a requirement for a steady multipoint binding of antigenic determinants to the lymphocyte, which is in turn needed for the induction of immune response.

In view of our interest in the role of conformation in immunology, we thought it would be worthwhile to find out whether the synthetic ordered polymer designated (Pro-Gly-Pro)$_n$, which we discussed earlier, needs cell cooperation for the induction of an efficient immune response. We compared it with a random polymer of a similar composition (Fuchs, Mozes, Maoz and Sela, 1974). In view of the results obtained, we extended the study to collagen and gelatin.

Heavily irradiated recipients were injected with syngeneic thymocytes, marrow cells or a mixture of both cell populations and were immunized with the above-mentioned antigens. An efficient immune response to the ordered collagen-like (Pro-Gly-Pro)$_n$ was obtained in the absence of transferred thymocytes. The thymus independence of (Pro-Gly-Pro)$_n$ was confirmed when thymectomized, irradiated mice were used as recipients. In contrast with these results, cooperation between thymus and marrow cells was necessary in order to elicit an immune response to the random (Pro^{66}, Gly^{34})$_n$ (Table 1).

Similarly, the immune response to the triple helical collagen was found to be independent of thymus, whereas for an effective response to its denatured product, gelatin, thymus cells were required (Table 2). These findings indicate that a unique

Table 1

THYMUS-MARROW CELL COOPERATION IN THE IMMUNE RESPONSE TO (Pro-Gly-Pro)$_n$, (Pro^{66}, Gly^{34})$_n$ AND Pro-Gly-Pro-OVALBUMIN (Fuchs et al., 1974)

Mouse strain	Immunogen	Assaying antigen	Assayed by*	Cells injected per recipient		
				10^8 thymocytes	2×10^7 marrow cells	10^8 thymocytes and 2×10^7 marrow cells
BALB/c	(Pro-Gly-Pro)$_n$	(Pro-Gly-Pro)$_n$-RNase	PFC	0**(2)	73 (30)	71 (24)
			HA	0 (2)	73 (30)	65 (23)
BALB/c	(Pro-Gly-Pro)$_n$	(Pro-Gly-Pro)$_n$-RNase	PFC	0 (2)	59 (22)	55 (22)
(Tx)			HA	0 (2)	59 (22)	48 (21)
SWR	(Pro-Gly-Pro)$_n$	(Pro-Gly-Pro)$_n$-RNase	PFC	15 (13)	53 (15)	64 (11)
			HA	0 (13)	57 (14)	44 (11)
C3H.SW	(Pro-Gly-Pro)$_n$	(Pro-Gly-Pro)$_n$-RNase	PFC		73 (11)	67 (9)
			HA		55 (11)	44 (9)
BALB/c	(Pro^{66}, Gly^{34})$_n$	(Pro^{66}, Gly^{34})$_n$-RNase	PFC		30 (27)	72 (21)
			HA		42 (19)	79 (19)
C3H.SW	(Pro^{66}, Gly^{34})$_n$	(Pro^{66}, Gly^{34})$_n$-RNase	PFC	11 (9)	32 (22)	67 (18)
			HA	18 (11)	27 (22)	76 (17)
BALB/c	Pro-Gly-Pro-ovalbumin	Pro-Gly-Pro-RNase	PFC		20 (25)	56 (27)
			HA		8 (25)	59 (27)
C3H.SW	Pro-Gly-Pro-ovalbumin	Pro-Gly-Pro-RNase	PFC		5 (19)	56 (9)
			HA		16 (19)	56 (9)

* PFC–Hemolytic plaque forming cell assay. Results higher than 100 plaques per spleen were considered positive.
HA–Passive micro hemagglutination assay. Hemagglutination titer at dilutions greater than 1 : 4 were considered positive.
** Percentage of syngeneic irradiated recipients producing detectable antibody titers. Numbers in parenthesis give the number of mice tested.

Table 2

THYMUS-BONE MARROW CELL COOPERATION IN THE IMMUNE RESPONSE TO COLLAGEN AND GELATIN (Fuchs et al., 1974)

Mouse strain	Immunogen and assaying antigen	Assayed by*	Cells injected per recipient		
			10^8 thymocytes	2×10^7 marrow cells	10^8 thymocytes and 2×10^7 marrow cells
SWR	RCM *Ascaris* collagen	PFC	21** (28)	74 (38)	70 (23)
		HA	19 (31)	70 (40)	57 (23)
SWR	Rat tail collagen	PFC	8 (13)	70 (10)	62 (13)
		HA	23 (30)	77 (35)	83 (30)
SWR	Rat tail gelatin	PFC	17 (23)	12 (32)	70 (23)
		HA	11 (26)	22 (32)	83 (23)

* PFC–hemolytic plaque forming cell assay. Results higher than 100 plaques per spleen were considered positive. HA–passive microhemagglutination assay. Hemagglutination titer at dilutions greater than 1:4 were considered positive.

** Percentage of syngeneic irradiated recipients producing detectable antibody titers. Numbers in parenthesis give the number of mice tested.

three-dimensional structure of immunogens possessing repeating antigenic determinants plays an important role in determining the need for cell to cell interaction in order to elicit an antibody response.

It is thus clearly demonstrated that changes in the steric conformation, unaccompanied by *any* changes in the primary structure of the protein, may lead to dramatic differences in the nature of the triggering of the immune response. These differences may also be correlated with different patterns of metabolism. It should be kept in mind that native collagen, due to its triple helical conformation, is rather resistant to proteolytic digestion by enzymes other than collagenase. The latter is probably not present in significant amounts in body fluids, as it is found in various organelles. On the other hand, gelatin is very susceptible to proteolytic digestion.

Concluding remarks

The common denominator of the various examples of immunochemical and some immunobiological studies I have discussed here is that proper antigen design may be helpful for a better understanding of the immune response on a molecular and cellular level.

I gave examples demonstrating the role of the steric conformation of the antigen in the definition of antibody specificity. These studies lead to the inevitable conclusion that an antigenic determinant is recognized at the biosynthetic site while the immunogenic macromolecule is still intact. Strong support for this view comes also from investigations, not mentioned here, on the inverse relationship between the net electrical charge of antigens and that of antibodies elicited.

The successful synthesis of molecules capable of provoking the formation of antibodies that can react with unique antigenic conformation-dependent determinants on native proteins may potentially lead to synthetic vaccines of the future. Any developments in this direction will have to take into account both chemical and genetical parameters, especially in view of the apparent close genetic link in several species between good immune response to certain antigens and the major histocompatibility locus.

The relative simplicity of the synthetic molecules facilitates the interpretation of the results obtained with them, and sometimes permits the detection of effects, such as genetic variations in immune response, which are not easily observed with complex natural antigens. A detailed analysis of the genetic control of the immune response at cellular level permitted a tentative assignment of various roles to the different cell types.

References

Arnon, R., Maron, E., Sela, M. and Anfinsen, C. B. 1971. *Proc. Natl. Acad. Sci. U.S.* **68**:1450.

Borek, F., Kurtz, J. and Sela, M. 1969. *Biochim. Biophys. Acta* **188**:314.

Crumpton, M. J. and Small, P. A., Jr. 1967. *J. Mol. Biol.* **26**:143.

Engel, J., Kurtz, J., Katchalski, E. and Berger, A. 1966. *J. Mol. Biol.* **17**:255.

Fuchs, S., Maoz, A. and Sela, M. 1974. *Israel J. Chem.* **12**:681.

Fuchs, S., Mozes, E., Maoz, A. and Sela, M. 1974. *J. Exp. Med.* **139**:148.

Maoz, A., Fuchs, S. and Sela, M. 1973a. *Biochemistry* **9**:4238.

Maoz, A., Fuchs, S. and Sela, M. 1973b. *Biochemistry* **9**:4246.

Maoz, A., Dym, H., Fuchs, S. and Sela, M. 1973c. *Eur. J. Immunol.* **3**:839.

Mitchison, N. A. 1971. *Eur. J. Immunol.* **1**:18.

Möller, G. and Michael G. 1971. *Cell. Immunol.* **2**:309.

Mozes, E. and Shearer, G. M. 1971. *J. Exp. Med.* **134**:141.

Pilz, I., Puchwein, G. Kratky, O., Herbst, M., Haager, O., Gall, W. E. and Edelman, G. M. 1970. *Biochemistry* **9**:211.

Pilz, I., Kratky, O., Licht, A. and Sela, M. 1973. *Biochemistry* **12**:4998.

Sarma, V. R., Silverton, E. W., Davies, D. R. and Terry, W. D. 1971. *J. Biol. Chem.* **246**:3753.

Schechter, B., Schechter, I., Ramachandran, J., Conway-Jacobs, A., Sela, M., Benjamini, E. and Shimizu, M. 1971a. *Eur. J. Biochem.* **20**:309.

Schechter, B., Schechter, I., Ramachandran, J., Conway-Jacobs, A. and Sela, M. 1971b. *Eur. J. Biochem.* **20**:301.

Schechter, B., Conway-Jacobs, A. and Sela, M. 1971c. *Eur. J. Biochem.* **20**:321.

Segal, D. M. 1969. *J. Mol. Biol.* **43**:497.

Shearer, G. M., Mozes, E. and Sela, M. 1972. *J. Exp. Med.* **135**:1009.

Sela, M. 1969. *Science* **166**:1365.

SELA, M. 1971. *Ann. N.Y. Acad. Sci.* **190**:181.

SELA, M. 1973. *Harvey Lectures 1971–1972.* **67**:213.

SELA, M., SCHECHTER, B., SCHECHTER, I. and BOREK, F. 1967. *Cold Spring Harbor Symp. Quant. Biol.* **32**:537.

SELA, M., MOZES, E. and SHEARER, G. M. 1972. *Proc. Natl. Acad. Sci. U.S.* **69**:2696.

TRAUB, W. and YONATH, A. 1965. *Israel J. Chem.* **3**:43.

TRAUB, W. and YONATH, A. 1966. *J. Mol. Biol.* **16**:404.

YAFFE, D. 1969. *Current Topics in Developmental Biology* **4**:37.

Receptor interactions and mitogenesis in lymphoid cells

GERALD M. EDELMAN

The Rockefeller University, New York, N.Y. 10021, U.S.A.

Our understanding of the origin of antibody diversity and of the molecular mechanisms of lymphocyte stimulation is still at a very early stage. Experiments stemming from the analysis of antibody structure suggest that these two outstanding problems of immunology can be attacked at the molecular level and that such an approach provides a framework for understanding cellular specificity in the immune response.

It appears that immunological specificity results from the interaction of a number of factors, including the specificity of the initial binding of the antigen, the affinity and avidity of the antibody receptors, and the cellular threshold for stimulation and clonal expansion. In order to understand the interaction of these factors in detail, it is necessary to study the structure of the lymphocyte membrane and particularly the mode of attachment of its receptors. An analysis of the structure and activity of mitogenic lectins has proven to be particularly helpful in carrying out this task. In this paper, I shall discuss studies on the structure and activity of the lectin, concanavalin A (Con A), particularly in relation to the problem of the structure and modulation of the lymphoid cell surface.

The primary and three-dimensional structure of concanavalin A have been determined. The data indicate that the native molecule is a tetravalent tetramer consisting of subunits with molecular weights of 26,000. The amino acid sequence, which was determined chemically, contains 237 residues. Interpretation of an electron density map at 2 Å resolution indicates that the predominant structural element is an extended polypeptide chain arranged in two anti-parallel pleated sheets or β-structures. Residues not included in the β-structures are arranged in regions of random coil. One of the pleated sheets contributes extensively to the interactions among the monomers to form both dimers and tetramers.

It has been found that Con A, when bound to the lymphocyte surface, can either induce redistribution of its receptors (cap and patch formation) or inhibit patch and cap formation of various receptors including those for Con A itself. At 37°C,

Con A forms few caps and inhibits receptor mobility. In contrast, incubation with Con A at 4°C followed by removal of unbound molecules and elevation of the temperature to 37°C resulted in extensive cap formation. The expression of these antagonistic activities is therefore highly dependent on the conditions under which cells are incubated with Con A.

Analysis of the subunit interactions reveals a simple structural basis of monomer-monomer interactions and dimer-dimer interactions and suggests means for altering both the valence and the surface specificity of the molecule. Chemical derivation of tetrameric concanavalin A with succinic anhydride or acetic anhydride converts the protein to a dimeric molecule without altering its carbohydrate-binding specificity. At low concentrations, the dose-response curves for the mitogenic stimulation of mouse spleen cells by native Con A and succinyl-Con A are similar. Above lectin concentrations of 10 μg/ml, however, the response to Con A is diminished while that for succinyl-Con A does not decrease until much higher doses are reached. We have attributed this difference mainly to the higher rate of cell death induced by the native Con A molecule. Con A also shows a greater capacity than succinyl-Con A to agglutinate sheep erythrocytes and to inhibit cap formation by immunoglobulin receptors on spleen cells. Moreover, at low concentrations, Con A induced its glycoprotein receptors to form caps, but succinyl-Con A did not induce cap formation.

Addition of antibodies directed against Con A to succinyl-Con A bound on cells restored the properties of agglutination, inhibition of cap formation by immunoglobulin receptors, and induction of cap formation by Con A receptors. Similar results have been obtained for acetyl-Con A. These data suggest that the altered biological activities of succinyl-Con A and acetyl-Con A are attributable mainly to their reduced valence. In addition, we have observed that colchicine and Vinca alkaloids reverse the inhibition of patches by tetrameric Con A. On the basis of all of these findings, we have constructed an hypothesis to explain the modulation of receptor mobility by Con A, colchicine and related drugs. It is proposed that binding of multivalent lectins alters the interaction of an assembly of colchicine-binding proteins with lectin receptors and other receptors, and reciprocally that the state of the colchicine-binding assembly alters the mobility and distribution of surface receptors on the cell membrane. This hypothesis opens the possibility that alterations in such assemblies are related to early signals in the mitogenic response.

References

EDELMAN, G. M., 1973. "Antibody Structure and Cellular Specificity in the Immune Response." Harvey Lecture (in press).

EDELMAN, G. M., CUNNINGHAM, B. A., REEKE, G. N. JR., BECKER, J. W., WAXDAL, M. J. and WANG, J. L., 1972. "The Covalent and Three-Dimensional Structure of Concanavalin A," *Proc. Natl. Acad. Sci. U.S.* **69**, 2580.

EDELMAN, G. M., YAHARA, I. and WANG, J. L., 1973. "Receptor Mobility and Receptor-Cytoplasmic Interactions in Lymphocytes." *Proc. Natl. Acad. Sci. U.S.* **70**, 1442.

GUNTHER, G. R., WANG, J. L., YAHARA, I., CUNNINGHAM, B. A. and EDELMAN, G. M., 1973. "Concanavalin A Derivatives with Altered Biological Activities." *Proc. Natl. Acad. Sci. U.S.* **70**, 1012.

YAHARA, I. and EDELMAN, G. M., 1972. "Restriction of the Mobility of Lymphocyte Immunoglobulin Receptors by Concanavalin A." *Proc. Natl. Acad. Sci. U.S.* **69**, 608.

YAHARA, I. and EDELMAN, G. M., 1973. "The Effects of Concanavalin A on the Mobility of Lymphocyte Surface Receptors." *Exp. Cell Res.*, **81**, 143.

T-cell recognition of cell surface antigens

MICHAEL FELDMAN, HARTMUT WEKERLE and IRUN R. COHEN
Department of Cell Biology, The Weizmann Institute of Science, Rehovot, Israel

Introduction

Immune reactions manifest themselves in two distinct ways: 1) the induction of antibodies; 2) the induction of cell mediated immunity. The diversity of antibodies was attributed to the diversity of the precursors of antibody producing B lymphocytes (Naor and Sulitzeanu, 1967; Ada and Byrt, 1969; Wigzell and Mäkelä, 1970). Whether the specificity of cellular immunity mediated by thymus derived lymphocytes (T cells) is similarly based on a state of diversity among T cells has been an open question. Studies were made of the conditions under which soluble antigens are recognized by B cells (Wigzell and Mäkelä, 1970; Walters and Wigzell, 1970). Yet, the conditions which determine the recognition of cell surface antigens by T cells were not known. There is evidence to suggest that the receptors of B cells are membrane-bound immunoglobulins (Mitchison, 1967; Sell, 1967a,b; Segal, Globerson, Feldman, Haimovich and Givol, 1969). The molecular nature of the receptors of T cells mediating graft reactions remained a controversial question.

Once recognition of cell-bound antigens takes place, lymphocytes undergo developmental processes leading to the final differentiation of effector cells, the lymphocytes which destroy the grafted cells. Is primary contact between the lymphocytes and the sensitizing cell antigens sufficient to switch on the entire sequence of differentiation processes culminating in the production of killer cells?

All systems of graft reaction are triggered by the recognition of "foreign" transplantation antigens. The state of natural tolerance, manifested by lack of reactivity towards autologous tissue, has been attributed to the elimination from the developing lymphoid system of cells capable of recognizing self antigens (Burnet, 1959; Burnet, 1963). This notion has not been subjected to rigorous experimental testing. Therefore, it is possible that lymphocytes which possess cell receptors for self-antigens do exist. If such lymphocytes are indeed present, then an obvious question suggests itself: What prevents such lymphocytes from reacting *in vivo* against self-antigens?

We have outlined a series of open questions with regard to lymphoid stimulation in cell mediated immunity. In order to approach them we have developed and studied in our laboratory an experimental system for the *in vitro* induction of cellular immunity (Ginsburg, 1968; Berke, Ax, Ginsburg and Feldman, 1969; Berke, Ginsburg and Feldman, 1969; Berke, Yagil, Ginsburg and Feldman, 1969; Berke, Ginsburg, Yagil and Feldman, 1969).

The *in vitro* system for the induction of cellular immunity

Most of our studies are based on a xenogeneic rat anti-mouse cellular immune reaction which takes place when normal rat lymphocytes are exposed in cell culture to monolayers of mouse embryo fibroblasts. During the first hours of culture a small fraction of the lymphocytes (about 3–5 %) firmly adhere to the sensitizing fibroblasts. This adherence reflects specific recognition and will be described in greater detail below. For the next 48 to 72 hours there appears to be little or no replication of lymphocytes while about 50–80 % of the original lymphocyte population dies and disappears. A stage of lymphocyte transformation to blast-like cells (BC) and rapid proliferation of the BC begins after the 72 hours of incubation. As they divide, the initially adherent lymphocytes spontaneously detach from the sensitizing fibroblast monolayer. An additional contact with the original sensitizing antigens is then required to activate the BC to lyse target fibroblasts. The target fibroblast cultures are labeled with radioactive chromium (^{51}Cr) and the extent of their lysis is determined by measuring the release of ^{51}Cr. Thus, the *in vitro* system appears to comprise four stages: 1) recognition, 2) induction, 3) proliferation, and 4) lysis. For operational purposes we call the first three stages the "sensitization phase". Sensitization cultures are usually maintained for 5 days. The lytic or "effector phase" is measured by transferring the BC to target fibroblast cultures for 18 to 44 hours.

We have recently modified the *in vitro* system to achieve sensitization of mouse lymph nodes against allogeneic fibroblasts (Cohen, Globerson and Feldman, 1971a; Altman, Cohen and Feldman, 1973). The mouse allosensitization system allowed us to use antibodies to characterize the origin of the lymphocytes participating in the reaction. We found that treatment of C3H lymph node cells with antibodies to C3H theta (θ) antigen abolished the induction of allosensitization (Altman, Cohen and Feldman, 1973). Thus, the *in vitro* induction of cell-mediated immunity measures T-cell function.

Specificity

The antigenic specificity of lytic reactions following sensitization of rat lymphocytes on fibroblasts of H-2^b or H-2^k phenotypes was measured by testing the relative lytic effect on target cells of either H-2 type. All experiments demonstrated strain specificity (Berke, Clark and Feldman, 1971). The lytic effect on target cells of H-2 phenotype identical with that of the sensitizing monolayer was always significantly greater than on unrelated phenotypes. Cross reactivity of lysis was usually less than 50%. Sensitization against a genetic hybrid resulted in a similar level of lysis of target cells from both parental strains. The same results of lysis of parental strains were obtained with lymphocytes sensitized on a mixture of cells from the two mouse strains.

The strain specificity of lymphocytes sensitized *in vivo* was always found to be less than that demonstrated by *in vitro* sensitized cells. Thus, *in vitro* sensitization of rat lymphocytes on mouse fibroblast monolayers is quantitatively more efficient and qualitatively more specific than the *in vivo* sensitization with identical genetic combinations.

Diversity of T lymphocytes

The diversity of antibodies seems to be based on underlying diversity among antibody-producing cells. This is deduced from (a) the specific binding of antigen to lymphocytes (Naor and Sulitzeanu, 1967), (b) the elimination of such lymphocytes after binding of highly radioactive antigens (Ada and Byrt, 1969), and (c) the specific adherence of antigen-binding cells to antigen-coated bead columns (Wigzell and Mäkelä, 1970). The reacting cells in all these cases were bone marrow derived lymphocytes (B cells). No such specific binding and hence no diversity could be demonstrated in unstimulated T cells involved in cell mediated immunity.

To test whether the specificity of cell mediated immune reactions is based on a diversity of cells, i.e., on the existence of lymphocytes each equipped with receptors for different antigens, the following experimental approach was applied (Lonai, Wekerle and Feldman, 1972; Wekerle, Lonai and Feldman, 1972): normal rat lymph node cells were plated on monolayers of mouse fibroblasts for a short incubation time usually varying between 30 minutes and 2 hours. During this time a certain number of lymphocytes adheres firmly to the fibroblasts. To test whether the adhering cells are those which possess specific receptors for the strain-specific antigens of the fibroblasts, the nonadhering lymphocytes were separated from the adherent ones and transferred to fresh monolayers. Both types of cultures, i.e., those containing the adherent and those containing the supernatant lymphocytes, were then cultured further to allow reactive lymphocytes to undergo sensitization. On the fifth day of sensitization the lymphoid cells were harvested from the sensitizing monolayers and equal numbers of viable lymphocytes were transferred onto ^{51}Cr-labeled target monolayers to assay cytotoxic activity. The result was that the cell population recovered from the adherent lymphocytes was significantly richer in BC and produced a two-fold greater cytolytic effect than did an equal number of the initially nonadherent lymphocytes (Lonai, Wekerle and Feldman, 1972). The nonadherent cells manifested a marked reduction in lytic potency. By repeating the separation procedure three consecutive times, through transfer of the nonadhering cells to fresh monolayers for further separation, we were able to eliminate all lymphocytes capable of reacting against the antigens of the absorbing monolayer. We then tested the capacity of the nonadherent lymphocytes to react against fibroblasts of mouse strains different from the ones used for separation. The result was that the lymphocytes which did not adhere to H-2^d fibroblasts were unable to undergo significant sensitization against fibroblasts of the H-2^d type. Such lymphocytes were, however, sensitized by H-2^k fibroblasts, and the extent of lysis obtained was similar to that produced by unseparated Lewis lymphocytes sensitized on H-2^k fibroblasts. Thus, mouse fibroblasts can be used as cellular immunoadsorbents to separate lymphocytes possessing cell receptors for mouse strain specific antigens. Therefore, the specificity of cell-mediated immune lysis seems to be based on the existence of diverse populations of lymphocytes.

Conditions determining antigen recognition

Antigen recognition by B lymphocytes has been shown to take place at temperatures of about 4°C (Wigzell and Mäkelä, 1970; Walters and Wigzell, 1970). Experiments were made to test whether the same is true for T cell mediated cellular immune reactions (Wekerle, Lonai and Feldman, 1972). The separation procedure of rat lymphocytes on mouse fibroblasts was attempted at 0° or 4°C, as compared to control cultures incubated at 37°C, which showed a marked separation effect. The result was that at 0° or 4° the adherent and nonadherent lymphocytes developed the same cytolytic activity after sensitization. Thus no specific adherence occurred at 0° or 4°C.

These results suggest that antigen-specific adherence of normal T lymphocytes requires metabolic energy. This was confirmed by assaying the separation effect at 37°C in the presence of different metabolic inhibitors. Dinitrophenol, at a concentration of 0.001 mol, and sodium azide at a concentration of 0.01 mol, prevented the specific separation. Lymphocyte viability was not affected at these concentrations. It is therefore concluded that the recognition of cell surface antigens by T cells is energy-dependent. The energy dependence of the specific adherence might be associated with surface migration and concentration of membrane receptors. Such energy-requiring processes have been demonstrated to occur during "cap formation" when lymphocytes are treated with bivalent anti-immunoglobulin antibodies (Taylor, Duffus and Petris, 1971). It is also possible that energy is required for the exposure of partly hidden receptors. To investigate this possibility, we used neuraminidase of *Vibrio cholerae* which had been shown to expose surface proteins of lymphoid and other cells (Schlesinger and Gottesfeld, 1971; Currie and Bagshawe, 1968). We tested whether pretreatment of Lewis lymph node cells with neuraminidase would affect specific adherence to fibroblast antigen. The treated cells showed an adherence effect at 0°C, which was even slightly higher than that of untreated lymphocytes incubated at 37°C (Wekerle, Lonai and Feldman, 1972). It thus appears that the process of antigen recognition is associated with the exposure of cryptic receptors.

Is antigen recognition sufficient to trigger T-cell mediated immunity?
We investigated whether recognition *per se* of cell surface antigens by lymphocytes is sufficient to trigger sensitization of T cells (Wekerle, Kölsch and Feldman, 1974). We assayed sensitization of normal rat lymphocytes against monolayers of mouse fibroblasts by measuring the acquisition of cytotoxicity against target monolayers syngeneic to the sensitizing monolayer, following 5 days of culturing on living or glutaraldehyde-fixed fibroblasts. Monolayers of either living or glutaraldehyde-treated fibroblasts were used as cellular immunoadsorbents to assay recognition.

Glutaraldehyde treatment of mouse fibroblasts did not seem to alter the cell surface antigens detectable by alloantibodies. To test whether lymphocytes recognize surface antigens of glutaraldehyde treated monolayers, the capacity of the latter to specifically adsorb lymphocytes possessing receptors for the fibroblast antigen was measured. The nonadhering rat lymphocytes, seeded on glutaraldehyde treated C3H fibroblasts, lost their capacity to become sensitized against fresh C3H cells, but retained their capacity to react against BALB/c fibroblasts. Thus, recognition of C3H antigen did take place. However, rat lymphocytes which did recognize such monolayers did not undergo sensitization. This was not due to the lack of feeder effect by the glutaraldehyde-fixed monolayers, since the addition of living rat fibroblasts, which sustained the survival of the lymphocytes, did not result in sensitization. Not even when living fibroblasts were added to syngeneic fixed monolayers did sensitization of the rat lymphocytes against the monolayer cells occur. Yet, addition of living fibroblasts to fixed allogeneic monolayers resulted in sensitization against the living fibroblasts but not against the fixed cells. Thus, rat lymphocytes seem to recognize the antigens of fixed monolayers, but cannot respond to them.

Recognition of self-antigens

Natural tolerance to self antigens could be explained either by the absence of self reactive lymphocytes, or by factors which inhibit the immune reactivity of lymphocytes which bear receptors for self-antigens.

As demonstrated above, the system of *in vitro* sensitization provides a means of inducing the sensitization of lymphocytes which bear preformed receptors specific for the sensitizing antigens. This system was therefore used to test experimentally whether or not there exist in normal animals lymphocytes with receptors that specifically recognize self-antigens (Cohen, Globerson, and Feldman, 1971b; Cohen and Wekerle, 1972a, 1973a).

The result was that inbred Lewis or Wistar rat lymph node cells could be induced to transform when cultured with syngeneic embryonic fibroblasts. The achievement of autosensitization in these cultures was demonstrated by the ability of the lymphocytes to cause immunospecific lysis of syngeneic target cells. Injection of these lymphocytes into newborn syngeneic rats produced splenomegaly to a degree consistent with a graft versus host reaction (GvH) (Cohen, Globerson and Feldman, 1971b). This finding was repeated using inbred mouse lymphocytes which had been autosensitized *in vitro.*

To learn whether receptors for self antigens are present on the surface of specific lymphocytes before the induction of autosensitization, syngeneic fibroblasts were used as immunoadsorbents (Cohen and Wekerle, 1972). BN rat lymph node cells were adsorbed to BN fibroblasts and the nonadherent lymphocytes were tested for their ability to become sensitized against either BN or allogeneic Lewis fibroblasts. BN lymphocytes that did not adhere to BN fibroblasts were inhibited in their ability to become sensitized against these fibroblasts. Nevertheless, such lymphocytes could be sensitized against Lewis fibroblasts to a degree equal to that of unadsorbed BN lymphocytes sensitized directly against Lewis fibroblasts. The specificity of adherence to syngeneic fibroblasts indicates that a fraction of normal peripheral lymphocytes bear receptors for self-antigens, and that these lymphocytes can become autosensitized.

The immunospecificity of autosensitization against syngeneic fibroblasts argues against the possibility that the reaction is the result of sensitization against extraneous antigens present in the culture medium. However, it is conceivable that autosensitization is induced by embryonic fibroblast antigens which were not accessible to the developing thymus-dependent immune system. To investigate this possibility, the induction of autosensitization against thymus reticulum cells obtained from adult rats was studied (Cohen and Wekerle, 1973b). Lymphocytes which mediate cellular immune reactions differentiate within the thymus. Hence, they should have access to reticulum cell antigens during their development. Suspensions of cells from Lewis or BN thymus glands were prepared. The cells of each individual gland were cultured separately. Using this procedure the thymus lymphocytes were preserved while the thymus reticulum cells grew into a confluent monolayer beneath the lymphocytes. The lymphocytes were collected from the reticulum cells after 5 days of culture. The degree of autosensitization was assayed by injecting 10^7 Lewis or BN lymphocytes into the right foot pad of each of six syngeneic rats. The left feet were injected with 10^7 unsensitized lymph node cells. The average weights of the right popliteal nodes were 4.5 (Lewis) and 3.8 (BN) times heavier than the average weights of the left popliteal nodes injected with control lymphocytes. The specificity of this *in vivo* GvH reaction was tested by assaying the relative cytolysis of Lewis or BN target fibroblasts produced by lymphocyte suspensions obtained from the GvH lymph nodes. The cytolysis of syngeneic target fibroblasts appeared to be immunospecific (Cohen and Wekerle, 1972a, 1973a).

These findings indicate that thymus lymphocytes may be autosensitized *in vitro* against reticulum cells present in the same thymus glands, and that autosensitization can be expressed both *in vivo* and *in vitro*. We may therefore conclude that self-tolerant lymphocytes exist in normal animals. These experiments appear to challenge the theories which base natural tolerance upon elimination of potentially reactive lymphocytes. It is likely that self tolerance *in vivo* is maintained by some regulatory mechanism which inhibits autosensitization of tolerant lymphocytes. The finding that *in vitro* autosensitized lymphocytes produce GvH reactions *in vivo* in syngeneic animals indicates that some expression of autosensitization cannot be completely suppressed *in vivo* once induction has occurred. We found that self-recognition can be inhibited by pretreatment of the lymphocytes with autologous serum. This inhibition is specific: allogeneic or xenogeneic sera do not inhibit self-recognition, nor does autologous serum inhibit the recognition of foreign antigens. Pretreatment of the fibroblasts with autologous serum, however, does not affect self recognition (Wekerle, Cohen and Feldman, 1973; Cohen and Wekerle, 1973b). The fact that only pretreatment with autologous serum of lymphocytes, but not of the syngeneic fibroblasts, prevents autosensitization suggests that the serum factors are not autoantibodies. They are likely to be soluble H-2 antigens shed off normal cells. These soluble antigens can be recognized, and thus block receptors for self, but they cannot trigger sensitization.

References

ADA, G. L. and BYRT, P. 1969. *Nature* **222**:1291.

ALTMAN, A., COHEN, I. R. and FELDMAN, M. 1973. *Cell. Immunol.* **7**:134.

BERKE, G., AX, W., GINSBURG, H. and FELDMAN, M. 1969. *Immunol.* **16**:643.

BERKE, G., GINSBURG, H. and FELDMAN, M. 1969. *Immunol.* **16**:659.

BERKE, G., YAGIL, G., GINSBURG, H. and FELDMAN, M. 1969. *Immunol.* **17**:723.

BERKE, G., GINSBURG, H., YAGIL, G. and FELDMAN, M. 1969. *Israel J. Med. Sci.* **5**:135.

BERKE, G., CLARK, W. R. and FELDMAN, M. 1971. *Transplantation.* **12**:237.

BURNET, M. 1959. *The Clonal Selection Theory of Acquired Immunity*. Nashville, Tenn.: Vanderbilt University Press.

BURNET, M. 1963. *The Integrity of the Body*. Cambridge, Mass.: Harvard University Press.

COHEN, I. R., GLOBERSON, A. and FELDMAN, M. 1971*a*. *J. Exp. Med.* **133**:821.

COHEN, I. R., GLOBERSON, A. and FELDMAN, M. 1971*b*. *J. Exp. Med.* **133**:834.

COHEN, I. R. and WEKERLE, H. 1972. *Science* **176**:1324.

COHEN, I. R. and WEKERLE, H. 1973*a*. In: *Microenvironmental Aspects of Immunity* (B. D. JANKOVIC and K. ISAKOVIC, eds.). New York: Plenum Publishing Corp., pp. 589–595.

COHEN, I. R. and WEKERLE, H. 1973*b*. *J. Exp. Med.* **137**:224.

CURRIE, G. A. and BAGSHAWE, K. D. 1968. *Brit. J. Cancer* **22**:843.

GINSBURG, H. 1968. *Immunol.* **14**:621.

LONAI, P., WEKERLE, H. and FELDMAN, M. 1972. *Nature, New Biology* **235**:235.

MITCHISON, N. A. 1967. *Cold Spr. Harb. Symp. Quant. Biol.* **32**:431.

NAOR, D. and SULITZEANU, D. 1967. *Nature* **214**:687.

SCHLESINGER, M. and GOTTESFELD, S. 1971. *Transplant. Proc.* **3**:1151.

SEGAL, S., GLOBERSON, A., FELDMAN, M., HAIMOVICH, J. and GIVOL, D. 1969. *Nature* **223**:1374.

SELL, S. 1967*a*. *J. Exp. Med.* **125**:289.

SELL, S. 1967*b*. *J. Exp. Med.* **125**:393.

TAYLOR, R. B., DUFFUS, W. P. H. and PETRIS, S. 1971. *Nature, New Biology* **233**:225.

WALTERS, C. S. and WIGZELL, H. 1970. *J. Exp. Med.* **132**: 1233.
WEKERLE, H., KOLSCH, E. and FELDMAN, M. 1974. *Eur. J. of Immunol.* **4**: 246.
WEKERLE, H., COHEN, I. R. and FELDMAN, M. 1973. *Nature, New Biology* **241**: 25.
WEKERLE, H., LONAI, P. and FELDMAN, M. 1972. *Proc. Nat. Acad. Sci. USA* **69**: 1620.
WIGZELL, H. and MÄKELÄ, O. 1970. *J. Exp. Med.* **132**: 110.

Structural analogies between the immune system and the nervous system

Niels Kaj Jerne

Basle Institute for Immunology, 487 Genzacherstrasse, CH-4058 Basle, Switzerland

These two systems stand out among all other organs (§1) of our body by their ability to respond adequately to an enormous variety of signals (§2). Both systems display dichotomies. The cells of both systems can receive as well as transmit signals. In both systems the signals can be either excitatorv or inhibitory (§3). The two systems penetrate most other tissues of our body, but they seem to avoid each other; a so-called blood-brain barrier prevents lymphocytes from coming into contact with nerve cells. The latter are sessile in brain, spinal cord and ganglia, forming a network of neurons. The ability of the axon of one neuron to form synapses with the correct set of other neurons requires something akin to epitope recognition.

Lymphocytes are a hundred times more numerous than nerve cells (§4). As they move about freely, lymphocytes can interact either by direct encounters or through the antibody molecules they release. These elements can recognize as well as be recognized and thus form a network (§5). As in the Nervous System, the modulation of the network by foreign signals represents its adaptation to the outside world (§6). Both systems thereby learn from experience and build up a memory which is sustained by reinforcement but cannot be transmitted to our offspring (§7). These striking analogies in the phenotypic expression of the two systems may result from similarities in the sets of genes that encode their structure and control their development and function.

The following notes further substantiate the statements enumerated above, as far as the Immune Systems is concerned.

§1. The cells of the immune system—the lymphocytes—are scattered throughout the body. Large accumulations of these cells occur in bone-marrow, thymus, spleen, lymph nodes and lymphatic tissues associated with the alimentary tract. Lymphocytes move about as single cells and circulate in the blood, the lymphatic vessels and in tissue fluids. The adult human body contains about 2×10^{12} lymphocytes and 2×10^{20} antibody molecules. The latter have been produced and secreted by lymphocytes. This total "organ" weighs almost one kilogram. In response to signals, lymphocytes can proliferate. Some become plasma cells which can produce and secrete 2,000 antibody molecules per second. All antibody molecules produced by a single cell, as well as by other cells of a given clone of lymphocytes, are identical. Antibody molecules belong to a variety of classes, subclasses and allotypes. Molecules of the same class and type can differ in the structure of the "variable" region of the molecule, which is determined by the amino acid sequences of the two polypeptide chains that combine to form that region. Two functional features of a "variable" region of an antibody molecule are: a) the combining site, by which the molecule can attach to an antigen; and b) the idiotype, which denotes antigenic patterns that can be recognized and become attached to combining sites of other antibody molecules. One individual is believed to produce millions of molecules differing in their "variable" regions, i.e., differing in the specific patterns of their combining sites and idiotypes. A '"resting" lymphocyte displays, as receptor in its surface membrane, the antibody molecules it will produce more abundantly when responding to a fitting signal.

§2. The signals to which lymphocytes respond involve the recognition by an antibody combining site of a molecular pattern ("epitome"). This is similar to the recognition of a substrate by the active site of an enzyme. Recognizable "epitopes" are displayed by the surfaces of all organic molecules, particles and cells (antigens) occurring in nature, which implies that the immune system can potentially respond to innumerable different epitopes. There is not a one-to-one recognition, however. A given epitope will be recognized with different degrees of precision by a variety of combining sites of the immune system, and a given combining site will recognize a variety of different epitopes. A measure of the degree of precision (or "fit") is the association constant of the reversible antibody-antigen attachment. The recognition, by a combining site of a receptor molecule on a lymphocyte, of an epitope displayed by an antigen particle leads to the reversible attachment of the antigen particle to the lymphocyte. It is not known what further signals are required for the lymphocyte to be stimulated to proliferate and to expand its synthetic apparatus. In certain cases it seems that another lymphocyte ("helper cell") must recognize another epitope on that same antigen particle in order to effect stimulation.

§4. Dichotomies

a) *Classes of lymphocytes*. The population of lymphocytes can be divided into two functional classes: T-cells and B-cells. Both classes arise from stem cells in the bone-marrow. The T-cells have passed through the thymus before entering the peripheral immune system. Only B-cells can develop into antibody secreting cells, and they are therefore thought to function mainly through the antibody molecules they release. The T-cells appear not to secrete antibody molecules. They are the cells that cause the phenomenon of cell-mediated immunity, i.e., those forms of immunity and sensitivity that can be transmitted from donor individuals to recipients only by the transfer of lymphocytes and not by the transfer of serum. Certain T-cells can kill target cells (e.g., cancer cells) by direct cell contact if these target cells display epitopes that the T-cells can recognize. T-cells can also function as the "helper cells" mentioned earlier. Furthermore, T-cells have been shown, in certain situations, to act as "suppressor"-cells toward B-cells.

b) *Direction of signals*. All antibody molecules, whether freely suspended in the body fluids or functioning as receptors on a lymphocyte, display combining sites as well as idiotypes. The latter are epitopes that can be recognized by combining sites of other antibody molecules. Every antobody molecule, and every lymphocyte displaying receptor molecules, can thus both recognize and be recognized. These properties derive solely from the structure and diversity of the variable regions of antibody molecules.

a) *Stimulation and suppression*. Many antigens which evoke antibody formation when injected into an animal will induce paralysis or tolerance when given in large amounts (high-zone tolerance) or in amounts below the antibody evoking theshold (low-zone tolerance). Thus, the presentation of epitopes to the fitting combining sites of the immune system can result in either excitation or inhibition of lymphocytes. The inhibitory effect is shown by the failure of the system subsequently to respond to otherwise stimulatory epitope concentrations. Both effects can also be demonstrated when presenting idiotypic epitopes. Thus the injection of a homogeneous myeloma protein (antibody produced by neoplastic cells) can lead either to the formation of anti-idiotypic antibody, or to tolerance. When, on the other hand, anti-idiotypic antibody is injected into an animal, the immune system is no longer capable of producing antibody molecules displaying the corresponding idiotype. No cases are so far known of a stimulatory effect of anti-idiotypic antibodies. For the sake of symmetry, this might well be expected to occur.

§4. The number of neurons in the human nervous system is often stated to be of the order of 10^{10}. If this is correct, the immune system disposes of a hundredfold more cells. Moreover, many lymphocytes are short-lived and are continuously replaced by new cells, both arising from lymphocyte proliferation and from stem cells.

§5. Network

a) *Formal network*. As the free antibody molecules and the lymphocyte receptor molecules of the immune system probably display millions of different idiotypes each of which can be recognized by a set of different combining sites, it is inconceivable that the combining sites of a given immune system should not recognize the idiotypes occurring in that same system. In all likelihood every given combining site will recognize, within this immune system, a set of different idiotypes, and every given idiotype will be recognized by a set of different combining sites.

b) *Functional network.* As mentioned earlier, the injection of molecules representing a given idiotype can lead either to anti-idiotypic antibody formation or to tolerance. Conversely, the injection of a large dose of anti-idiotypic antibody molecules has been shown to suppress the formation of antibodies displaying that idiotype, whereas low concentrations of anti-idiotypic antibody molecules appear to provoke formation of antibodies of the corresponding idiotype. The concentration of antibody molecules in human blood corresponds to about 10^{17} combining sites and 10^{17} idiotypes mer ml. If, for a moment, we assume that the total system has 10^7 different idiotypes and 10^7 different combining sites at its disposal, then each given one would be present, on the average, in a concentration of 10^{10} per ml blood. Antigen concentrations of this order of magnitude have, in several situations, been shown to evoke immunological responses.

§6. The immune system as a network can be viewed in the abstract, independently from outside influences. Questions can be asked with respect to the eigen-behavior of this isolated system, and as to the stability of the degree of diversity it displays. Obviously, the system is designed to deal with signals emanating from outside the system. Here we can distinguish between epitopes presented by antigens produced by or occurring in other organs and tissues of the same individual ("self" antigens), and antigens that penetrate into the individual from the outside world ("foreign" antigens). The immune system must obviously adapt to the self-antigens. This modulation occurs in early ontogeny. (It may break down later in pathological situations.) From its ability to respond to invading foreign antigens (including those that may occur on mutant cells arising in the individual) the system apparently derives the value for which it has been selected in evolution. Foreign epitopes may provoke the system because they are normally absent and therefore have not contributed to the state of the network.

§7. The transient provocation by a given antigen leaves a trace in the immune system, which may either be an enhanced or a diminished ability to respond to this antigen at a later occasion. This "memory," which can be of lifelong duration (cf. immunity to measles, to given influenza strains), is usually thought to be the result of the presence of specific "memory" lymphocytes, which may persist because of the network modulation provoked by the initial encounter with that antigen. Clearly, such modulations cannot be inherited: children must be vaccinated.

Dissipative structures in neural nets

J. D. COWAN

Department of Biophysics and Theoretical Biology
The University of Chicago, Chicago, Illinois 60637

0. Introduction

To understand how brains work is certainly among the most difficult of all intellectual problems. A current opinion would have it that "we have only the most superficial understanding of the functioning of the nervous system" (Handler, 1974). This may well be the case, nonetheless there have been very considerable advances in neuroanatomy, neurophysiology, and neurochemistry in the past twenty years or so, which, together with the advent and development of cybernetics and artificial intelligence studies, suggest that new levels of insight may soon be reached, about how brains work. In this paper I shall discuss some aspects of the mathematical theory of neural nets, as an example of cybernetic studies of how brains might work. It is of interest that the theory I shall discuss speaks to the existence of "dissipative structures" in neural activity, similar to those considered by Katzir-Katchalsky, as a possible basis for memory mechanisms (Katzir-Katchalsky et al., 1974).

It is now some twenty years since Sholl in his pioneering monograph on the organization of the cerebral cortex (Sholl, 1956) concluded that the functional units of nervous activity were not specific circuits, but were instead randomly connected aggregates of nerve cells, whose pattern of joint activity conveyed information. Similar conclusions based on behavioral studies were arrived at by the psychologist Hebb (1949). Since then microelectrode studies of single nerve cells have led to a differing view, most clearly stated by Barlow (1974), that there exist hierarchies of neural aggregates, each abstracting information from patterns of joint cellular activities, in such a way that the proportion of cells becoming active per unit time decreases with increasing position in the hierarchy. Thus the activity of individual cells high in a polysynaptic hierarchy is functionally significant, whereas low in the hierarchy it is the pattern of joint activity that is required. Are these two views of neural organization fundamentally different, or is there in fact a spectrum of functionally significant neural activities, ranging from individual firing patterns to those of large aggregates? In this paper recent attempts will be described (Wilson and Cowan, 1972, 1973; Feldman and Cowan, 1975a, b) that seek to answer these and related questions by way of a mathematical theory of interactions within and between neural aggregates. It turns out that the most important determinant of local as opposed to global specificity is the degree of correlation present in stimuli, and in the pattern of activity within and between aggregates, the latter correlation depending on anatomical factors. In all cases, equations can be formulated that adequately describe the dynamic spatio-temporal activity. Such equations generally turn out to be coupled nonlinear integral equations, of the form

$$\mathbf{F}(\mathbf{r}, T) = \boldsymbol{\phi}[A \times \mathbf{F}(\mathbf{r}, T) \otimes h_1(T) \pm \mathbf{P}(\mathbf{r}, T) \otimes h_1(T) - \boldsymbol{\theta}(\mathbf{r}, T)] \tag{0.1}$$

$\mathbf{F}(\mathbf{r}, T)$ represents the mean activity of nerve cells comprising a net or tissue. T is measured in units of the nerve cell membrane time constant μ. A is a matrix specification of the coupling parameters in the net, i.e., the strengths and signs of neural synapses, $\boldsymbol{\theta}(\mathbf{r}, T)$ represents refractory effects built up in nerve cells following excitation, $\mathbf{P}(\mathbf{r}, T)$ is any external excitation to the cells, $h_1(T)$ is a linear filter with time constant μ equal to the membrane time constant, and $\boldsymbol{\phi}$ is a nonlinear function representing the steady state relationship between the activity $\mathbf{F}$ and the trigger-zone voltage $(A \times \mathbf{F} + \mathbf{P}) \otimes h_1 - \boldsymbol{\theta}$ that drives the net.

Such equations can be given either a discrete or a continuous representation, depending upon how they are to be used. In the discrete case, the equation for the ith component $F_i(\mathbf{r}, T) = F_i$ represents the behavior of the ith cell, and ϕ represents the smooth dependence of firing rate on trigger-zone voltage or current, seen for example in motoneurons (Kernell, 1968; Feldman and Cowan, 1975a,b). In the continuous case the equation for $F_i(\mathbf{r}, T)$ represents the behavior of the ith cell type in the tissue, and ϕ represents the smooth relationship between F_i, the mean proportion of cells of the ith type becoming activated per unit time at $\mathbf{r}$, as a function of the mean trigger zone voltage or current built up in the cells of the ith type at $\mathbf{r}$.

A convenient representation of ϕ in the discrete case is given by the function (Feldman and Cowan, 1975a),

$$\phi[y] = [r + \exp(-v(y - \theta_\infty))]^{-1} - [r + \exp(v\theta_\infty)]^{-1} \tag{0.2}$$

and in the continuous case by the function (Wilson and Cowan, 1972, 1973),

$$\phi[y, F] = \left[1 - \int_{T-r}^{T} F\,dt\right] \times$$
$$\times\ [[1 + \exp(-v(y - \theta_\infty))]^{-1} - [1 + \exp(v\theta_\infty)]^{-1}] \tag{0.3}$$

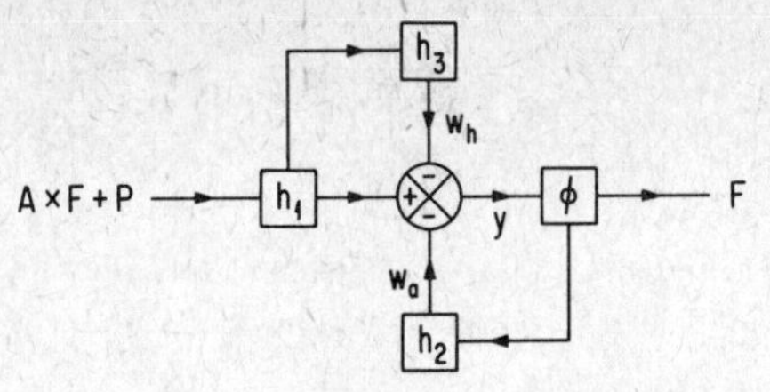

Figure 1
The structure of the neural model, showing both afterhyperpolarization and accommodation.

where $y = (A \times \mathbf{F} + \mathbf{P}) \otimes h_1 - \boldsymbol{\theta}$, r is the absolute refractory period of cells in the net or tissue, and v and θ_∞ are parameters determining the shape of ϕ. In the discrete case these parameters are related to the variability of cellular properties such as membrane time constants and resting thresholds (Kernell, 1966), and in the continuous case to the variability of properties such as the number and distribution of synaptic endings per cell. More complex refractory effects resulting from afterhyperpolarization (Kernell, 1968) and accommodation (Fitzhugh, 1969) are represented by the function $\boldsymbol{\theta}$. Thus

$$\boldsymbol{\theta} = \mathbf{W}_h + \mathbf{W}_a \tag{0.4}$$

where $\mathbf{W}_h$, the afterhyperpolarization, is given by the temporal convolution

$$\mathbf{W}_h = \mathbf{F} \otimes h_2 \tag{0.5}$$

where h_2 is a first order linear filter with time constant μ_2 and where $\mathbf{W}_a$, the accommodation, is given by the temporal convolution

$$\mathbf{W}_a = (A \times \mathbf{F} + \mathbf{P}) \otimes h_1 \otimes h_3 \tag{0.6}$$

where h_3 is another filter with time constant μ_3.*

Such equations can be depicted in a useful black box form, as shown in Fig. 1. A shorter symbolic representation of a special form of such equations, limited to spatially uniform systems, has recently been developed (Stein et al., 1974a,b), which obviously extends to the spatially nonuniform case. Figure 2 shows such representation.

Finally, it should be noted that in the continuous representation of neural activity, the ith component of the vector $A \times \mathbf{F}$, $(A \times \mathbf{F})_i = \sum_j A_{ij} F_j$, becomes a sum of spatial integrals, i.e., the term $A_{ij} F_j$ becomes the integral

$$\int_{-\infty}^{\infty} K_{ij}(\mathbf{r}, \mathbf{r}')\, F_j\!\left(\mathbf{r}', T - \frac{|\mathbf{r} - \mathbf{r}'|}{v_j}\right) d\mathbf{r}'$$

* Other ways of incorporating relative refractory effects are possible, particularly in the continuous representation (Wilson and Cowan, 1972).

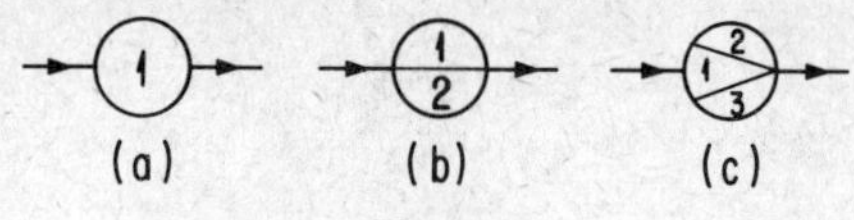

Figure 2
Symbolic representation of neural model: (a) absolute refractoriness only, (b) afterhyperpolarization added, (c) accommodation added to (b).

where $K_{ij}(\mathbf{r}, \mathbf{r}')$ is a kernel or distribution function denoting the mean number of synaptic endings on a given cell of the ith class at $\mathbf{r}$, per cell of the jth class a distance $|\mathbf{r} - \mathbf{r}'|$ away, and where v_j denotes the velocity of propagation of impulses along axons of cells of the jth type.*

1. Absolute refractory nets

Although the effects of incorporating afterhypolarization and accommodation are most interesting, in that spatio-temporal oscillations and propagating waves of great complexity are seen in such nets, and although such effects are of relevance for information processing models of neural activity, particularly in the cerebellum and associated motor systems (Wall and Cowan, in preparation), only the phenomena seen in absolute refractory nets will be fully discussed in this paper. Thus equation (0.1) simplifies to

$$\mathbf{F}(\mathbf{r}, T) = \phi[A \times \mathbf{F}(\mathbf{r}, T) \otimes h_1(T) \pm \mathbf{P}(\mathbf{r}, T) \otimes h_1(T)] \tag{1.1}$$

a nonlinear integral equation of Hammerstein type (Cochran, 1972). Furthermore, only the time averaged activity

$$\langle \mathbf{F}(\mathbf{r}, T) \rangle = \int_{T-1}^{T} F(\mathbf{r}, T')\, dT' \tag{1.2}$$

will be considered. That is, only activity smoothed with a time constant equal to the membrane time constant (μ msec) will be considered. This process eliminates all temporal integrals from consideration, and reduces equation (1.1) to the nonlinear integro-differential equation,

$$\frac{\partial \langle \mathbf{F}(\mathbf{r}, T) \rangle}{\partial T} = -\langle \mathbf{F}(\mathbf{r}, T) \rangle + \phi[A \times \langle \mathbf{F}(\mathbf{r}, T) \rangle \pm \langle \mathbf{P}(\mathbf{r}, T) \rangle] \tag{1.3}$$

This equation is much more readily integrated by numerical methods than is Eq. (1.1) (Wilson and Cowan, 1972, 1973). An equation more or less of this form, for a net

* In what follows, v_j, which is 0 (m/sec), will be neglected, since it produces transit time effects of negligible importance for the problems considered in this paper.

comprising cells of a single homogeneous type, was first introduced by Beurle (1956), and in fact even earlier there are precursors to be found in the work of Rapaport (1950, 1951) and Shimbel (1950, 1951) and Allanson (1956). Since 1956, related equations have been investigated by Caianiello (1961), Smith and Davidson (1962), Griffith (1963, 1965), and Harth et al. (1970), among others. An early version of the present work, concerning equations for large scale activity in nets comprising at least two distinct cell types, one excitatory, the other inhibitory, is to be found in Cowan (1968, 1970).

Nets with excitation only

It will be convenient to consider first the properties of a net comprising a single excitatory cell type. Thus Eq. (1.3) reduces to

$$\frac{\partial\langle E(\mathbf{r}, T)\rangle}{\partial T} = -\langle E(\mathbf{r}, T)\rangle + \phi[A \times \langle E(\mathbf{r}, T)\rangle \pm \langle P(\mathbf{r}, T)\rangle] \tag{1.4}$$

The coupling matrix A consists of only the single term A_{EE}. In what follows the continuous representation will be employed. Thus the term $A \times \langle E(\mathbf{r}, T)\rangle$ becomes the spatial integral

$$\int_{-\infty}^{\infty} K_{EE}(\mathbf{r}, \mathbf{r}')\langle E(\mathbf{r}, T)\rangle\, d\mathbf{r}'$$

where $K_{EE}(\mathbf{r}, \mathbf{r}')$ is a kernel specifying the coupling between cells within the net, and ϕ is defined as in Eq. (0.3), so that Eq. (1.4) reduces to

$$\frac{\partial\langle E(\mathbf{r}, T)\rangle}{\partial T} = -\langle E(\mathbf{r}, T)\rangle + [1 - r\langle E\rangle]\mathscr{S}[N] \tag{1.5}$$

where $\mathscr{S}[N]$ is the "logistic" function defined in (0.3), and where N, the membrane voltage, is defined as

$$N(\mathbf{r}, T) = \alpha\left[\int_{-\infty}^{\infty} K_{EE}(\mathbf{r}, \mathbf{r}')\langle E(\mathbf{r}, T)\rangle\, d\mathbf{r}' \pm \langle P(\mathbf{r}, T)\rangle\right] \tag{1.6}$$

where α has the dimension of volts.

To obtain exact solutions of Eqs. (1.5) and (1.6) is difficult. However certain special cases can be solved that provide the necessary insights into the resulting phenomena.

Case 1: Spatially uniform net

The simplest case occurs when $K_{EE}(\mathbf{r}, \mathbf{r}')$ is constant over the entire net. That is to say, any pair of cells in the net is connected; so that all cells have the same mean number of synapses, K_{EE}. Furthermore if $\langle P(\mathbf{r}, T)\rangle$ is also uniform over the net, Eqs. (1.5) and (1.6) are independent of $\mathbf{r}$, and so is the activity $\langle E\rangle$, which satisfies the equation

$$\frac{\partial\langle E(T)\rangle}{\partial T} = -\langle E(T)\rangle + [1 - r\langle E\rangle]\mathscr{S}[N(T)] \tag{1.7}$$

where

$$N(T) = \alpha K_{EE}\langle E(T)\rangle \pm \alpha\langle P(T)\rangle \tag{1.8}$$

Simple topological analysis can therefore be used to investigate the stability of steady states in the net (Allanson, 1955; Harth et al., 1970). It turns out that even with $\langle P\rangle$ equal to zero, if the mean number of synapses per cell from recurrent excitatory connections K_{EE} is sufficiently large, then non-zero stable steady states of activity are possible. Indeed if K_{EE} is very large, zero activity, $\langle E\rangle = 0$, is an unstable steady state. Such a state can be excluded by limiting K_{EE} and ν_E, the variability of K_{EE}, so that $K_{EE}\nu_E < \theta$. Non-zero stable steady states can be excluded by ensuring that the excitation built up by such states is always sub-threshold, i.e., setting K_{EE} so that $\alpha K_{EE}[1 + r]^{-1} < \theta_\infty$. Provided that $K_{EE}\nu_E$ is not too large, the graph of $\phi = [1 - r\langle E\rangle]\mathscr{S}[N(\langle E\rangle)]$, plotted as a function of $\langle E\rangle$ is as shown in Fig. 3.

It will be seen that there is a systematic change in the topology of the singular points of $\phi(\langle E\rangle)$, i.e., in the points at which $\partial\langle E\rangle/\partial T = -\langle E\rangle + \phi = 0$ when $\langle P\rangle = 0$, $\langle E\rangle = 0$ is the only stable steady state. As $\langle P\rangle$ increases, the stable steady state value of $\langle E\rangle$ slowly increases, until at $\langle P\rangle = \langle P\rangle_{\text{threshold}}$, two stable steady states appear, separated by an instability. This corresponds to the beginning of the so-called non-thermodynamic branch in the development of a dissipative structure (Prigogine and Nicolis, 1971). Eventually for superthreshold values of $\langle P\rangle$, a single non-zero stable steady state exists. Figure 4 shows the graph of such states, plotted as a function of $\langle P\rangle$. It will be seen that a simple hysteresis loop can be generated by raising

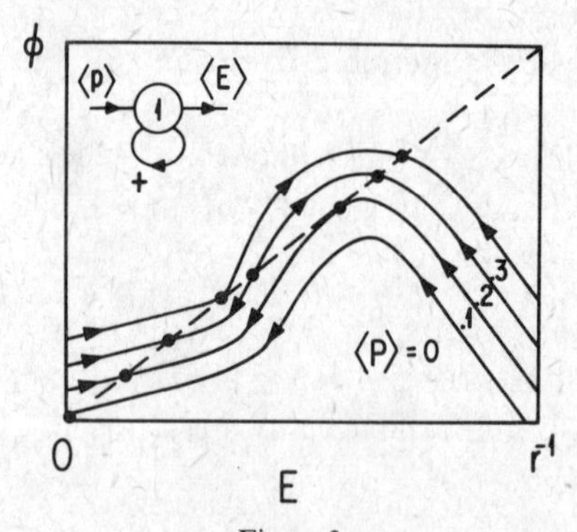

Figure 3
Isoclines of Eq. (1.7). Arrows indicate the sign of $\partial E/\partial t(\leftarrow\ -ve,\ \rightarrow\ +ve)$. Closed circles ● indicate steady states ($\partial E/\partial t = 0$). Neural net configurations shown in upper left-hand corner.

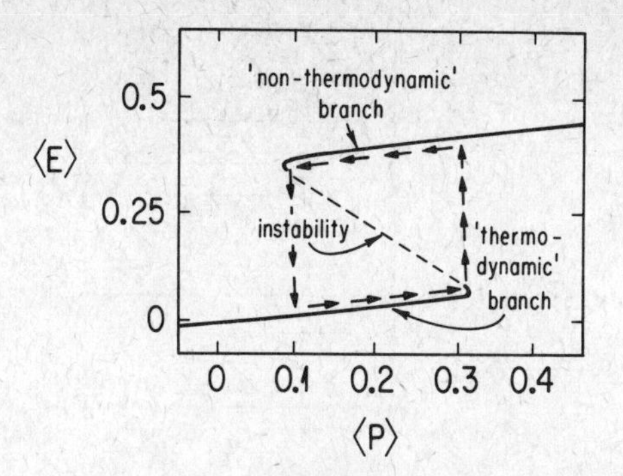

Figure 4
Steady state values of $\langle E \rangle$ as a function of $\langle P \rangle$. Hysteresis loop generated if $\langle P \rangle$ is varied slowly between the values 0.1 and 0.3.

$\langle P \rangle$ slowly from zero to 0.3 units, at which point the solution of Eq. (1.7) bifurcates, and the stable steady state value of $\langle E \rangle$ switches from nearly zero to nearly 0.5.

If now $\langle P \rangle$ is slowly lowered, $\langle E \rangle$ will slowly diminish along the non-thermodynamic upper branch, until the critical valve $\langle P \rangle = 0.1$ is reached, where the system switches back to the lower branch. Thus a simple "flip-flop" has been shown to exist in a neural net comprising only excitation cells (Harth et al., 1970). In other terms, the net may be said to exhibit the "cusp" catastrophe (Thom, 1972). In summary, then, the spatially uniform net under uniform stimulation has been shown to exhibit simple hysteretic switching.

Case 2: Spatially non-uniform net

If $\langle P \rangle$ is spatially non-uniform, however, so that Eqs. (1.5) and (1.6) obtain, the situation seems more complicated. Intuitively, however, it is to be expected that any locally induced activity will propagate throughout the net, "switching on" the net as it propagates. Such a situation was first considered by Beurle (1956). Here an approximate calculation will be outlined that serves to show the validity of the above. It is assumed that $\mathcal{S}[N(\langle E \rangle)]$, the proportion of cells activated per unit time by the excitation $\langle E \rangle$, is simply proportional to $\langle E \rangle$. Thus Eqs. (1.5) and (1.6) can be approximated by the logistic equation

$$\frac{\partial \langle E \rangle}{\partial T} = \langle E \rangle + [1 - r\langle E \rangle] m \langle E \rangle \tag{1.9}$$

the solution of which is

$$\langle E \rangle = \frac{M}{2r}[1 + \tanh(\beta(T - T_0) + K_0)] \tag{1.10}$$

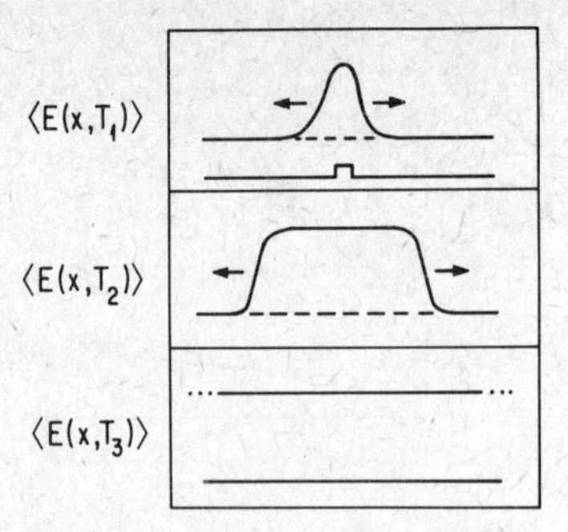

Figure 5
Generation of travelling wavefronts (solitons) in a net of excitatory neurons. The region (initially) stimulated is indicated beneath the top graph.

where $M = 1 - 1/m$ is the proportion of cells used in the passage of a wavefront, and where $\beta = (m - 1)/2$. If this solution represents a wavefront propagating with velocity v, then

$$\langle E(\mathbf{r}, T)\rangle = \frac{M}{2r}\left[1 + \tanh\left(\beta\left(T - \frac{|\mathbf{r}|}{v}\right) + K_0\right)\right] \tag{1.11}$$

assuming a constant stimulus $\langle P(\mathbf{r})\rangle = \delta(\mathbf{r}, \mathbf{r}_0)$ generating the initial pulse of $\langle E\rangle$. Figure 5 shows three stages in the propagation of such fronts.

As expected, since the refractory period r is small, and since there is no inhibition in the net, cells can fire many times during the passage of a front, or indeed support the passage of many fronts. In fact fronts propagate through the net in the manner of solitons (Scott et al., 1973), in a fashion similar to the propagation of action potentials along poisoned axons, the recovery of which has been impeded (A. Huxley, unpublished). Evidently there is a critical value of m, for such propagation, related as before to K_{EE} or more generally to $K_{EE}(\mathbf{r}, \mathbf{r}')$, and to v_E. If m is too small the activity will decay to $\langle E\rangle = 0$, whereas if m is sufficiently large, the whole net will switch on. There is no intermediate level at which stable propagation can occur, or at which stable spatio-temporal activity (dissipative structures) of any higher complexity can occur. It is this paucity of stable structures that leads one to the introduction of inhibition.

2. Nets with both excitation and inhibition

Turning now to the case of nets incorporating both excitatory and inhibitory cells, it will again be convenient to consider first the simpler case of spatially uniform nets.

Spatially uniform nets

The relevant equations for this case are a simple extension of Eqs. (1.7) and (1.8), i.e.,

$$\frac{\partial\langle E(T)\rangle}{\partial T} = -\langle E(T)\rangle + [1 - r\langle E\rangle]\,\mathscr{S}_E[N_E(T)] \tag{2.1}$$

$$\frac{\partial\langle I(T)\rangle}{\partial T} = -\langle I(T)\rangle + [1 - r\langle I\rangle]\;\mathscr{S}_I[N_I(T)] \tag{2.2}$$

where

$$N_E(T) = \alpha[K_{EE}\langle E\rangle - K_{EI}\langle I\rangle \pm \langle P\rangle] \tag{2.3}$$

$$N_I(T) = \alpha[K_{IE}\langle E\rangle - K_{II}\langle I\rangle \pm \langle Q\rangle] \tag{2.4}$$

and where $\langle E\rangle$ and $\langle I\rangle$ are respectively the proportions of excitatory and inhibitory cells in the net becoming activated per unit time.

As before, in the spatially uniform case, "phase plane" analysis can be carried out on the equations, to determine the nature of any singular points. Figure 6, for example, shows the locus of points of constant $\langle E\rangle$ and $\langle I\rangle$, plotted on an $\langle E\rangle$–$\langle I\rangle$ phase plane. It will be seen that there exist three possible steady states, separated by two unstable steady states. Figure 7 shows the resulting steady states of $\langle E\rangle$ plotted as a function of $\langle P\rangle$, the excitatory stimulus, for zero external inhibitory stimulation $\langle Q\rangle$.

It will be seen that there is now a stable steady state of intermediate activity $\langle E\rangle \sim 0.25$, accessible from either of the other two stable steady states by the variation of $\langle P\rangle$, with two hysteresis loops now possible. The actual details of the switching between such states, while complicated, shows the general properties found in all

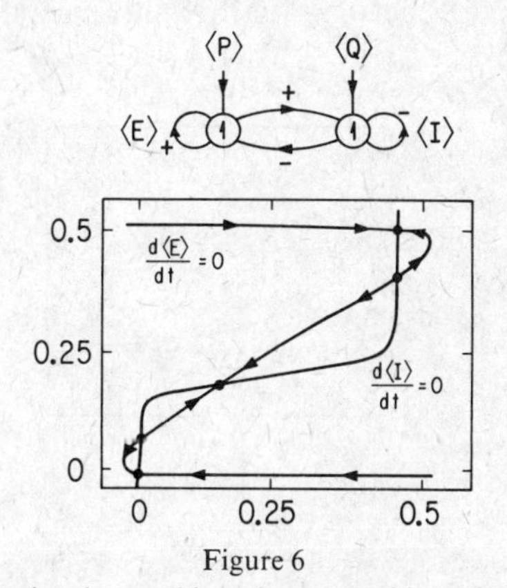

Figure 6
Steady states of $\langle E\rangle$ and $\langle I\rangle$ showing three stable and two unstable steady states (from Wilson and Cowan, 1972). Net configuration shown above graph.

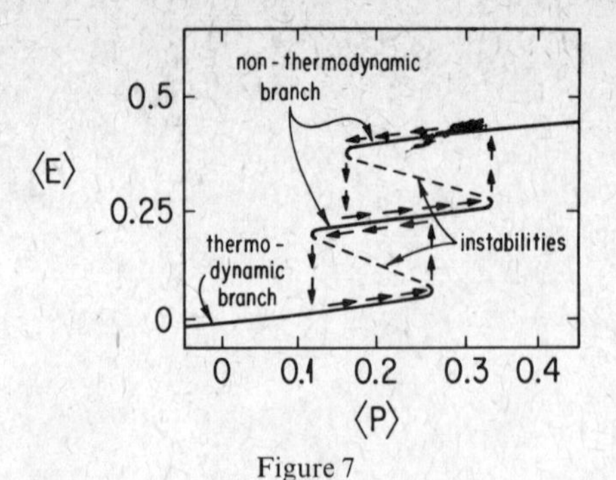

Figure 7
Steady state values of $\langle E \rangle$ as a function of $\langle P \rangle$ for $\langle \phi \rangle = 0$, showing multiple hysteresis loops.

switching systems. Thus there is a strength-duration curve (Fitzhugh, 1969) for the switching between steady states, as shown in Fig. 8.

The time course of the activity follows, more or less, the curve defined in equation (1.10), as shown in Fig. 9. Such a time course is very simple, showing no temporal oscillations. This follows from the absence of relative refractoriness. Much more complex transients can be obtained if such refractory properties are incorporated (Wilson and Cowan, unpublished; Stein et al., 1974a). It will be seen, however, that superthreshold transients do display the interesting property that even after the stimulus has ceased, the mean activity in the net keeps increasing until a stable steady state is reached. This, of course, is possible only because recurrent excitatory loops in the net provide a positive feedback which is strong enough to augment and then maintain a high level of activity. Such activity is limited only by inhibition or refractoriness. Given the nonlinearity that stems from the existence of a threshold for activation in cells, all the elements requisite for the development of dissipative structures are present in neural nets. Spatially uniform non-zero stable steady states

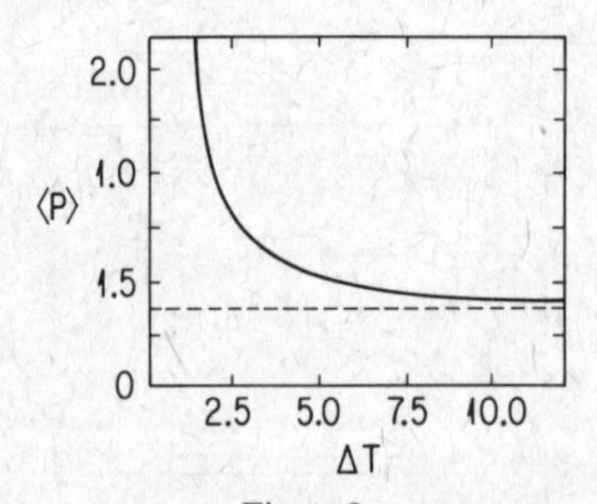

Figure 8
Strength–duration curve for the excitation of a net from one stable steady state to another. Axes indicate the stength $\langle P \rangle$ and duration ΔT of a rectangular pulse, just sufficient to cause the net activity to switch from a low state to the next higher stable steady state (from Wilson and Cowan, 1972).

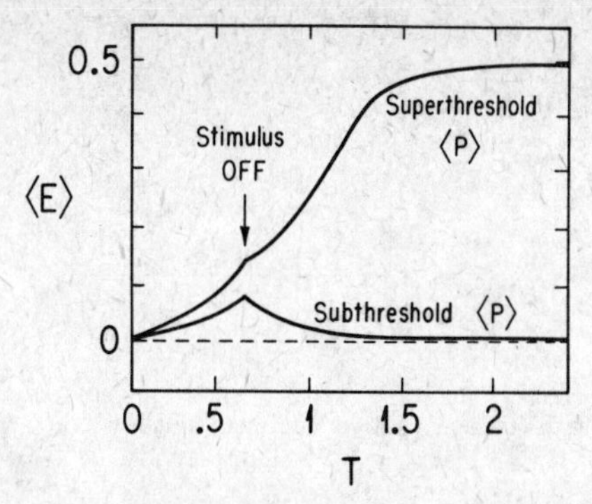

Figure 9
Transient response of net in response to rectangular pulse of excitation, showing sub- and superthreshold stimulation.

are perhaps the simplest of dissipative structures. However, there is another simple dissipative structure, the limit cycle (Minorsky, 1962) that is also readily generated by neural nets comprising both excitatory and inhibitory cells. Such cycles can occur given not only strong positive feedback, but strong negative feedback as manifested in the product $K_{EI}K_{IE}$ that specifies the mutual interaction between excitatory and inhibitory cells. In addition, a further requirement is that both v_E and v_I be large, i.e., that the variability in the number of synaptic endings per cell, or equivalently in cellular thresholds, be small. This means that highly synchronous activity is possible in a net, given constant $\langle P \rangle$. Because of the negative feedback provided by inhibition, such activity is oscillatory. Figure 10 shows the phase plane and Fig. 11 the associated stable limit cycle.

Such a limit cycle corresponds to the generation of bursts of neural activity in the net. It is of interest that these cycles are the analogue of predator-prey cycles generated by the Lotka–Volterra equations for ecological kinetics (May, 1973). Indeed it

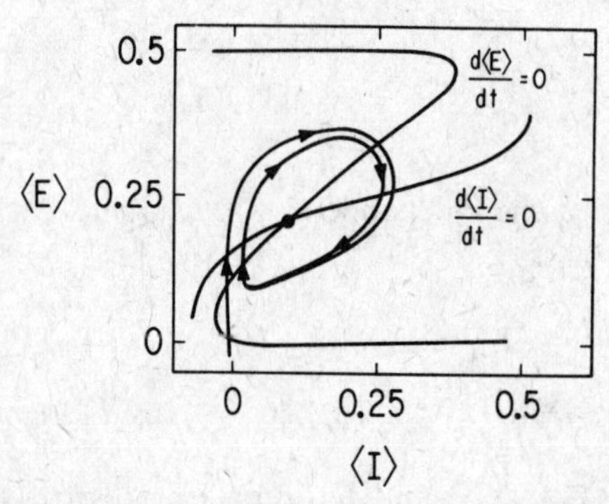

Figure 10
Phase plane showing a limit cycle trajectory in response to constant $\langle P \rangle$ (from Wilson and Cowan, 1972).

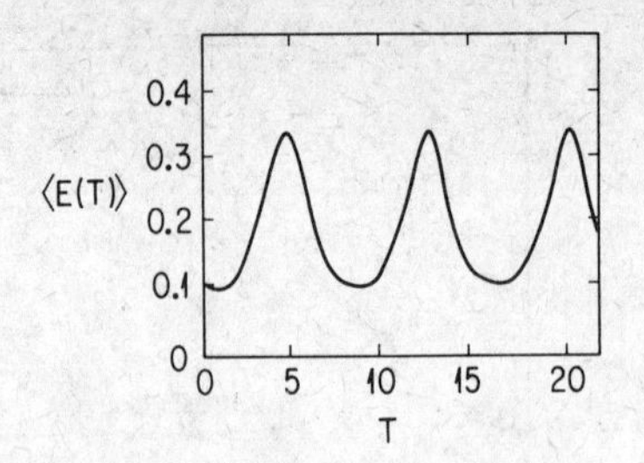

Figure 11
Limit cycle activity in a net with strong negative feedback, and low connection noise (from Wilson and Cowan, 1972).

was such an analogy that led the author to the original formulation of similar neural net equations, from which the current ones were developed (Cowan, 1965, 1968).

It is important to recognize that these equations and properties are structurally stable (Thom, 1972), in the sense that small parametric fluctuations will not change the nature of the stable states, a property of open nonlinear dissipative systems, but not of conservative systems.

Spatially non-uniform nets

With appropriate modifications, the responses of similar nets to spatially non-uniform stimuli are of the same character. It will be convenient to consider here only spatially homogeneous nets. That is, nets the kernels of which are functions only of $\mathbf{r} - \mathbf{r}'$, so that the spatial integrals introduced in Sec. 0 become the convolutions

$$\int_{-\infty}^{\infty} K_{ij}(\mathbf{r} - \mathbf{r}')\, F_j\left(\mathbf{r}', T - \frac{|\mathbf{r} - \mathbf{r}'|}{v_j}\right) d\mathbf{r}' =$$

$$= K_{ij}(\mathbf{r}) \otimes F_j\left(\mathbf{r}, T - \frac{|\mathbf{r} - \mathbf{r}'|}{v_j}\right) =$$

$$= K_{ij}(\mathbf{r}) \otimes F_j(\mathbf{r}, T) \tag{2.5}$$

and Eqs. (2.1)–(2.4) all become $\mathbf{r}$-dependent, with N_E and N_I defined, respectively, as (Wilson and Cowan, 1973):

$$N_E(\mathbf{r}, T) = \alpha[K_{EE}(\mathbf{r}) \otimes \langle E(r, T)\rangle - K_{EI}(\mathbf{r}) \otimes \langle I(\mathbf{r}, T)\rangle \pm \langle P(\mathbf{r}, T)\rangle] \tag{2.6}$$

$$N_I(\mathbf{r}, T) = \alpha[K_{IE}(\mathbf{r}) \otimes \langle E(\mathbf{r}, T)\rangle - K_{II}(\mathbf{r}) \otimes \langle I(\mathbf{r}, T)\rangle \pm \langle Q(\mathbf{r}, T)\rangle] \tag{2.7}$$

The four kernels K_{EE}, K_{EI}, K_{IE}, and K_{II} represent, respectively, recurrent excitation, lateral excitation of inhibitors, lateral inhibition of excitors, and recurrent inhibition. As defined in Sec. 0, these kernels represent the mean number of synaptic endings on a given cell type, from all cells of a specified type, a distance $|\mathbf{r}|$ away. Histological measurements indicate that such kernels, in general, are decreasing functions of $|\mathbf{r}|$, with space constants σ_{ij} ranging from a few to several hundred microns (Sholl, 1956). A convenient approximation to such kernels is shown in Fig. 12, the rectangular kernel of width $2\sigma_{ij}\mu m$ and amplitude $b_{ij}\mu m^{-1}$.

With this approximation, the convolution defined in Eq. (2.5) reduces to the simple integral

$$b_{ij} \int_{\mathbf{r}-\sigma_{ij}}^{\mathbf{r}+\sigma_{ij}} F_j(\mathbf{r}, T)\, d\mathbf{r} \tag{2.8}$$

It remains to determine the relative sizes of these space constants. In what follows it will be assumed that

$$\sigma_{EE} < \sigma_{IE} \tag{2.9}$$

and that

$$\sigma_{II} < \sigma_{EI} \tag{2.10}$$

i.e., that the range of lateral interactions between cell types is greater than that of recurrent interactions. The net effect of this assumption is to *localize* any activity in the net, since the lateral excitation of inhibitors, and the lateral inhibition of excitors will quench any propagating activity generated in the net. In effect, a locally addressed computing net can be built up from interconnected nerve cells, given the above conditions.

The responses of such a net are similar to those already described for the spatially uniform cases, except that the activity is localized. Thus if there is strong positive feedback in the net, b_{EE} high and high v_E and v_I, spatially inhomogeneous stable steady states result, as shown in Fig. 13.

As in the spatially uniform case, there is an excitation or switching condition. However, instead of a strength-duration curve, the condition is expressed as a strength–duration–area surface. Figure 13 shows spatially inhomogeneous stable steady states generated by brief excitatory stimuli of superthreshold dimensions.

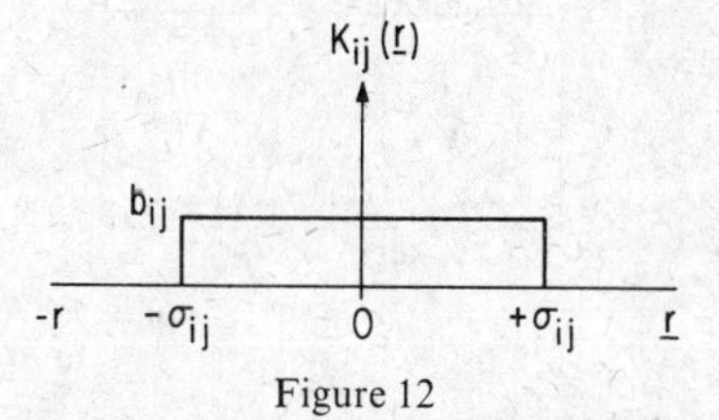

Figure 12
Approximation to the kernel $K_{ij}(\mathbf{r})$ specifying the mean number of synapses on cells of the ith class from all cells of the jth class a distance $|\mathbf{r}|$ away.

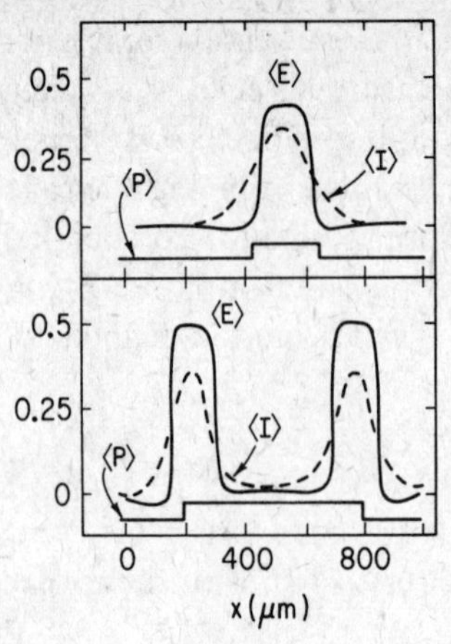

Figure 13
Spatially inhomogeneous stable steady states generated in response to two differing stimuli (from Wilson and Cowan, 1973).

In effect, the net functions as a spatially localized flip-flop, thus combining computer logic with a spatially coded addressing system. It will be seen that a narrow stimulus generates one localized stable pulse, whereas a broad stimulus generates two stable pulses, one at each end of the stimulus. This latter condition results from the well-known effects of lateral inhibition in the net (Ratliff, 1965). The pulses produced are stable, and can only be removed by restimulation of the net with stimuli of sufficiently lowered (or even negative, i.e., inhibitory) amplitude. In similar fashion, if there is strong negative feedback in the net, as well as strong positive feedback

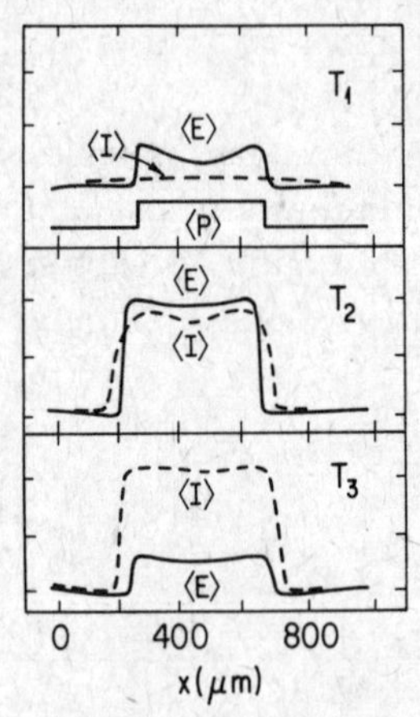

Figure 14
Three epochs in a spatially localized limit cycle in response to the maintained stimulus $\langle P(\mathbf{r}) \rangle$ shown in the upper frame (redrawn from Wilson and Cowan, 1973).

and low connection "noise", high v_E and v_I, then localized limit cycles are seen in response to maintained superthreshold stimuli. Figure 14 shows spatial profiles of the cycle at three epochs, the temporal profile being similar to that shown for the uniform case, in Fig. 11.

It would appear that no new states are present in the response of such nets to spatially non-uniform stimuli. Such is not the case. There are in fact at least two new types of response that are of significant interest. The first concerns the response of a net to maintained excitatory stimuli, the inhibitory neurons of which have themselves been disinhibited, by means of a low level spatially uniform stimulus $\langle Q \rangle = -Q_0$. Figure 15 shows the resulting effect.

It will be seen that the disinhibited net no longer can be addressed locally, and that the response to a localized excitatory stimulus consists of travelling waves of excitation closely followed by inhibition (strong negative feedback is assumed in this case. Weak negative feedback would give propagating fronts intermediate in structure between those shown in Fig. 5 for the purely excitatory net, and the waves shown in Fig. 15). It is important to note that the troughs generated behind such waves are the result of inhibition, and not of refractoriness. Such waves are therefore quite different from those considered by Beurle (1956), or from those simulated by Farley and Clark (1961).

The second new type of response is of a different character. It will be recalled that the kernels $K_{ij}(\mathbf{r})$ are all of relatively short range. If, in addition, the coupling between cell types is not too strong, and if the connection noise is fairly high, and if recurrent excitation is sufficiently strong, but not too strong, then in response to a brief stim-

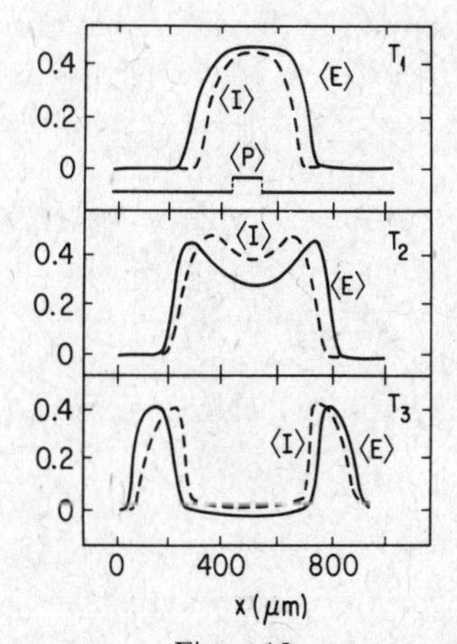

Figure 15
Generation of travelling waves in a disinhibited net with strong negative feedback (from Wilson and Cowan, 1973).

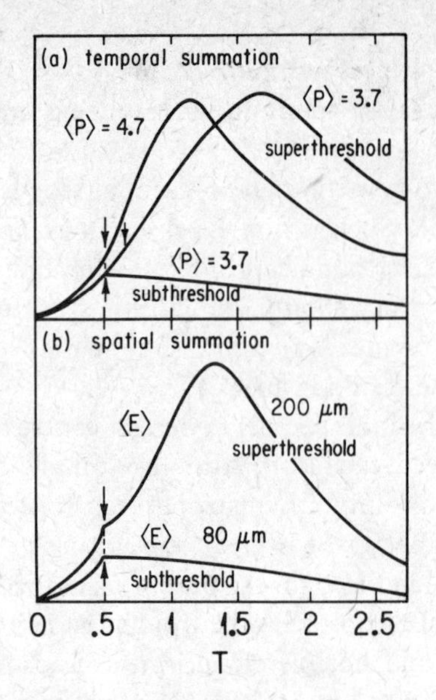

Figure 16

Active transients showing (a) temporal summation, and (b) spatial summation, ↑, stimulus OFF (redrawn from Wilson and Cowan, 1973).

ulus, instead of an all-or-none switching of the net activity, a prolonged but transient response is exhibited. Figure 16 shows such transients.

It will be seen that the net response continues to augment even after cessation of the stimulus, in a manner similar to that discussed in connection with the switching on of stable steady states. In fact there is a strength-duration-area condition for the generation of this type of response, which is referred to as an active transient in what follows. Temporal and spatial summation, i.e., the generation of superthreshold active transients in response to long lasting or big stimuli, are direct consequences of such a condition, and are demonstrated in Fig. 16. It will be seen that the *latency* of the peak response is related to stimulus intensity, all things being equal, a coding property of some significance. Such a transient cannot be generated in response to uniform stimuli, but only to relatively localized stimuli.

3. More complex nets

It will be evident that even the simplest possible case, spatially homogeneous nets comprising one or two absolutely refractory cell types, already exhibits complex dynamical patterns of activity. It is all too easy to add to the complexity.

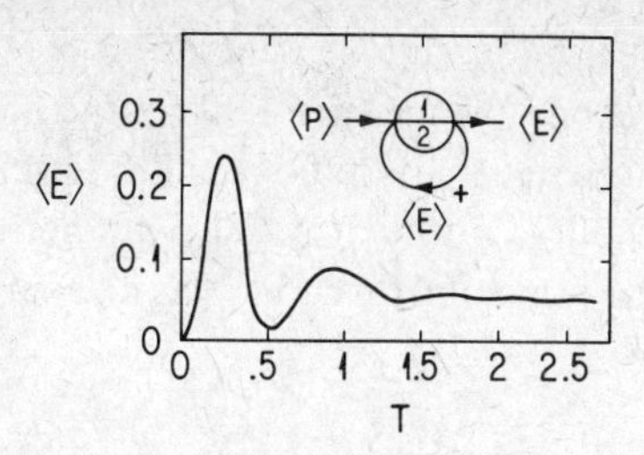

Figure 17
The transient response of an excitatory net with afterhyperpolarization, to a *maintained* stimulus $\langle P \rangle$. Net configuration shown in upper right hand corner.

Afterhyperpolarization

Afterhyperpolarization has already been cited as a factor responsible for the generation of oscillatory transients in spatially homogeneous nets. Figure 17 shows, for example, the response of an excitatory net to a maintained stimulus.

It will be seen that there is an oscillatory transient preceding the settling down of the net to a much lower steady state than would obtain in the absence of afterhyperpolarization.

More than two cells types

In similar fashion, if the variation in connectivity is not homogeneous, so that effective thresholds vary over the net, or if other cellular properties support a natural classification into many differing cell types, then much more complex spatio-temporal patterns obtain. Figure 18 shows for example the responses of a uniform net comprising six differing cell types to maintained stimuli (Feldman and Cowan, (1975b).

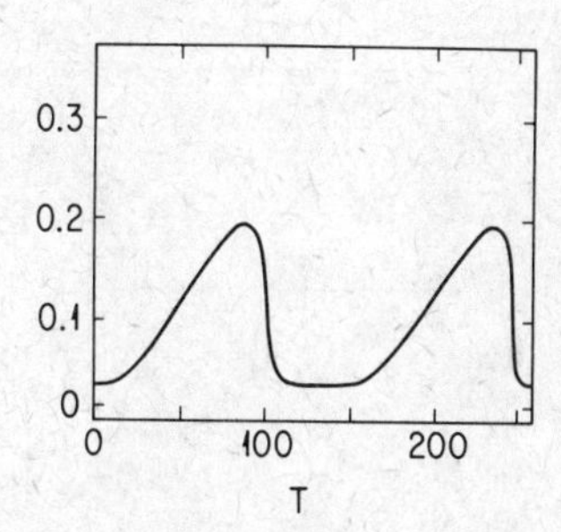

Figure 18
The limit cycle response of model medullary inspiratory neurons to maintained stimulation (from Feldman and Cowan, 1975b).

Such a net is intended as a model for the generation of stable respiratory rhythms in the brain stem. It will be seen that the time scale of the limit cycle shown in Fig. 18 is two orders of magnitude greater than that of the limit cycles previously considered. This is a consequence of the large number of cell types and the increased complexity of interactions in the net, and it may be thought of as an example of cooperative activity in the net (cf. Little, 1974).

Spatial inhomogeneity

Another source of additional complexity occurs if the kernels $K_{ij}(\mathbf{r}, \mathbf{r}')$ are no longer spatially translation invariant. Such a situation occurs in the retina and associated visual nets, and in fact many aspects of visual physiology can be seen to be associated with the properties of these kernels. For example, if the kernels are such that the various space constants increase as a function of retinal eccentricity, then it can be shown that the response of the visual system is more or less independent of the size of visual stimuli, although of course size itself is encoded in a more-or-less direct fashion into spatial coordinates in the visual system. One consequence of this is that spatially periodic stimuli of the form

$$\langle P(\phi) \rangle = L_0 \left[1 + \alpha \cos\left(\frac{2\pi n}{L} \phi \right) \right] \tag{3.1}$$

where ϕ is retinal eccentricity, are amplitude modulated by spatially inhomogeneous nets, as shown in Fig. 19.

It turns out that the peak modulation amplitude varies with the spatial frequency of the stimulus, i.e., spatial frequency is encoded into retinal eccentricity, in the visual system (cf. van Doorn et al., 1972). This has important consequences for theories of spatial filtering in the visual system (Cowan and Wilson, in preparation).

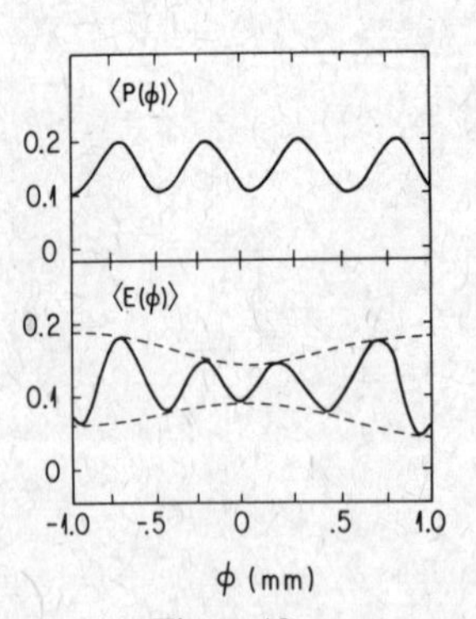

Figure 19
Response of a spatially inhomogeneous net to a spatial sine wave.

Synaptic plasticity

Finally, changes in net responses involving time scales still longer than those considered hitherto, require either accommodation or fatigue mechanisms, or changes in synaptic strengths brought about by prolonged or repeated stimulation. Given such mechanisms numerous aspects of neural adaptation to stimuli can be investigated using the theory of neural nets as presently outlined (Wilson, 1974 (in press); Cowan and Wilson, in preparation).

4. Discussion

It should now be obvious that the behavior of neural nets, as described above, bears many resemblances to that of switching circuits composed of transistor-like elements. In this respect neural nets are similar to chemical reactions (Schwartz, 1975), so it is not surprising that the dissipative structures seen in certain reactions (Zhabotinsky and Zaikin, 1973), and those analyzed in model systems (Prigogine and Nicolis, 1971; Eigen, 1971) should exhibit properties akin to those seen in neural nets. It was undoubtedly Aharon Katzir-Katchalsky's interest in such reactions that led him to investigate the possible role of dissipative structures in memory (Katchalsky et al., 1974). Early formulations of the theory of neural nets (McCulloch and Pitts, 1943) may be thought of as an attempt to model only the threshold logic underlying kinetics. McCulloch and Pitts in effect proposed a local binary coding system for information processing in the brain, without regard to any spatial or temporal manifolds that might exist in the brain and its environment. Current theories of neural nets, as exemplified in this paper, may be thought of as a step toward the inclusion of such manifolds in the neural code, containing as they do dynamical phenomena that lead to stable, spatio-temporal dissipative structures, in which the mean proportion of cells becoming activated per unit time, in a given neural region, may serve as the unitary element out of which a complex code for information processing could be constructed.

What of noise in such neural nets? Early work on redundant neural nets (von Neumann, 1956; McCulloch, 1959; Winograd and Cowan, 1963) and more recent work on redundant nets that decode ambiguous stimuli (Marr, 1969) may be seen as ways of using not local binary codes, but codes involving the joint pattern of activity of many cells, in a fashion related to the way in which the current dynamical theories employ spatial codes. Other attempts to treat noise in neural nets by the methods of statistical mechanics (Wiener, 1949; Cowan, 1968, 1970) may be thought of as attempts to analyze temporal aspects of neural coding by thermodynamic methods. The current theories combine and extend many of these notions, dealing as they do with thermodynamic and non-thermodynamic states of spatio-temporal activity, and with the switching and oscillation of such states. When extended to cover fluctuations by the methods used in non-equilibrium statistical mechanics (Graham,

1973), such theories may well provide a basis for realistic neural modelling. Of course the complexity of nets comprising many cell types may vitiate the utility of the methods of analysis outlined in this paper, and it may be that topological methods recently introduced for the analysis of certain complex dynamical systems (Thom, 1972) may prove to have an important role in such an analysis.

It remains to be seen if these theories will prove to be useful for the investigation of how brains work. Had he lived, it is certain that Aharon Katzir-Katchalsky would have played a leading role in such investigations.

Acknowledgments

The work reported in this paper was carried out in collaboration with Professor H. R. Wilson, Dr. J. L. Feldman, and others at The University of Chicago, with financial support provided in part by the Alfred P. Sloan Foundation, and the U.S. National Institutes of Health. Thanks are due to Mrs. Angell Pasley for considerable help in the preparation of this manuscript.

References

ALLANSON, J. T., 1956. In *Information Theory,* Third London Symposium (E. C. CHERRY, ed.). London: Butterworth and Company, p. 103.

BARLOW, H. B., 1973. *Perception* **1**:371–394.

BEURLE, R. L., 1956. *Phil. Trans. Roy. Soc. Lond. B 669,* **240**:55–94.

CAIANIELLO, E. R., 1961. *J. Theor. Biol.* **1**, 204.

COCHRAN, J. A., 1972. *The Analysis of Linear Integral Equations.* New York: McGraw-Hill.

COWAN, J. D., 1965. *Progress in Brain Research,* Vol. 17 (N. WIENER and J. P. SCHADE, eds.). Amsterdam: Elsevier.

COWAN, J. D., 1968. In *Neural Networks* (E. R. CAIANIELLO, ed.). Berlin: Springer.

COWAN, J. D., 1970. In *Mathematical Problems in the Life Sciences II* (M. GERSTENHABER, ed.). Providence, R. I.: Amer. Math. Soc.

DOORN, A. J. v., KOENDERINGK, J. J. and BOUMAN, M. A., 1972. *Kybernetik* **10**(4):223–230.

EIGEN, M., 1971. *Naturwissenschaften* **58**:465–523.

FARLEY, B. G. and CLARK, W. A., 1961. In *Information Theory,* Fourth London Symposium (E. C. CHERRY, ed.). London: Butterworth and Company, p. 242.

FELDMAN, J. L. and COWAN, J. D., 1975a. *Biological Cybernetics* **17**, 19–38.

FELDMAN, J. L. and COWAN, J. D., 1975b. *Biological Cybernetics* **17**, 39–51.

FITZHUGH, R., 1969. In *Biological Engineering* (H. P. SCHWAN, ed.). New York: McGraw-Hill, pp. 1–86.

GRAHAM, R., 1973. In *Quantum Statistics in Optics and Solid-State Physics.* Berlin: Springer-Verlag, pp. 1–97.

GRIFFITH, J. S., 1963. *Bull. Math. Biophys.* **25**:111–120.

GRIFFITH, J. S., 1965. *Bull. Math. Biophys.* **27**:187–195.

HANDLER, P., 1974. *American Scientist* **62**, 4.

HARTH, E. M., CSERMELY, T. J., BEEK, B. and LINDSAY, R. D., 1970. *J. Theor. Biol.* **26**, 93.

HEBB, D. O., 1949. *The Organization of Behaviour.* New York: Wiley.

KATZIR-KATCHALSKY, A., ROWLAND, V. and BLUMENTHAL, R., 1974. *Neurosciences Research Bull.* **12**, 1.

KERNELL, D., 1966. *Science* **152**: 1637–1640.
KERNELL, D., 1966. *Brain Res.* **11**: 685–687.
LITTLE, W. A., 1974. *Math. Biosciences* (in press).
MARR. D., 1969. *J. Physiol.* **202**: 437–470.
MAY, R., 1972. *Science* **177**, 900.
MCCULLOCH, W. S., 1959. In *Self-Organizing Systems* (M. YOVITTS and S. CAMERON, eds.). London: Pergamon Press, p. 264.
MCCULLOCH, W. S. and PITTS, W., 1943. *Bull. Math. Biophys.* **5**: 115–137.
MINORSKY, N., 1962. *Nonlinear Oscillations.* Princeton, N.J.: D. van Nostrand Co. Inc.
NEUMANN, J. V., 1956. In *Automata Studies* (C. E. SHANNON and J. MCCARTHY, eds.). Princeton, N.J.: Princeton University Press, pp. 1–69.
PRIGOGINE, I. and NICOLIS, G., 1971. *Quart. Rev. Biophys.* **4**(2 and 3): 107–148.
RAPAPORT, A., 1951. *Bull. Math. Biophys.* **12**, 109.
RAPAPORT, A., 1951. *Bull. Math. Biophys.* **13**, 85.
RATLIFF, F., 1965. *Mach. Bands.* London: Holden-Day.
SCHWARTZ, G., 1975. In *Functional Linkage in Macromolecular Systems* (F. O. SCHMITT, ed.). Cambridge, Mass.: M.I.T. Press.
SCOTT, A. C., CHU, F. Y. F. and MCLAUGHLIN, D. W., 1973. *Proc. IEEE,* pp. 1443–1483.
SHIMBEL, A., 1950. *Bull. Math. Biophys.* **12**, 241.
SHIMBEL, A., 1951. *Bull. Math. Biophys.* **13**, 319.
SHOLL, D. A., 1956. *The Organization of the Cerebral Cortex.* London: Methuen.
SMITH, D. R. and DAVIDSON, C. H., 1962. *J. Soc. Computing Mach.* **9**, 268.
STEIN, R. B., LEUNG, K. V., OGUZTORELI, M. N. and WILLIAMS, D. M., 1974. *Kybernetik* **14**: 223–230.
STEIN, R. B., LEUNG, K. V., MANGERON, D. and OGUZTORELI, M. N., 1974. *Kybernetik* **15**: 1–9.
THOM, R., 1972. *Structural Stability and Morphogenesis.* New York: Benjamin.
WIENER, N., 1949. *Cybernetics.* New York: M.I.T.-Wiley.
WILSON, H. R. and COWAN, J. D., 1972. *Biophysical J.* **12**: 1–24.
WILSON, H. R. and COWAN, J. D., 1973. *Kybernetik* **13**(2): 55–80.
WINOGRAD. S. and COWAN, J. D., 1963. *Reliable Computation in the Presence of Noise.* Cambridge, Mass.: M.I.T. Press.
ZHABOTINSKY, A. M. and ZAIKIN, A. N., 1973. *J. Theor. Biol.* **40**: 45–61.

The transduction of chemical signals into electrical information at synapses

ROBERT WERMAN
Department of Zoology, The Hebrew University of Jerusalem, Jerusalem, Israel

Introduction

I will describe some attempts to characterize the behavior of postsynaptic membranes when exposed to neurotransmitter. The analysis is based on, takes advantage of and is limited by the use of electrophysiological techniques. The results shown are those from my laboratory and are the result of work with a number of talented collaborators[1]. I have no desire to slight the works of other laboratories, but I would rather present a more personal scientific story here.

When a nerve impulse reaches a nerve terminal, depolarization in the presence of calcium ions leads to the release of a chemical, or transmitter. The transmitters that will be discussed here are acetylcholine, γ-aminobutyric acid and glycine. These chemicals then diffuse across a synaptic gap of 20 to 50 nm where they interact with specific receptors, probably proteins. By this means nerves transfer information to other nerve cells, to muscles and to glands. Synaptic transmission to nerves and muscle will be discussed in this paper but there is no reason to believe that transmission to glands is of a different character.

Characterization of the synaptic ionophore

In almost all cases the interaction of transmitter with its postsynaptic receptor leads to an increase in membrane conductance of an ion or of several ions found on both

[1] These collaborators include: Morrie Aprison, Martin Blank, Neville Brookes, Bob Davidoff, Rich Manalis, Kobi Mazliah, Ian McCance and Lea Ziskind. I thank them all for allowing me to show their published and, in some cases, unpublished work and for allowing me to skip over other, equally important, finidngs.

sides of the membrane. It is generally presumed that the ionic conductance increases as the result of a conformational change in the membrane which results in the opening of a narrow, hydrophillic channel in the membrane (Shohami and Ilani, 1973). Although it has been claimed that these channels may sometimes permit both cations and anions to traverse (Eccles, Eccles and Ito, 1964; Machne, Dunning and Kanno, 1973), I feel that this possibility is unlikely and certainly not proven in any given case (for example, Lux, Loracher and Neher, 1970). It seems more probable that the narrow channels involved generally contain fixed charges, thereby permitting ionic movement only of uniform and opposite charge.

Our first approach to the problem of characterizing the transmitter–receptor interaction was to ask whether it was possible to characterize the molecular mechanism involved in activation of a receptor by electrophysiological techniques. The solution of this problem turns out to be an analytically simple one (Werman, 1965) but rather more difficult, although not impossible, technically. First of all, we make use of the fact that the conductance increase will move the membrane potential towards a potential value that is determined by the sum of terms, each of which comprises the transport number of an ion whose conductance is increased, multiplied by the Nernst potential of that ion. Mathematically, this potential is singular and can also be described by the constant-field equation (Goldman, 1943; Hodgkin and Katz, 1949; Werman, 1965) which is based on several reasonable assumptions, only one of which is obviously challangeable, namely that the voltage drops uniformly across the thickness of the membrane. The latter equation looks like this:

$$E = \frac{RT}{F} \ln \frac{\sum_{1}^{c} P_c C_{c,o} + \sum_{1}^{a} P_a A_{a,i}}{\sum_{1}^{c} P_c C_{c,i} + \sum_{1}^{a} P_a A_{a,o}} \quad (1)$$

where P is a permeability, c refers to cations, a to anions, C to cation activity, A to anion activity; i, inside the membrane and o, outside the membrane[2].

It is obvious that all processes which produce an increase in ionic conductance of only one ion lead to the same constant-field, or reversal, potential. Two ion processes are similarly underdetermined: all such processes with the same ratio of ionic permeabilities will, in that case, lead to the same reversal potential. However, when three (or more) ions are involved, one can take advantage of the added indeterminancy. Thus, it can be shown that the three ion processes evoked by each of two

[2] Physical chemists find it difficult to understand why this complicated and incompletely justified equation is used by biologists. It might help to recall that a living cell is a small structure and that it is not usually possible for the investigator to determine concentrations of ions inside the cell. The constant field equation allows all manipulations to be made on the outside and, within the time limits and external concentration realm where ionic concentrations on the inside are not affected by external changes, the equation is very useful.

stimulants are identical with one another if the reversal potentials are equal before and the new reversal potentials after changing the concentration of one of the involved ions are again equal (Werman, 1965, 1966). This observation is based on our ability to describe the identity of each of the reversal potentials as a straight line relating two permeability ratios: an infinite number of such possibilities can be fitted to the data. But each set of reversal potentials, before and after changing the concentration of one ion is described by a separate straight line. There is clearly at most one point, the intersection of the two lines, which is common to both conditions, indicating a single solution. Thus, the criterion establishes the singularity of the process among an infinite array of possibilities, providing a very powerful tool for characterization of a membrane process by electrophysiological means[3].

This powerful tool provides a delicate instrument for determining whether a compound suspected of being transmitter has the same action as the physiologically released transmitter (Werman, 1966). Indeed, the criterion has been successfully used in the demonstration that glycine is an inhibitory transmitter in the mammalian spinal cord (Werman, Davidoff and Aprison, 1968; Werman, 1972a,b). It is now clear that the three-ion test[4] only characterizes the ionophore process. It is theoretically possible that two different receptors may utilize the same ionophore but when the actions of GABA and glycine were compared on the goldfish Mauthner cell, where both produce increased chloride conductance, both receptors as well as both ionophores appeared to be distinct (Mazliah, 1973; Mazliah and Werman, 1974).

The three-ion test can be used to ask a number of interesting questions about synapses. For example, is the phenomenon of neural facilitation indeed entirely presynaptic? Does the ionophore change in the presence of transmitter? Does transmitter desensitization involve changes in the ionophores? The phenomena of facilitation and desensitization will be explained in relation to the experiments answering the questions.

Fig. 1 shows an experiment in which the reversal potentials of the synaptic potential and the potential produced by local iontophoresis of acetylcholine were compared at the neuromuscular junction of a frog sartorius muscle fiber. In this case, the reversal potentials were -3.0 and -3.3 mV, respectively, when sucrose replaced 50% of the sodium. The difference is within the experimental variability of the

[3] The case of more than three ions is also easily treatable: the criterion is now changed to a comparison of reversal potentials before and after $n-2$ changes of concentration of different involved ions. That case and the case where concentrations on both sides of the membrane are changed are treated by me elsewhere (Werman, 1965).

[4] In almost all synapses studied to date there appear to be only one or two ions participating. In one case, the frog neuromuscular junction, in addition to Na^+ and K^+ ions (Takeuchi and Takeuchi, 1960), Ca^{++} conductance is also increased but under normal conditions its contribution appears minimal (Takeuchi, 1963; Katz and Miledi, 1969). In general the experimental procedure is to introduce one or two ions which will mimic the actions of the native ions whose conductance is increased and then to compare reversal potentials before and after one concentration change.

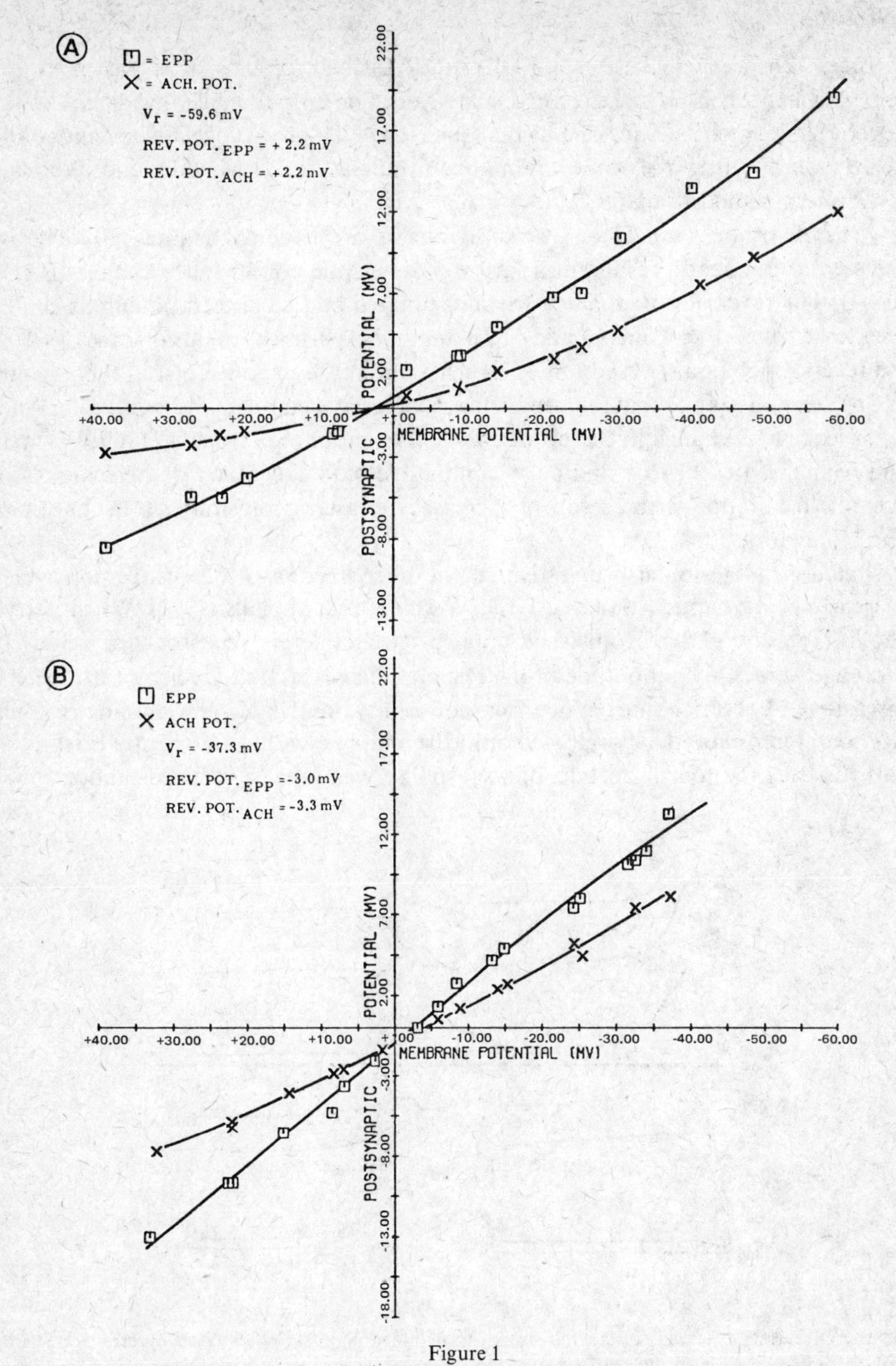

Figure 1

Identity of reversal potentials of the endplate potential (EPP) and iontophoresed acetylcholine (ACH). Glycerol-treated frog sartorius muscle. A. Lithium chloride replaces one half external sodium chloride. B. Sucrose replaces one half external sodium chloride. From Manalis and Werman (unpublished).

method. When the sucrose was replaced by LiCl, three ions, K^+, Na^+ and Li^+ took part in the conductance change. The reversal potentials were moved 5 mV in the depolarizing potential and were again equal. Thus the physiological transmitter and acetylcholine utilize the same ionophore (Manalis, 1969; Manalis and Werman, 1969; Werman and Manalis, 1970).

When two shocks separated by a brief interval are given to the presynaptic nerve, the second frequently results in a larger postsynaptic conductance change (Fig. 2, line 1). The reversal potentials of the unfacilitated and facilitated potentials can be seen to be equal (Fig. 2, lines 4 and 5). In another experiment in the absence of Li^{++}, the reversal potentials were both -7.4 mV, and in the presence of Li^{++} both were 2.8 mV, demonstrating that the conductance increase accompanying facilitation is not a consequent of change in the properties of the ionophore (Manalis, 1969; Werman and Manalis, 1972). This finding is compatible with the generally held view that synaptic facilitation is the result of increased release of transmitter (Otsuka, Endo and Nonomura, 1962).

Manalis and I similarly investigated the phenomenon of desensitization which is clearly postsynaptic (Manalis, 1969; Werman and Manalis, 1970). When acetylcholine (or one of its agonists) is applied to cholinergic receptors for relatively prolonged times, the amplitude of the response decreases and the receptors become insensitive to acetylcholine. When four consecutive pulses of acetylcholine of equal size were iontophoretically released onto the endplate within sufficiently brief intervals (0.5 sec) the fourth acetylcholine response was greatly reduced (about 25%),

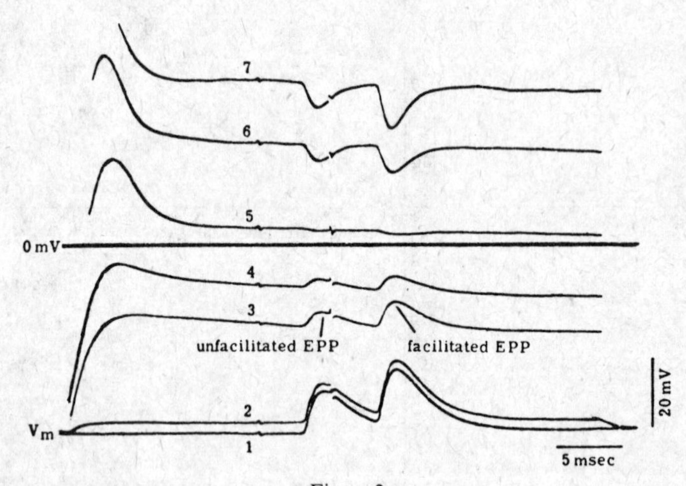

Figure 2
Identity of reversal potentials of facilitated and unfacilitated endplate potentials. Glycerol-treated frog sartorius muscle. Sweep 1 is at resting potential and progressively larger depolarizing steps are given in sweeps 2 to 7. Both zero potential and the reversal potential are seen to lie between sweeps 4 and 5. From Manalis and Werman (unpublished).

indicating about a 38 % reduction in conductance change. Again the reversal potentials of the first and the fourth responses were found to be identical, indicating that the process of desensitization does not involve a change in the nature of the ionophore.

Manalis and I (Manalis, 1969; Manalis and Werman, 1969; Werman and Manalis, 1970) also investigated the problem of weak agonism. There are certain congeners of acetylcholine which are only capable of producing a very weak response even when applied in very high concentrations. We asked whether part or all of this difference in maximum depolarization produced by different agonists was related to changes in the molecular characteristics of the ionophore. Table 1 shows that the strong agonists carbachol, succinylcholine and nicotine, the augumenting agent, dexamethonium, and the weak agonists, heptyltrimethylammonium and edrophonium all utilize the same molecular mechanisms for conductance increase. In each experiment both acetylcholine and the agent to be tested were applied by iontophoresis onto the same endplate and the reversal potentials were compared. Iontophoresis of decamethonium ions does not produce depolarization but augments and prolongs acetylcholine action. Decamethonium was thus always tested simultaneously with acetylcholine. The results are expressed as the ratio of the driving forces (the reversal potential less the resting potential), that for the drug to that for acetylcholine (Manalis, 1969; Manalis and Werman, 1969; Werman and Manalis, 1970).

The five percent reduction of driving force in the case of edrophonium can hardly explain the reduction of the maximum response to edrophonium to less than one-seventh of that produced by acetylcholine. It is probable that in the case of weak agonists, only a small proportion of the agonist-receptor complexes produce the necessary conformational change for increased ionic conductance. Those ionophores that are activated, however, do not appear to substantially differ from the ionophores activated by strong agonists.

Table 1

DRIVING FORCES FOR POTENTIALS PRODUCED BY AGONISTS OF ACETYLCHOLINE COMPARED WITH DRIVING FORCE OF ACETYLCHOLINE AT THE SAME SYNAPSE (FROG NEUROMUSCULAR JUNCTION)

Compound	Driving force ratio
Acetylcholine (2 doses)	0.99
Carbachol	1.01
Succinylcholine	1.02
Nicotine	1.00
Decamethonium (acetylcholine)	1.01
Heptyl-trimethylammonium	1.00
Edrophonium	0.95

Stoichiometry of transmitter–receptor interactions

It should now be clear that the immediate electrical consequent of activation of a synaptic receptor is a change—usually an increase—in ionic conductance. This fact and the fundamental law of electricity which states that conductances in parallel are additive gives the electrophysiologist a powerful tool for measuring the interaction between a transmitter and its receptor. If one assumes that the activation of single receptors results in a mean conductance change that is normally distributed, the activation of n receptors should produce n times the unitary conductance change. Indeed, by an elegant technique, Katz and Miledi (1971) have demonstrated that activation of the acetylcholine receptor in the frog neuromuscular junction produces, at 22°C, a one msec increase in conductance of 10^{-51} mhos. This conductance is one order of magnitude less than the maximal conductance through a simple 0.6 nm channel (Hille, 1970).

The conductance increases produced can be measured as a function of transmitter concentration and therefore treated the same way as the product in an enzyme-substrate reaction. Thus, in general,

$$g' = f(A) \tag{2}$$

where A is the concentration of transmitter and g' the measured change in ionic conductance. More specifically, assuming that n molecules of transmitter are needed to activate a receptor and that K_i is the association constant of the ith step,

$$g' = g'_{\max} \frac{A^n}{A^n + K_n A^{n-1} + \ldots + K_n K_{n-1} \ldots K_2 K_1} \tag{3}$$

It can be shown (Werman, 1969) that, in general, for this function,

$$\lim_{A \to 0} \frac{d \log g'}{d \log A} = n \tag{4}$$

Thus the slope of the log-log plot of conductance change as a function of concentration is a monotonically decreasing function which decreases from n to 0 (Werman, 1969). Now the function we are talking about is different from a binding function which gives the amount of bound ligand in relation to the total number of sites. The binding function is given by:

$$y = \frac{nA^n + (n-1)\, A^{n-1} K_n + \ldots + A K_n K_{n-2} \ldots K_3 K_2}{n(A^n + K_n A^{n-1} + \ldots + K_n \ldots K_2 A + K_n \ldots K_1)} \tag{5}$$

This function can be determined directly by chemical but not by electrophysiological means and does not tell how many sites are activated. The electrophysiological data, on the other hand, measure the analogue of the rate of formation of the product of the enzyme–substrate reaction, which can be assumed to be linearly related to the average number of sites activated per unit time.

In order to determine the stoichiometry of a transmitter–receptor interaction using the approach indicated by equation (4), Brookes and Werman (1973) used the flexor tibae muscle of the locust which is a small bundle of 10–20 fibers, 20 to 50 μ in diameter, with a length of about 7 mm. Fig. 3 shows the form of measurement used to obtain our data. A sawtooth inward current is delivered across the membrane of the muscle fiber and the voltage change is recorded by a second microelectrode placed intracellularly close to the current electrode. The currents were 1.6 sec in duration (which is very long in comparison with the membrane time constant) and were repeated every 10 sec. The response is seen to be linear. When γ-aminobutyric acid (GABA), the inhibitory transmitter to arthropod muscles, is flowed across the preparation, a transient hyperpolarization is seen accompanied by a sustained increase in ionic conductance. This preparation exhibits no desensitization but the large conductance change results in a redistribution of chloride ions, the current carrier in this case, and a return of the voltage to baseline levels (and even slight depolarization) with maintained GABA application. The conductance increase, however, is not reduced. Since the voltage changes under steady-state conditions were negligible (and minor even during the transient), possible membrane field effects (Magleby and Stevens, 1972) can be ignored.

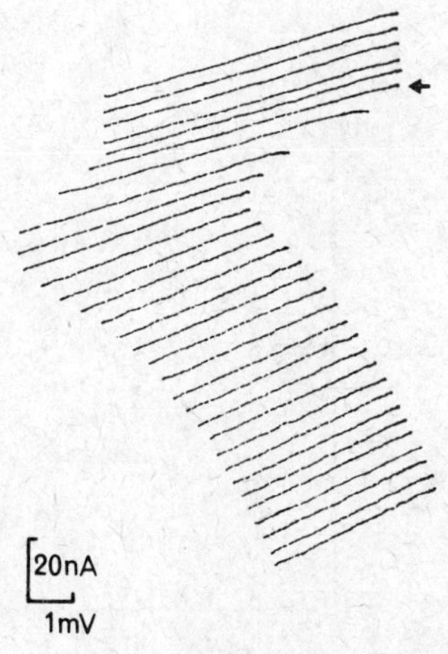

Figure 3

X–Y plotter recording of rate of increase of membrane conductance of locust flexor tibae fiber on application of GABA. The resting potential is at the right and the pen lifts when the peak hyperpolarization to an inward current step is reached. The pen was shifted down manually between current ramps which were given at intervals of 10 sec. From the arrow downward the preparation was perfused with 0.24 mM GABA. The response consists of a transient hyperpolarization and a sustained conductance increase (change in slope). From Brookes, Blank and Werman (1973).

The pooled data from 19 experiments with GABA in concentrations from 4–10 × 10^{-5} M are shown in Fig. 4. The values obtained with increasing concentration of GABA are fitted by a regression line with a log-log slope of 3.15 ± 0.06 (standard error of the estimate). It can be seen that there is some hysteresis in the system and that the curve obtained on descent is shifted to the left and the slope somewhat reduced (2.45 ± 0.11). It can also be seen that there is no apparent reduction in slope even at levels 1% of the maximum conductance change obtainable. It will be shown later that this lack of curvature differentiates the behavior of the system from the possibility that all combinations of GABA with the receptor—one to one, two to one, and three to one—are capable of increasing membrane conductance. Thus we can conclude that at least three (and probably no more) molecules of GABA are needed to activate the inhibitory receptor in this system.

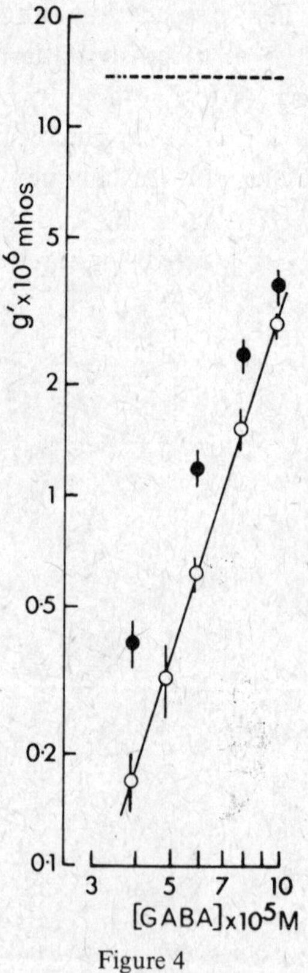

Figure 4

Experimental limiting slope of log-log plot of conductance increase, g', against GABA concentration in locust muscle. ○, mean points (±S.E.M., 9 experiments) measured on cumulative ascent of the concentrations; ●, mean points (10 experiments) on descent. The slope of the regression line is 3.15 ± 0.06 (S.E. of estimate). The dashed line is extrapolated mean maximum response. From Brookes and Werman (1973).

In order to examine the effect of higher doese of GABA on the system and at the same time avoiding drastic increases in the membrane conductance and irreversible changes, we replaced 75% of the external chloride ions with proprionate ions. Such a change reduced the unitary conductance to 18% of its control level on the average and the range of GABA concentrations that could be examined was now increased to 50×10^{-4} M and beyond. When a maximum conductance response is obtainable, a Hill plot can be constructed which provides further stoichiometric information (Fig. 5). Examination of Fig. 5 shows that there is deviation of the experimental points from a straight line above a level of 60% activation of receptors. If the experimental points were all to fall on a straight line, the reaction would be one of infinite cooperativity described by the equation:

$$g' = g'_{\max} /(1 + K/A^n) \qquad (6)$$

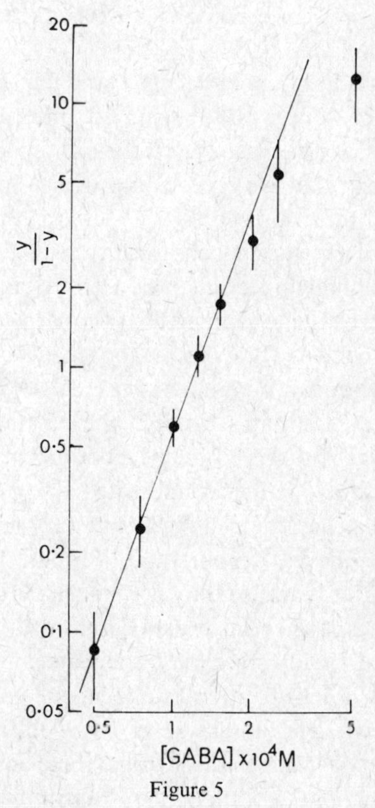

Figure 5

Experimental Hill plot for GABA in locust muscle (proprionate saline). The regression line was computed from the lower five points (slope = 2.78 ± 0.05, mean of nine experiments). From Brookes and Werman (1973).

or in the Hill form,

$$\frac{g'/g'_{\max}}{1 - g'/g'_{\max}} = \frac{A^n}{K} \tag{7}$$

The deviation illustrated indicates positive cooperativity and the Hill plot illustrated in Fig. 6 can be approximately fitted if $n = 3$ and $K_1 = 5K_2 = 25K_3$. Thus, the equation describing the GABA interaction at room temperature is:

$$\frac{g'}{g'_{\max}} = \frac{A^3}{A^3 + K_3A^2 + K_3K_2A + K_3K_2K_1} \tag{8}$$

where the concentration giving a 50% activation of receptors ($A_{0.5}$) is 1.1×10^{-4} M and $K_1 = 4.5 \times 10^{-4}$ M, $K_2 = 9.0 \times 10^{-5}$ M and $K_3 = 1.8 \times 10^{-5}$ M. This reaction is one of relatively great positive cooperativity and its Hill plot shows a 10% deviation from a straight line of 3 when the Hill factor equals 0.34 as compared to a 10% deviation for the zero cooperativity case when the Hill factor is only 0.001 (Brookes and Werman, 1973).

It is important to note that both the Hill (Fig. 5) and the log-log (Fig. 4) plots show that there is no reduction of slope with lowering of concentrations of GABA to the range where they produce as little as 1% of the maximal response. This finding indicates that it is not likely that we are dealing with a mixed state, whereby the interaction of both two or three molecules of GABA or even of one, two or three molecules of GABA can all produce a conductance change. If any significant contribution to the change in conductance was produced by binding of fewer than three molecules of GABA, it can be demonstrated that the slope would again decrease at lower concentrations approaching the value of the smallest number of ligand molecules that are capable of producing a conductance change (Werman, 1971). Furthermore, this finding clearly differentiates between the function of state measured by our experimental techniques and the binding function. The latter function, even in the presence of the positive cooperativity indicated by the upper limb of the Hill plot, is expected to show a reduction in slope—going towards a value of one—that is already marked at the 1% level (Brookes and Werman, 1973).

When a cholinergic receptor is attacked in the same manner as the GABA receptor, we run into a new problem. The phenomenon of desensitization described above does not permit one to make the steady-state measurements which are necessary for the analysis. It turns out that desensitization is dependent on the ionic environment and in one case, that of the molluscan neuron H (hyperpolarizing, chloride conductance increase) response, appropriate changes in the ionic environment can eliminate desensitization entirely without affecting the transmitter induced conductance increase (Ziskind and Werman, 1975a). When this is done, the necessary steady-state measurements could again be made (Ziskind and Werman, 1975b).

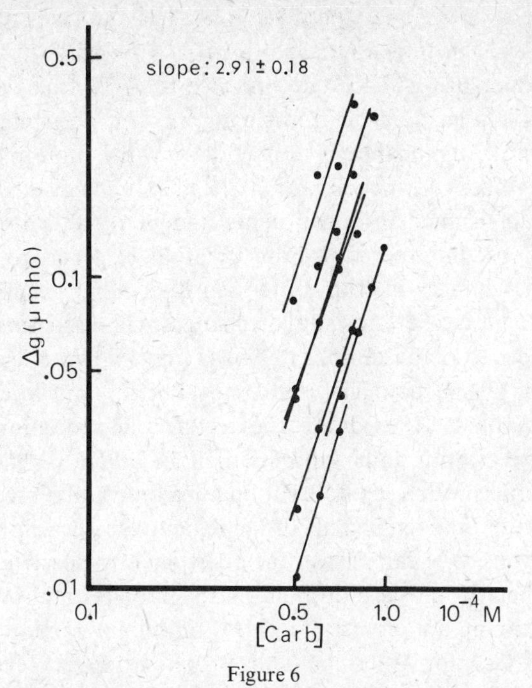

Figure 6

Log-log plot of conductance increase as a function of carbachol concentration in molluscan neurons. Measurements were made in seven H cells after all Na^+ and Ca^{++} ions were replaced by an equimolar concentration of Mg^{++} ions in order to eliminate desensitization. The mean regression slope was 2.91 ($\pm$0.18 S.D.). From Ziskind and Werman (1975b).

In Fig. 6, the mean log-log slope of the conductance change as a function of cholinergic agent concentration, in this case carbachol, in seven H cells in the absence of desensitization was 2.91 ($\pm$0.18 S.D.) indicating that three or more molecules of carbachol were necessary to activate the receptor. It is of interest that the log-log slope obtained in the presence of desensitization was 1.56 which is very close to the slope obtained for acetylcholine at vertebrate neuromuscular junctions where desensitization is also present (Werman, 1969; Ziskind and Werman, 1975b). This may indicate that three molecules (certainly at least two are involved) are also necessary to activate that cholinergic synapse.

There is no evidence that all synapses require three—or even more than one—molecules of transmitter in order to activate the synaptic conductance change. In one study in our laboratory of GABA and glycine receptors in the same vertebrate neuron, both receptors showed a requirement for four molecules of ligand. We were also able to demonstrate that both the receptors and ionophores are microscopically distinct but macroscopically distributed over the same region of the cell (Mazliah, 1973; Mazliah and Werman, 1974).

It is also possible that polymerization or depolymerization of synaptic receptor subunits may occur under different environmental conditions. We have found that under appropriate environmental conditions it is possible to change the log-log slope of the conductance—GABA dependence to 4 (Werman and Brookes, 1973) and even 2 (Brookes, personal communication). The possibility of such changes having physiological significance is in accord with the more plastic view of membranes that is beginning to invade biology: a membrane as part of a living system may be capable of change in constituents and in organization. One must keep the possibility of subunit reorganization in mind, it seems to me, in attempting to understand the changes in transmitter sensitivity that occur with degeneration of the presynaptic nerve. Perhaps similar changes will also prove to be involved in the cellular counterpart of the complicated behaviors known as learning and memory.

On the basis of pharmacological (Khromov-Borisov and Michaelson, 1966) and chemical (DeRobertis, 1971) evidence, the acetylcholine receptor in vertebrates has been postulated to contain four subunits. If it be assumed that each molecule of acetylcholine interacts with a separate subunit in a four subunit receptor, than at least three subunits must be occupied in order to activate the cholinergic receptor in molluscan H neurons. One can answer the question as to possible preference of some subunits over others by the use of antagonists. Brookes and Werman studied the effects of picrotoxin, an antagonist of GABA, on the interaction of GABA with the locust inhibitory receptor (Werman and Brookes, 1969; Werman and Brookes, 1971). We were able to show that the conductance *decrease* produced by varying doses of picrotoxin when the GABA concentration is held constant is fitted by a curve implicating one molecule of picrotoxin in the interaction. This finding indicates that picrotoxin probably blocks only one site and that the GABA receptor under these conditions apparently does not include more than three potential sites of interaction.

Thermodynamic parameters of drug–receptor interaction

Since the equilibrium constants of the drug–receptor interaction may be obtained by electrophysiological analysis, it is also possible to obtain the temperature dependence of the interaction. Figure 7a demonstrates that cooling greatly increases the affinity of the locust muscle inhibitory receptor to GABA (Brookes and Werman, 1973). A 9.5°C change in temperature shifts the curve about one-half of a log unit. The slopes of the curves obtained at 15°, moreover, do not differ significantly from the slopes at higher temperatures, and we felt justified in using the concentration giving a half-maximum response, $A_{0.5}$, as a measure of a mean dissociation constant for the reaction (cf. Werman and Brookes, 1973, for discussion of this point). A Van 't Hoff plot of the temperature dependence obtained in this manner is shown in Fig.7b. The standard enthalpy change, H^0, for the overall interaction of GABA with receptor was calculated from the slope of the regression line and was found to

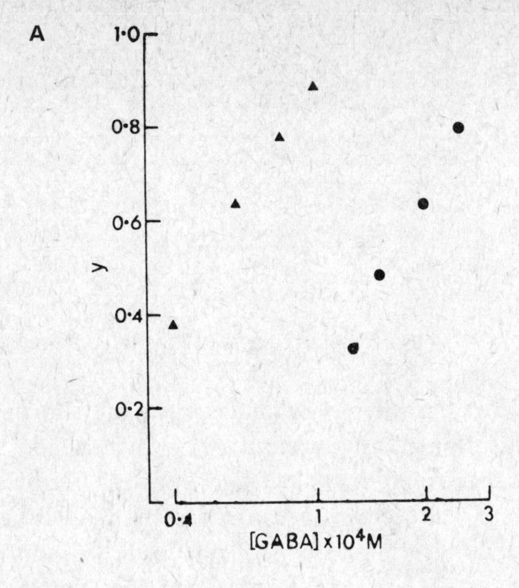

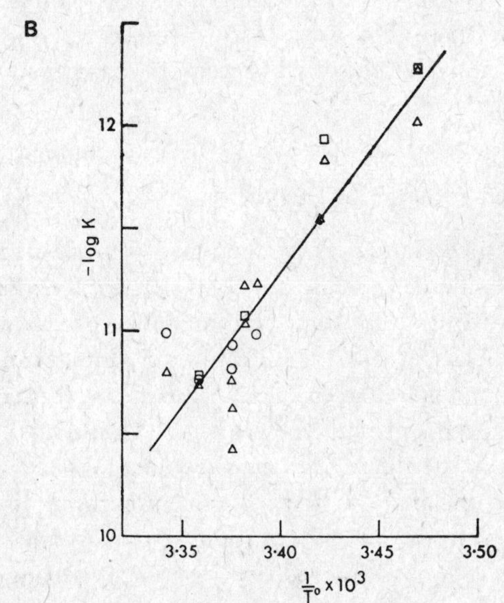

Figure 7

Temperature dependence of the GABA-receptor interaction in locust muscle. From Brookes et al. (1973). (A) Log-log dose response curves at 15° and 24.5°. Responses are expressed as fractions of extrapolated maximum conductance increase from different preparations at 15° (▲) and 24.5° (●). (B) Van't Hoff plot for interaction of GABA and receptor. The negative logarithm of the estimated dissociation constant K is plotted against the reciprocal of absolute temperature. K was estimated in different experimental procedures: △ from ascending concentrations; ○ descending concentrations; □ intermediate washout and recovery between concentrations. The three sets of data appear to form a single population. The regression slope was $12.95\ (\pm 1.36) \times 10^3$.

Table 2
THERMODYNAMIC PARAMETERS FOR OVERALL INTERACTION BETWEEN GABA AND RECEPTOR
($H^0 = -59.3$ KCAL/MOLE)

T, °C	$A_{0.5}$, M × 10^4	F^0, kcal/mole	S^0, e.u./mole
15	0.45	−16.1	−150
19	0.63	−15.7	−149
23	1.52	−14.5	−151

be −59.3 (±6.2) kcal/mole. The derived thermodynamic parameters are found in Table 2 (from Brookes et al., 1973).

It is of interest that the large thermodynamic values are similar to the old heat data on protein denaturation (cf. Dowben, 1969), which would suggest that we might be dealing with a conformational change in a protein (see below, on kinetics). Moreover, the values are quite close to some known multi-subunit interactions such as the interaction of three molecules of adenosine 5′-monophosphate with fructose 1,6 diphosphatase (Taketa and Pagell, 1965) or 4 molecules of O_2 with hemoglobin (Roughton, Otis and Lyster, 1955). This correspondence is certainly consistent with the hypothesis that the GABA receptor is a multi-subunit (oligomeric) receptor.

Kinetics of the transmitter–receptor interaction

Analysis of the conductance change as a function of time is straightforward (see Fig. 3). When the conductance increase to GABA in locust muscle was plotted against time, there were two time constants, an initial fast time constant and a later slow time constant (Brookes et al., 1973). Moreover, the second time constant appeared to be more prominent when the GABA dosage is reduced. A semilog plot of the kinetics in one experiment with four doses of GABA at 23°C is shown in Fig. 8. It can be seen from the figure that the slow component is almost the same for the three lowest doses and that the first component becomes progressively more prominent with increasing dose. The time constants of the fast and slow components of the experiment in Fig. 8 were analyzed by a graphical peeling process and are shown in Table 3.

The equation assumed was:

$$1 - g'/g'_{\max} = a_0 \exp(-t/\tau_a) + b_0 \exp(-t/\tau_b) \quad (9)$$

where a_0 and b_0 are the initial fractions showing fast and slow kinetics respectively and τ_a and τ_b are the fast and slow time constants respectively. The analysis shows that the fast component is independent of concentration and that increase in GABA concentration increases the fraction initially exhibiting fast kinetics.

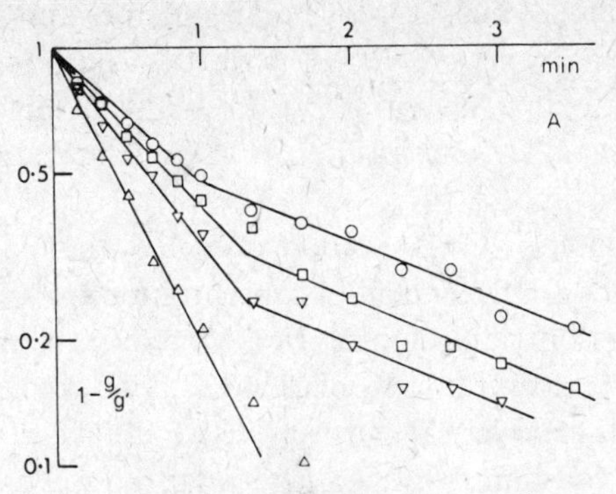

Figure 8

Kinetics of onset of GABA action in locust muscle at 23°C. The instantaneous conductance increase g was measured at multiples of 10 sec after application of GABA until equilibrium conductance increase, g', was reached. Each GABA application was of 7 min duration. The GABA concentrations used were: ○ 100 μM; □ 150 μM; 190 μM; △ 240 μM ($A_{0.5}$ = 140 μM). All responses were recorded from the same fiber. From Brookes et al. (1973).

The first few applications of GABA to a preparation elicit responses which are progressively faster and the effect is more marked at higher concentrations. This phenomenon is called "sensitization" and is a purely kinetic one: the steady-state conductance change is not altered. Figure 8 was obtained after sensitization of the preparation and the effects of the smallest concentration were reproducibly repeated at the end of the experiment. Analysis of the kinetics of repeated applications of GABA to a "naive" preparation indicates that there is increasing participation of a fast component and decreasing participation of a slow component with repeated GABA applications (Brookes et al., 1973).

Table 3

TIME CONSTANTS OF FAST AND SLOW EXPONENTIAL COMPONENTS OF RESPONSE TO GABA AT 23°C

GABA, M × 10^4	g'/g'_{max}	a_0	τ_a, sec	τ_b, sec
1.0	0.19	0.34	—*	182
1.5	0.51	0.49	32	169
1.9	0.78	0.62	32	180
2.4	0.92	—**	33	—**

* Deviation from linearity prevented determination of a reliable value (see Fig. 8).

** The slow process was not well defined at this concentration. a_0 is at least 0.9 (see Fig. 8).

Surprisingly, we found that decreasing the temperature increased the rate of GABA action (Brookes et al., 1973). Even when care is taken to keep the GABA concentration close to that giving a 50% response at different temperatures, in other words using a smaller concentration at lower temperatures, there is increasing participation of a fast process.

The exponential onset of GABA action resembles that of a first order reaction. Since the rate constant is independent of concentration the kinetics cannot represent the effect of a macroscopic diffusion barrier. Moreover, the relative insensitivity of the rate constant to temperature (Brookes et al., 1973) would appear to preclude a process utilizing metabolic energy, such as active uptake of GABA. It would also appear that the rate constant described here is far too slow to represent the rate of binding of GABA to receptor. Therefore, the kinetic described is most likely that of a conformational change of the receptor, the change permitting an increased chloride conductance. This argument is certainly true for the slow component of the biphasic kinetic and may well also pertain to the faster component.

Conclusions

I have attempted to show that use of electrophysiological techniques can tell us much about the nature of the transmitter–receptor interaction. The major theoretical tools are quite simple, and involve the (potential) uniqueness of the conductance change produced by the interaction and the measurement of algebraically summing conductances. The technical problems are a bit more difficult but are, in general, within the firing range of the experimentalist.

The picture that results from these studies of the transmitter-receptor interaction is that of several molecules of transmitter interacting with subunits of a receptor and leading to a highly specific ionic conductance change (usually an increase). The nature of the ionic environment or binding of some congeners of the transmitter as ligands can introduce factors (steric hindrance?) which reduce the likelihood of the conformational change. Thus, weak agonism and desensitization may be related phenomena with different time constants (Werman and Manalis, 1970). An immediate test of this conclusion would be to see whether conditions which remove desensitization also convert weak agonists to strong agonists.

Finally, recent promising work on the isolation of synaptic receptors suggests that the ionophore may be an integral part of the same molecule as the receptor (DeRobertis, 1971). This suggests the possibility of using the electrophysiological data to verify the isolation of the receptor and also its recovery in a form that has the same properties as the *in situ* receptor. Since chemists characterize the isolated receptors by binding data it would be necessary to recalculate the appropriate binding data from the function of state in order to compare *in vitro* and *in situ* behaviors of the receptor. A more direct test might involve reconstructing the receptor into a lipid bilayer and determining a) the function of state of the activated receptor and b) the ionic selectivity of the ionophore. The data could then be directly compared with the electrophysiologically obtained data (Werman, 1969).

References

BROOKES, N., BLANK, M. and WERMAN, R., 1973. "The kinetics of the conductance increase produced by γ-aminobutyric acid at the membrane of locust muscle fibers," *Mol. Pharmacol.* **9**:580–589.

BROOKES, N. and WERMAN, R., 1973. "The cooperativity of γ-aminobutyric acid action on the membrane of locust muscle fibers," *Mol. Pharmacol.* **9**: 571–579.

DEROBERTIS, E., 1971. "Molecular biology of synaptic receptors," *Science* **171**:963–971.

DOWBEN, R. M., 1969. General Physiology, New York; Harper & Row, pp. 230–231.

ECCLES, J. C., ECCLES, R. M. and ITO, M., 1964. "Effects produced on inhibitory postsynaptic potentials by the coupled injections of cations and anions into motoneurons," *Proc. Roy. Soc.* **B160**:192–210.

GOLDMAN, D. E., 1943. "Potential, impedance and rectification in membrane," *J. Physiol.* **27**:37–60.

HILLE, B., 1970. "Ionic channels in nerve membranes," *Prog. Biophys. Molec. Biol.* **21**:1–32.

HODGKIN, A. L. and KATZ, B., 1949. "The effect of sodium ions on the electrical activity of the giant axon of the squid," *J. Physiol.* **108**:37–77.

KATZ, B. and MILEDI, R., 1969. "Spontaneous and evoked activity of motor nerve endings in calcium Ringer," *J. Physiol.* **203**:689–706.

KATZ, B. and MILEDI, R., 1971. "Further observations on acetylcholine noise," *Nature* **232**:124–126.

KHROMOV-BORISOV, N. V. and MICHAELSON, M. J., 1966. "The mutual disposition of cholinoreceptors of locomotor muscles and the changes in their disposition in the course of evolution," *Pharmacol. Rev.* **18**:1051–1.

LUX, H. D., LORACHER, C. and NEHER, E., 1970. "The action of ammonium on postsynaptic inhibition of cat spinal motoneurons," *Exp. Brain Res.* **11**:431–447.

MACHNE, S., DUNNING, B. B. and KANNO, M., 1973. "pH dependence of reversal potential for acetylcholine-induced currents in neurons," *Am. J. Physiol.* **225**:601–605.

MAGLEBY, K. L. and STEVENS, C. F., 1972. "The effect of voltage on the time course of end-plate currents," *J. Physiol.* **223**:151–171.

MANALIS, R. S., 1969. "Reversal potential measurements for potentials produced by a group of cholinergic compounds iontophoretically applied to the frog endplate." Doctoral thesis, Indiana University, Bloomington, 142pp.

MANALIS, R. S. and WERMAN, R., 1969. "Reversal potentials for iontophoretic potentials produced by several cholinometics," *Fed. Proc.* **12**:292.

MAZLIAH, J., 1973. "The influence of glycine and GABA on Mauthner cells in the goldfish." Master's thesis, Hebrew University, Jerusalem, 74pp.

MAZLIAH, J. and WERMAN, R. "The actions of glycine and GABA compared on the same cell in vertebrates," *Israel J. Med. Sci.* **10**:566, 1974.

OTSUKA, M., ENDO, M. and NONOMURA, Y., 1962. "Presynaptic nature of neuromuscular depression," *Jap. J. Physiol.* **12**:573–584.

ROUGHTON, F. J. W., OTIS, A. B. and LYSTER, R. L. J., 1955. "Determination of the individual equilibrium constants of the four intermediate reactions between oxygen and sheep haemoglobin," *Proc. Roy. Soc.* **B144**:29–54.

SHOHAMI, E. and ILANI, A., 1973. "Model hydrophobic ion exchange membrane," *Biophys. J.* **13**:1242–1260.

TAKETA, K. and POGELL, B. M., 1965. "Allosteric inhibition of rat liver fructose 1,6 diphosphatase by adenosine 5′-monophosphate," *J. Biol. Chem.* **240**:651–662.

TAKEUCHI, A. and TAKEUCHI, N., 1960. "On the permeability of end-plate membrane during the action of transmitter," *J. Physiol.* **154**:52–67.

TAKEUCHI, N., 1967. "Effects of calcium on the conductance change of the endplate membrane during the action of transmitter," *J. Physiol.* **167**:141–155.

WERMAN, R., 1965. "The specificity of molecular processes involved in neural transmission," *J. theoret. Biol.* **9**:471–477.

WERMAN, R., 1966. "Criteria for identification of a central nervous system transmitter," *Comp. Biochem. Physiol.* **18**:745–766.

WERMAN, R., 1969. "An electrophysiological approach to drug-receptor mechanisms," *Comp. Biochem. Physiol.* **30**:997–1017.

WERMAN, R., 1971. "The number of receptors for calcium ions at the nerve terminals of one endplate," *Comp. Gen. Pharmacol.* **2**:129–137.

WERMAN, R., 1972a. "Amino acids as central neurotransmitters," *Res. Publ. A.R.N.M.D.* **50**:147–180.

WERMAN, R., 1972b. "CNS cellular level: membranes," *Ann. Rev. Physiol.* **34**:337–374.

WERMAN, R. and BROOKES, N., 1969. "Interaction of γ-aminobutyric acid with the postsynaptic inhibitory receptor of insect muscle," *Fed. Proc.* **28**:831.

WERMAN, R. and BROOKES, N., 1971. "The stoichiometry of transmitter-receptor interactions," *Experientia* **27**:1120.

WERMAN, R., DAVIDOFF, R. A. and APRISON, M. H., 1968. "The inhibitory action of glycine on spinal neurons in the cat," *J. Neurophysiol.* **31**:81–95.

WERMAN, R. and MANALIS, R. S., 1970. "Reversal potential measurements for strong and weak agonists of acetylcholine at the frog neuromuscular junction," *Israel J. Med. Sci.* **6**:320–321.

ZISKIND, L. and WERMAN, R., 1975a. "Sodium ions are necessary for cholinergic desensitization in molluscan neurons," *Brain Res., 88* (in press).

ZISKIND, L. and WERMAN, R. 1975b. "At least three molecules of carbamylcholine are needed to activate a cholinergic receptor," *Brain Res., 88* (in press).

Chemotaxis in bacteria as a simple sensory system

D. E. KOSHLAND, JR.
Department of Biochemistry, University of California, Berkeley, California, 94720

There are certain common features in the response to chemical stimuli over a wide range of species. For example, the specificities of the responses to a restricted group of chemicals (Moncrieff, 1967) suggest the kind of specificity identified with protein molecules. Moreover, the attractant need not be metabolized in order to provide the stimulus (Weibull, 1960). Since biological patterns tend to recur in nature, the existence of a common biochemical response to stimuli is not unreasonable. The studies on chemotaxis reported were initiated with the hope that analogies to higher neural processes might exist but with the understanding that such relationships would need careful analysis.

Chemotaxis in bacteria was discovered in the 1880's by Engelmann (1902) and Pfeffer (1888) who demonstrated that microorganisms were attracted to chemicals and would migrate up a gradient to the position of highest concentration. Gabricevsky (1900) demonstrated that the bacteria responded positively to some chemicals, negatively to others, and were indifferent to a third group. This phenomenon has been studied by many workers in the subsequent years, most particularly by the discerning studies of Adler and his co-workers on *E. coli* (1969). Adler demonstrated that there were two general classes of chemotaxis mutants: one specific in which mutation eliminated response to a particular chemical, and a second in which the bacteria were unable to respond to any chemical attractants. The first mutants were presumably identified with the individual receptor molecules for each chemoattractant, the latter with the general apparatus for chemotaxis itself. Recently Hazelbauer and Adler (1971) have studied mutants which indicate that the galactose-binding transport protein, isolated by Wu, Boos and Kalckar (1969), also serves as the receptor in chemotaxis.

The approach of my laboratory to sensory systems was similar to those employed in our research group for a number of years on isolated enzyme systems (Koshland and Neet, 1968). However, if such a biochemical approach to chemotaxis were to succeed, added quantitative tools would be needed and the initial studies have been directed to the development of the needed procedures. Three apparati have been developed, each for a specific purpose. The first, a "migration velocity apparatus," measures the movement of a mass of bacteria as a statistical average in much the same way that the gross diffusion of a solute in a liquid is described (Dahlquist, Lovely, and Koshland, 1972). The second, which we call a "tracker," is a device which measures the movement of individual bacteria in a defined spatial gradient (Lovely, Macnab, Dahlquist, and Koshland, in press). The third is a "temporal gradient apparatus," which allows us to follow the movements of an individual bacterium in a temporal gradient in the absence of a spatial gradient (Macnab and Koshland, 1972). Using these quantitative devices, some genetics and some biochemistry, we are beginning to learn something about the chemotactic phenomenon.

Migration velocity apparatus and Weber's law

The apparatus (Dahlquist et al., 1972) designed for measuring the migration of the bacteria as a statistical ensemble is shown schematically in Fig. 1. A linear density gradient of glycerol (0.5 to 3%) is used to stabilize a solution in an observation cell of 10 ml volume. The bacterial density is determined by monitoring the intensity

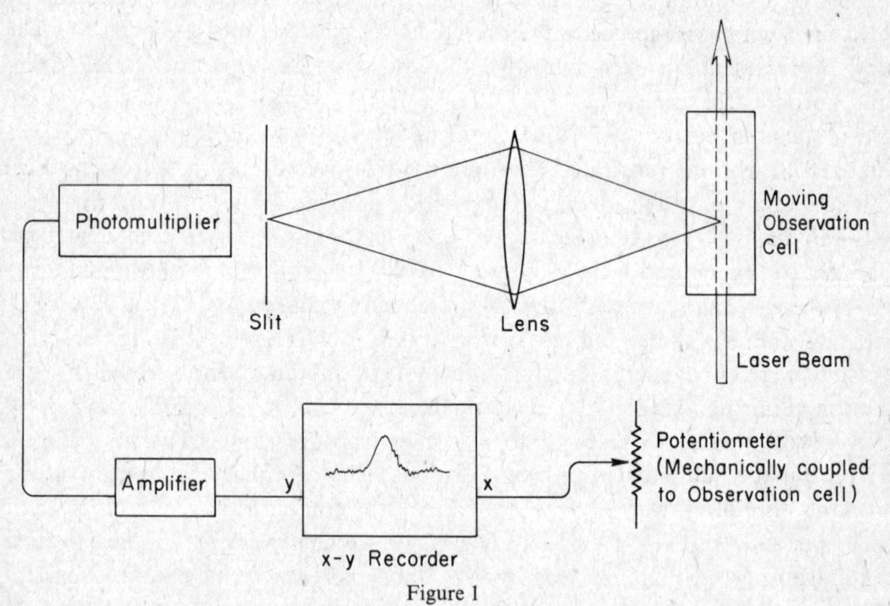

Figure 1

Colony migration apparatus. A schematic representation of the apparatus designed to detect the migration of a bacterial population in a defined gradient.

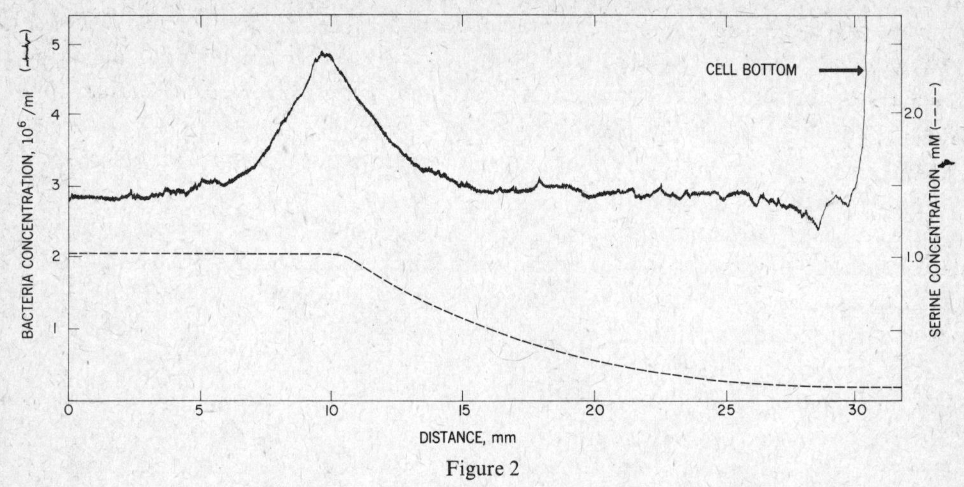

Figure 2

The response of *S. typhimurium* to an exponential gradient of L-serine. The plateau concentration of serine employed was 10^{-3} M and the decay distance of the gradient was 6.4 mm. The dashed line represents the bacteria redistribution after 15 min.

of light scattered by the bacteria. The light, supplied by an He–Ne laser, shines up through the bottom of the observation cell. A photomultiplier tube measures the intensity of scattered light. At less than 10^7 bacteria per ml the intensity of the right angle scattered light was found to be proportional to the bacterial concentration. The concentration of bacteria at various positions of the cell can therefore be determined by moving the observation cell vertically by means of a screw drive. The cell position is recorded on the x axis and the signal from the photomultiplier is recorded on the y axis of an x-y plotter. The distribution of bacteria in the cell can be scanned in 1–3 min and this distribution is then followed as a function of time. By suitably mixing chambers before the introduction of the liquid to the column, various distributions of attractant and/or bacteria can be constructed in the observation cell.

There is a complicated response to linear gradients, but a relatively simple response to an exponential gradient

$$\frac{\mathrm{d}c/c}{\mathrm{d}x} = \frac{\mathrm{d}\ln c}{\mathrm{d}x} = \text{constant}$$

The result of superimposing such a distribution of serine on an initially uniform bacterial distribution is shown in Fig. 2. In this case the bacteria accumulate at the top of the gradient as a well defined peak. The bacterial concentration on either side of this peak remains fairly constant, however, unlike the response to a linear increase which shows a distinct trough adjacent to the peak. Thus, bacteria appear to be moving through the gradient region at a steady rate. The average velocity of the bacteria is determined by the proportional changes, i.e., ratios, in concentrations. Thus the bacteria to a fair approximation follow the Weber–Fechner law (Thompson, 1967) over very narrow ranges. It is of interest that this Weber–Fechner law, which states that the just noticeable increase in stimulus is proportional to the intensity of the stimulus, has been observed in psychic phenomena including many responses in higher organisms (Thompson, 1967).

The similarity does not end there. Weber's law is only roughly correct in higher species and significant deviations occur for the bacteria as well. As discussed above, the bacteria move with uniform velocity through a cell with an exponential gradient extending over a rather limited concentration range. But when the plateau concentration range was varied over many orders of magnitude, the velocities of migration no longer agreed with Weber's law (Dahlquist et al., 1967). The deviations observed were well beyond experimental error and the bacteria deviate by amounts similar to those observed in mammals in phenomena such as the estimation of light intensity, the observations of sounds, and many other behavioral characteristics (Thompson, 1967).

The "tracker" and the mean free path of bacterial motion

In order to understand the kinetics of chemotaxis and later to study behavioral mutants, we needed a precise description of the motion of an individual bacterium. A device for describing this motion in quantitative terms had been developed by Berg (1971) and it is a fine one for some purposes. However, the Berg device did not allow us to observe the movements of the bacterium in defined stable gradients of attractant over long periods of time. The apparatus (Lovely et al., 1974) described in Fig. 3 was devised to solve this problem. A cuvette, essentially identical to the one in the colony migration apparatus, is mounted on a microscope stage and bacteria are placed therein in a defined gradient. The long-range objective of a Leitz microscope is focused to see the bacteria in this vessel using dark field optics. The long working distance of the objective (7 mm) allows one to see the bacteria well away from the walls of the vessel. A "joy stick" allows one to activate motors which move the observation cell in the x and z directions and a pedal activates motors which move the cell in the y direction. Using a crosshair in the eyepiece of the microscope and the optical changes in the bacterium, an individual bacterium can be kept in focus by operation of the "joy stick" and the pedal. The motion of the observation cell and hence of the bacterium is recorded on computers and from this a record of the motion in xyz coordinates can be reconstructed.

In order to correlate the movement of a colony and the movement of an individual bacterium it is important to have a statistical number of the latter. After tracking a reasonable number of bacteria, the net velocity upwards of a colony of such bacteria was calculated and found to be 2.9 μm/sec (Lovely et al., 1974). The velocity

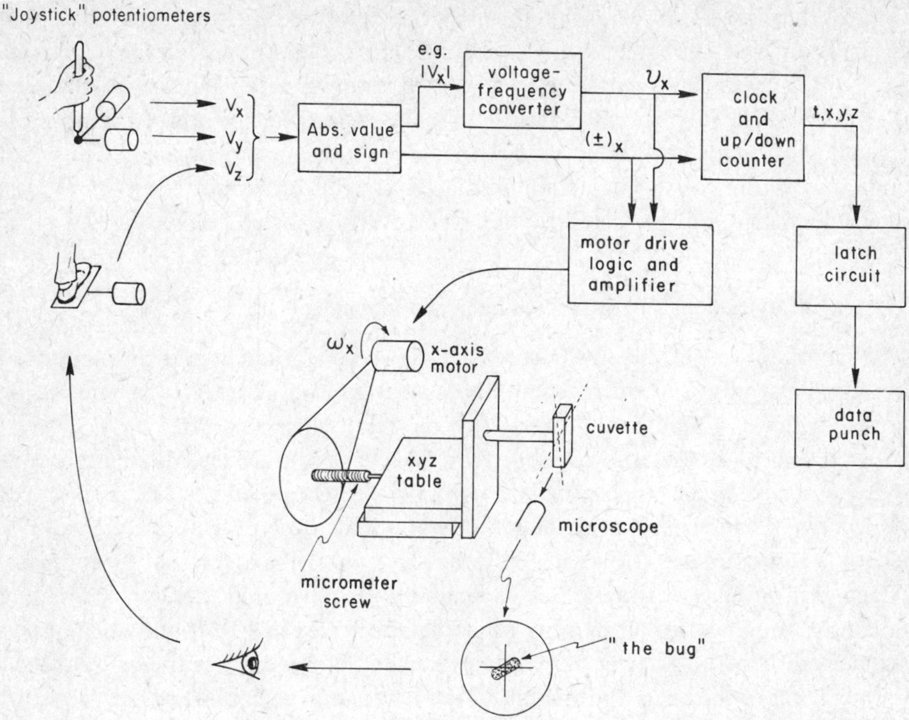

Figure 3

Bacterial tracking apparatus. Schematic illustration of device to follow three dimensional movements of bacteria in the presence and absence of gradients. Observation cell can be filled with bacteria in a defined gradient as described for the colony migration apparatus. An individual bacterium is kept in focus by movement of this observation cell using the "joy stick" for the *x-y* directions and the foot pedal for the *z* direction. Joy stick and pedal activate stepping motors which are connected to punch tape apparatus to record data for calculating coordinates.

of the colony as determined by the colony migration apparatus was 2.8 μm/sec (Dahlquist et al., 1972). This good agreement is pleasing and suggests that both instruments are measuring true characteristics of the bacterial population. Just as in perfect gas theory, the collective properties of a mass can be deduced from the motions as an individual.

Observations under the microscope show that *Salmonella* and *E. coli* appear to travel in roughly straight lines and then turn abruptly. Sometimes they appear to be tumbling head over tails for a brief period and then swim off in a new direction at random. These qualitative observations were converted to a quantitative analysis by Berg and Brown (1972), who found that a) the length of the runs (distance between tumbles or "twiddles") was Poissonian, b) the angle of the turns averaged 62° with a Poissonian distribution, c) the length of an average run was increased by travelling up a gradient of attractant, d) the length of an average run going down a gradient of attractant was about the same as normal. The results supported the widely held assumption that bacteria migrate by a "biased random walk," but they contradicted the view that the migration followed a "shock reaction," i.e., increased tumbling on going down a gradient (Weibull, 1960). In fact the movement depended on decreased tumbling on going up the attractant gradient, a conclusion supported by our tracking experiments and also indicated by the experiments reported below.

The temporal gradient apparatus and bacterial "memory"

At this stage we have shown that the bacteria sense a gradient and that they travel up a gradient of attractant by modulating their mean free path. It remains to find how they detect a gradient and how they modulate their mean free path.

In the first place the criteria needed for a bacterium to detect a gradient are very severe. *Salmonella* are approximately two micrometers in length. The drop in concentration of serine in one of the defined gradients described above over a 2 μm length is approximately 1 part in 10^4. A bacterium sensing a gradient by comparing the concentrations of attractant at its head and its tail would therefore be required to contain an analytical apparatus which detects 1 part in 10^4. If one calculates the statistical fluctuations in the more dilute gradient, however, of individual molecules in the region of the head and the tail of the bacterium, even making very conservative estimates, the fluctuations are approximately 1 part in 10. Thus, the statistical fluctuations would indicate that the bacteria cannot detect a gradient whereas the experimental results indicate that they can. How can one proceed further?

To distinguish between spatial sensing and temporal sensing mechanisms, Macnab devised temporal gradient apparatus (Macnab and Koshland, 1972). This apparatus uses a rapid mixing device to plunge the bacteria from one uniform concentration of attractant (c_i) into a final uniform concentration (c_f) where they are observed microscopically. Control experiments established that neither the mixing apparatus nor the absolute concentration of attractant affected the motility pattern, e.g., they swim similarly in uniform distribution at 10^{-5} M attractant, or 0.5×10^{-5} M, or 2×10^{-5} M attractant. If the bacteria use instantaneous spatial sensors at their "heads" and "tails" all they "see" after mixing is a uniform distribution of attractant. They should thus swim normally. If, on the other hand, bacteria operate by a temporal sensing mechanism and their "memory" span is greater than the mixing time of the instrument, they should show an altered motility pattern. The latter is what was observed.

If the bacteria are subjected to no gradient, $c_i = c_f$, then motility is "normal." Normal motility is defined as the motility observed in a wild-type culture in a non-gradient situation. The *Salmonella* swim in a fairly coordinated manner, slight changes in direction occurred often and occasionally a bacterium would tumble and then start swimming in a new direction. A stroboscopic multiple exposure of such behavior is shown in the middle portion of Fig. 4. If the bacteria were subject to a positive gradient of attractant, i.e., $c_f > c_i$, supercoordinated swimming was observed, i.e., the bacteria swam for considerably longer distances before tumbling (cf. upper portion of Fig. 4). If on the other hand the bacteria were subjected to a negative gradient of attractant, i.e., $c_f < c_i$, the motion was very uncoordinated, i.e., the bacteria swam only short distances before tumbling and starting out in a new direction (lower portion of Fig. 4). These patterns were not the result of the mixing process itself since bacteria treated in the same way in the absence of a temporal gradient ($c_i = c_f$) showed normal motility. Thus, the bacteria must have a time-dependent mechanism which allows them to remember the concentration of their immediate past (c_i) even though they are presently in a uniform gradient of attractant at a new concentration (c_f). Controls established that the absolute levels of attractant concentration did not affect motility, i.e., the bacteria gave the same normal motility in uniform distributions of attractant at concentrations equal to both c_i and c_f.

The conclusion that a time-dependent process is involved is further supported by the observation of a relaxation process, i.e., the abnormal motility a few seconds after mixing gradually reverts to normal motility during the passage of time. A few *minutes* after plunging into a different concentration ($c_i \neq c_f$) the bacteria which had been subjected to the rapid mixing were indistinguishable from normal bacteria. It therefore follows from these experiments that the bacteria are not sensing an immediate spatial gradient but in fact have some device for comparing their previous environment with their present one, i.e., they have some type of primitive "memory" apparatus.

Gradients of repellent caused a similar behavior with a precisely inverse pattern, i.e., the bacteria tumbled more frequently on going to higher concentrations ($c_f > c_i$) and tumbled less frequently on going to lower concentrations ($c_f < c_i$) (Tsang et al., 1973).

These results fitted excellently with the tracking studies with one possible exception: our studies indicated a time dependent effect in both negative and positive gradients. Berg's tracking studies showed a quantitative alteration from the normal pattern in the positive direction only. Actually there is no conflict since the temporal studies showed an asymmetry in the response also. Large positive concentration jumps give long relaxation times in the range of many minutes for steep gradients ($c_f \gg c_i$) whereas large negative jumps ($c_f \ll c_i$) gave much shorter responses of the order of seconds. Thus the temporal studies are consistent with a more pronounced effect in the positive direction. Apparently the motility pattern is altered by a temporal sensing device which can enhance or suppress tumbling, but suppression of tumbling during movement in favorable directions is more significant quantitatively than the enhanced tumbling in unfavorable directions.

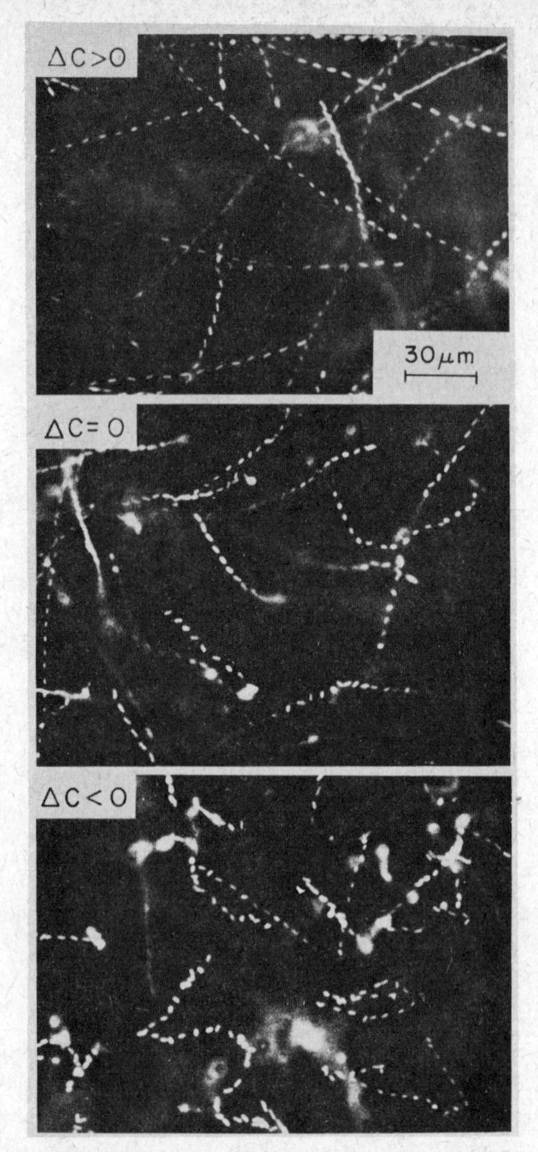

Figure 4

Motility tracks of *S. typhimurium*, taken in the time interval 2 → 7 sec after subjection of bacteria to a sudden (200 msec) change in attractant (serine) concentration in the temporal gradient apparatus. *Upper*: $c_i = 0$, $c_f = 0.76$ mM. Smooth, linear trajectories. *Middle*: $c_i = c_f = 0$ (control). Some changes in direction; bodies often show "wobble" as they travel. Bright spots indicate tumbling or nonmotile bacteria. *Lower*: $c_i = 1$ mM, $c_f = 0.24$ mM. Poor coordination; frequent tumbles and erratic changes in direction. (Photomicrographs were taken in dark-field with a stroboscopic lamp operating at five pulses sec^{-1}. Instantaneous velocity of bacteria in straight line trajectories is of the order of 30 $\mu m\ sec^{-1}$.)

Ribose binding protein

A number of proteins are apparently involved in the transport of metabolites across the bacterial membrane (Pardee, 1968). As mentioned above, one of these, the galactose binding protein of *E. coli*, has been identified (Hazelbauer and Adler, 1971) with chemotaxis. Further evidence has been obtained by the isolation of the ribose binding protein of *Salmonella typhimurium* (Aksamit and Koshland, 1972). Using the shock procedure of Heppel (1967) and Anraku (1968), R. Aksamit in our laboratory has purified a ribose binding protein which is a monomer and has a molecular weight of 29,000 (Aksamit and Koshland, 1972).

The specificity of this protein for binding of ribose derivatives is the same as for the chemotactic response (cf. Table 1). Both show a very high, if not almost absolute, specificity for ribose. The dissociation constant of ribose from the pure protein ($K_d = 0.6 \times 10^{-7}$ M) is approximately the same at the midpoint of the chemotactic response curve as determined in the temporal gradient. The actual dissociation equilibrium constant of allose, a structural analog of ribose, is 10^{-4} M and agrees well with the optima for the chemotactic response. Thus, a correlation between the properties of the purified receptor protein and the behavioral response of a living species has been obtained in this sensory system.

Table 1
SPECIFICITY OF RIBOSE-BINDING PROTEIN AND CHEMOTAXIS OF *SALMONELLA TYPHIMURIUM*

Compound	Chemotaxis response	Binds to ribose-binding protein
Ribose	Positive	Positive
D and L arabinose	None	None
D-xylose	None	None
D-lyxose	None	None
α-D-ribose-1-P	None	None
α-D-methylribofuranose	None	None
1,4-anhydroribitol	None	None
D-glucose	None	None
D-galactose	None	None
D-mannose	None	None
D-fucose	None	None
L-rhamnose	None	None
α-D-ribose-5-P	None	None
2-deoxy-D-ribose ribitol	None	None
Allose	Positive	Positive

The light effect

We have discovered that visible light at high intensity specifically disrupts normal motility by causing tumbling, the effect being reversible if short pulses are used. This "light effect" not only provides a convenient tool for generating tumbles at will, but probides clues to the control of motility (Macnab and Koshland, 1974).

The phenomenon of light-induced tumbling was first observed when a xenon arc source was used in conjunction with an oil-immersion condenser of high numerical aperture. The absence of the effect with a dry microscope condenser of lower numerical aperture indicated a strong intensity dependence. Heat or ultraviolet radiation damage was not the cause of the tumbling, since IR and UV blocking filters provided no protection.

By the use of narrow-band interference filters a very crude action spectrum was obtained. The maximum effect occurred in the region 350 to 450 nm, no measurable effect occurring at 530 nm or beyond. The intensity dependence of the phenomenon was studied, both with white light and monochromatic light at a variety of wavelengths. In all cases, if light at a given intensity was just sufficient to be fully effective (all bacteria responding by tumbling) then light at half that intensity was much less effective (20–40 % of bacteria responding) and light at quarter the intensity had no detectable effect.

The response to the light stimulus was instantaneous, as judged by visual observation. The duration of the resulting tumble increased with increasing light intensity. The effect of short pulses (< 1 sec) was reversible; normal motility resumed after the tumble.

Tumbles induced by light stimuli were indistinguishable from those which occur intermittently in normal motility, and whose frequency of occurrence is altered in chemotaxis as a result of gradient sensing. The interaction of a "favorable" chemotactic stimulus (one suppressing tumbling) with the "unfavorable light" stimulus was examined. Bacteria were subjected to a concentration jump from 0 to 1 mM serine (a strong attractant) in the temporal gradient apparatus and observed immediately thereafter. The bacteria were all swimming smoothly for several minutes *even when exposed to light pulses which would normally have elicited tumbling*. Thus, light as a stimulus causing tumbling can be neutralized by a chemotactic stimulus causing smooth motility, just as has been shown previously for opposing chemotactic stimuli, i.e. attractants and repellents.

The action spectrum, though crude, is compatible with that of a flavin. Riboflavin, flavin mononucleotide, flavin adenine dinucleotide, etc. all display absorption maxima at around 370 and 440 nm, a minimum at around 405 nm, and virtually no absorption above 520 nm. Other bacterial chromophores such as quinones and cytochromes have quite different spectra. This suggests that the light effect in *Salmonella* involves excitation of a flavin, although a more refined action spectrum and independent chemical evidence will be necessary before this can be stated with certainty. If we assume (as is reasonable for a non-photosynthetic system) that the absorbing molecular species normally functions in its ground electronic quantum state, excitation to a higher state would in general result in loss of function. Furthermore, high light intensity would be necessary to secure appreciable population of the excited state because of its short lifetime. Since flavins are involved in electron transport, the light effect may be caused by an alteration of electron transport and it is interesting that the effect is most pronounced during aerobic metabolism. Interruption of the energy supply has been suggested as the basis for phototactic and aerotactic responses (Links, 1955; Clayton, 1958). However, since tumbling is an active process it seems more likely that an *alteration* of flagellar operation is occurring. Most importantly however the results indicate that electron transport is involved in the signalling system and gives us a clue to the generation of the behavioral response.

Relation of chemotaxis to higher neural systems

The signalling processes in a monocellular organism cannot have the complex circuitry of higher species with a central nervous system but it could quite possibly use similar biochemical pathways and a similar pattern of recording information. The bacterium under consideration is shown in Fig. 5. The information received at receptors distributed over the membrane is transmitted to the flagella. The distance the signal must travel is much shorter than in higher species, but the stimulus must be transmitted and must involve a behavioral response. In the case of *Salmonella*, a response to specific chemicals is apparently analogous to the specific responses in higher organisms, e.g., odor in man and pheromones in insects. The high specificity of the response, e.g., to ribose only, indicates a protein molecule is the receptor and such a molecule has been isolated. The response does not require the metabolism of the attractant as is also true in higher species. An exactly parallel and inverse relationship is seen in repellents. The analogy to pheromones in insects and to taste and odor in mammals is impressive. The bacteria obey Weber's law roughly over short ranges of stimulus and deviate significantly from it over wide ranges, again as seen in higher species.

The bacteria utilize a time-dependent mechanism which can be termed a "memory" in the sense that it allows the bacterium to compare its past with its present. This memory is very short, but it requires a biochemical mechanism which allows an integration over time. Quite obviously, this is a long way from the memory of man with its vast storage and retrieval mechanisms. Yet the relation of the bacterial system to higher species may be important, just as the analogy in protein synthesizing pathways between bacteria and higher species is close but not identical.

The level of some chemicals is clearly critical for the signal to the flagella. More-

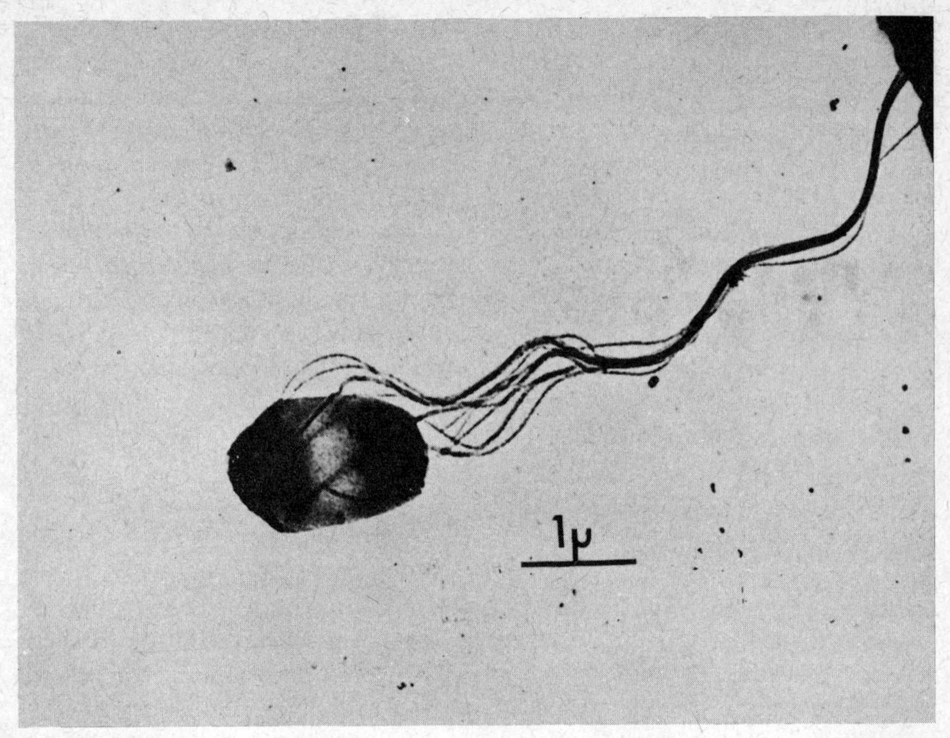

Figure 5
An electron microscope picture of a *Salmonella*. The receptors are located in the periplasmic space in outer membrane. The binding of a chemoattractant generates a signal which is transmitted in some manner to produce a behavioral response in the flagella.

over the inputs to this level from a variety of attractants and repellents act as a summation process. Such levels could logically be related to equivalent levels in a neuron. In that case the state of polarization depends on additive and subtractive relationships of stimuli from chemicals, dendrites and hormones. Moreover the signal in the bacteria is transmitted from the receptor to the flagella just as a signal in higher species must be transmitted from a receptor to the motor apparatus. The distances are greater in the latter case but again the chemistry may not be dissimilar.

Finally, if new interneuron connections are to be initiated or reinforced by the synthesis of new proteins or are to be deleted by hydrolysis of old proteins and decreased protein synthesis, some change in chemical level is indicated. Small molecules such as cyclic AMP, inducers or compressors are required to initiate protein synthesis or inhibit protein breakdown. It is logical therefore that a change in the level of some compound X could itself trigger other more permanent chemical changes as well as being involved with the firing of an individual neuron. Learning and memory will therefore be products of a common biochemical event.

The behavior patterns of the single cell bacteria will obviously be simpler than the more complex processes in higher species but the preliminary indications are that the similarities are significant. The monocellular system offers the advantages of simplicity. Hopefully it can provide insight into biochemical mechanisms of communication which are common to all species.

Acknowledgments

I am grateful for financial support from the National Science Foundation (grant GB7057) and the U.S. Public Health Service (grant AM-GM-10765).

References

ADLER, J., 1969. *Science* **166**:1588–1597.
AKSAMIT, R. and KOSHLAND, D. E., Jr., 1972. *Biochem. Biophys. Res. Commun.* **48**:1348–1353.
ANRAKU, Y., 1968. *J. Biol. Chem.* **243**:3116–3122.
BERG, H. C., 1971. *Rev. Sci. Instruments* **42**:868.
BERG, H. C. and BROWN, D. A., 1972. *Nature* **239**:500.
CLAYTON, R. K., 1958. *Arch. Mikrobiol.* **29**:189–212.
DAHLQUIST, F. W., LOVELY, P. and KOSHLAND, D. E., Jr., 1972. *Nature New Biology* **236**:120–123.
ENGELMANN, T. W., 1902. *Arch. ges. Physiol. Pflügers* **57**:375–390.
GABRICEVSKY, G. N., 1900. *Z. Hyg.* **35**:104–122.
HAZELBAUER, G. L. and ADLER, J., 1971. *Nature New Biology* **230**:101–104.
HEPPEL, L. A., 1967. *Science* **156**:1451–1455.
KOSHLAND, D. E., Jr. and NEET, K. E., 1968. *Ann. Rev. Biochem.* **37**:359–410.
LINKS, J., 1955. Thesis, Leiden.
LOVELY, P., MACNAB, R., DAHLQUIST, F. W. and KOSHLAND, D. E., Jr., 1974. *Rev. Sci. Instruments* **45**:683.
MACNAB, R. M. and KOSHLAND, D. E., Jr., 1972. *Proc. Nat. Acad. Sci. U.S.* **69**:2509–2512.
MACNAB, R. M. and KOSHLAND, D. E., Jr., 1974. *J. Mol. Biol.* **84**:399.
MONCRIEFF, R. W., 1967. *The Chemical Senses.* London: Leonard Hill.
PARDEE, A. B., 1968. *Science* **162**:632–637.
PFEFFER, W., 1888. *Untersuch Botan. Inst. Tübingen* **2**:582–589.
THOMPSON, R. F., 1967. *Foundations of Physiological Psychology.* New York: Harper & Row.
TSANG, N., MACNAB, R. and KOSHLAND, D. E., Jr., 1973. *Science* **181**:60–63.
WEIBULL, C., 1960. *The Bacteria,* Vol. I (I. C. GUNSALUS and R. Y. STANIER, eds.). New York: Academic Press, pp. 153–202.
WU, H. C. P., BOOS, W. and KALCKAR, H. M., 1969. *J. Mol. Biol.* **41**:109–120.

Workshop on stability and origin of biological information
The first Aharon Katzir-Katchalsky conference

GENERAL DISCUSSION

Rehovot, Israel, June 23–28, 1973

F. O. SCHMITT (Chairman): Mr. President, Ladies and Gentlemen: This is the final session of our symposium. Being self-organizing, like DNA, its purpose is to provide an opportunity for open discussion by all participants. You may address your questions or discussion toward anyone of the panel, or you may care to discuss something of your own interest. No set speeches have been arranged; the program is to be entirely spontaneous.

As a chairman's privilege, I should like at the outset to say a few words about the relation between Aharon and the Neurosciences Research Program (NRP) which I have the honor to chair. The organization is unique in being a tightly knit group of Associates who are experts in fields cognate to neuroscience and who are mutually dedicated to the development of the more theoretical aspects. NRP Associates meet regularly at Stated Meetings to debate the issues and to hear new evidence on various subjects arranged by the in-house Staff. The group also carries out Work Sessions on subjects near the cutting edge of neuroscience.

Aharon joined the group in the year of its founding, 1962. At Stated Meetings he thrilled the Associates with his highly original lectures on the bearing of biophysical chemistry, particularly membrane properties and phenomena of irreversible reactions, on life science, and especially on neuroscience.

A few years ago he lectured before the NRP Associates on network thermodynamics, a subject which he was later to develop with Oster and others at the University of California at Berkeley. The lecture resulted in much discussion by other Associates, particularly Drs. Eigen, Onsager and McConnell. At the next meeting he related new developments in the field and suggested that the subject may have application to the neural nets of brain.

The subject was considered as a possible topic for a Work Session in its own right, although Paul L. Penfield, Jr., of M.I.T., author of the book on *Tellegen's Theorem and Electrical Networks* (Cambridge, Mass: The M.I.T. Press, 1970), pointed out certain limitations presented by lack of data related to the brain, as compared with electrical networks for which the theory was designed. Meanwhile, Aharon had become interested in dissipative structures and the possible role of such processes in brain function. His interest was stimulated by the fact that during activity the brain may manifest certain steady, D.C. potentials, of the order of a millivolt, as a global phenomenon, relatively independent of the neuronal organization of the brain. At an earlier Intensive Study Program, Vernon Rowland, of Case-Western Reserve University, had pointed out that small shifts of steady potentials may occur if the animal correctly performed an act for which it was previously trained; if it did not perform the act correctly, everything else being the same, the shift of potential did not occur. It was also known that certain neurophysiological processes and also behavior are influenced by polarization of the brain by appropriately applied potentials.

Therefore a Work Session, to be co-chaired by Professors Katzir-Katchalsky and Rowland, was devised at which the physiological aspects of steady potentials in the brain and the field of dissipative structures and of the thermodynamics of irreversible processes were jointly discussed. The participants looked forward to the possibility of bringing to the attention of neuroscientists phenomena and theories quite different from those comprising classical neurophysiology, which deals primarily with self-propagating action-potential waves, synaptic phenomena, and the specificity of the "wiring diagram" of the neuronal networks subserving the particular physiological function under consideration at the time. Aharon was, of course, fully aware of the more classical neurophysiological approach, but the Work Session was designed to explore the possible role of the dissipative structures and irreversible processes, especially at membrane surfaces, as a general phenomenon within the brain, holistically considered. The Work Session was an exciting one, thanks primarily to Aharon's dynamic personality and leadership in new fields of exploration of biophysical chemistry applied to biology. We are grateful to Dr. Robert Blumenthal, of the National Cancer Institute, a student of Aharon, who helped plan the Work Session and who, as a labor of love, has co-authored the small monographic *NRP Bulletin* entitled "Dynamic Patterns of Brain Cell Assemblies," published by the NRP (A. Katchalsky, V. Rowland, and R. Blumenthal, Vol. 12, No. 1, 1974) and covering basic principles discussed at the Work Session.

I cannot refrain from recounting that, unlike most Work Sessions, which are attended primarily by experts in the field under discussion, some ten additional NRP Associates elected to attend this one. For half a day, following the two-and-a-half-day scheduled Work Session, Aharon recapitulated the upshot of the Work Session and dwelt on still other more tangential subjects, all very dear to his heart.

The air was electric with the scientific challenge, the power, and the charm of his presentation. These remarks are contained in part in the published *NRP Bulletin*'s record of the Work Session, but the memory of that discourse will live on in the hearts and minds of the individuals who were privileged to hear his spontaneous, unscheduled lecture on that morning of May 19, 1972.

We shall now start the discussions of that which happened in the last few days of this memorial symposium in honor of Aharon.

Thus far, Professor John Ross, Chairman of the Chemistry Department at M.I.T., has had no opportunity to discuss certain aspects relevant to our general topic. I shall therefore now call on Professor Ross.

J. Ross: During our meeting we were much concerned with far-from-equilibrium phenomena and for the purpose of the remainder of the discussion it may be useful to list a variety of nonlinear phenomena and a variety of coupling mechanisms which may lead to the requisite nonlinearity.

First, we have stable reaction mechanisms, in which case fluctuations from the macroscopic progress of the reaction decay in time. Close to chemical equilibrium all reaction mechanisms are stable. Far-from-equilibrium fluctuations in a stable reaction mechanism may decay either monotonically or in an oscillatory manner. Instability in a reaction mechanism is said to occur if a fluctuation from the macroscopic progress may grow in time either with or without an oscillatory component. A limit cycle system is one in which oscillations of concentrations or other thermodynamic variables occur, these oscillations being stable in amplitude and frequency but not necessarily in phase.

Nonlinear reaction mechanisms may show the possibility of multiple steady states. Examples may be cited of the possibility of both reversible and irreversible transitions between stable branches of a multiple steady-state system. If reversible transitions occur then hysteresis may be observed provided that the fluctuations within the stable branches of steady states are small compared to the separation of the stable branches.

Chemical systems may be maintained at far-from-equilibrium conditions by continuous feeding of reactant into the system and continuous removal of products. We may think of this process as an asymmetric mass flow from reactants to products. In an analogous way, and yet one which may be much easier to achieve experimentally, the system may be closed to mass flow but not isolated, so that energy may cross the boundary of the system. In that case an asymmetric energy flow achieves the purpose of attaining far-from-equilibrium conditions. This may be done by shining onto the system light of a frequency absorbed by a reactant only, and after absorption of light the energy is quickly converted into heat by a radiationless transition. The change in temperature causes changes in relative concentrations of reactants and products if the enthalpy change of the reaction is not zero. Another similar mechanism is through joule heating in a reaction in which, say, the reactants are ions and the products are neutral species. A passage of current through the system achieves the goal of maintaining the system far from equilibrium.

Another topic not yet mentioned is the effects of periodic external forces on nonlinear mechanisms; i.e., at marginal stability resonance response is possible, and for such systems the analogy with electronic components is striking.

There was some discussion about fluctuations but none on the interesting questions of the behavior of fluctuations near instabilities or of any similarities in such behavior to that observed at an equilibrium critical point. The study of this question, theoretically and experimentally, possibly by light scattering techniques, seems most interesting.

Before proceeding with the listing of different types of phenomena, let me say a few words about different types of mechanisms which can give us such processes. In order to obtain the possibility of multiple stationary states or oscillations, for instance, we know that the macroscopic equations of reaction must be nonlinear. In a homogeneous reaction mechanism this may be achieved by auto or cross catalysis. There may, in addition, be coupling of a homogeneous reaction mechanism to other thermodynamic properties such as temperature or pressure, provided that the variation of rate coefficients with these variables is taken into account. If the reaction mechanism occurs within an enclosure then very simple types of mechanisms coupled with nonlinear permeation coefficients across the boundary of the enclosure are sufficient. The varieties of feedback loops which lead to nonlinearities are extensive. We talked a great deal about enzyme reactions and, just as an example, if the enzyme activity is a function of pH and the enzyme acts as a catalyst for a reaction in which hydrogen ions are a product, then the possibility for autocatalysis exists. Although we heard almost nothing about electrochemical reactions, of which many show oscillatory behavior, I believe that the importance of that topic is such that in future conferences we shall concentrate on it extensively.

If we couple a scalar reaction mechanism to vectorial processes such as diffusion, thermal conduction or mass flow, then various spatial phenomena may occur. Again it is useful to distinguish between heterogeneous and homogeneous catalysis and the occurrence, respectively, of local and global instabilities leading to spatial structures (dissipative structures). Furthermore I would like to distinguish between intrinsic and extrinsic spatial structures: An intrinsic structure is one in which the characteristic length scale is determined primarily by the time and length scales of the reaction and transport processes, whereas in an extrinsic case the characteristic length scale of the spatial pattern is determined by the size of the system. In biology the word epimorphic is analogous to intrinsic, and morpholactic to extrinsic. Next we need to distinguish between time-invariant structures, temporally oscillatory structures and chemical waves. In the case of the "Belusov–Zhabotinsky" reaction all of these classifications have been observed.

From the discussion we have had it seems to me useful to give you a very simple example of how a spatial structure may arise. Let us consider two non-interacting gases A and B, in a container. In order to study thermal diffusion in this system, we could assume that A would go preferentially to the hotter regions. Now let us shine on the system light of the frequency which is absorbed by A but not by B. Consider next a fluctuation in temperature which occurs in a small region within the system. If the temperature is higher within that small region, due to a fluctuation, then by thermal diffusion the mole fraction of A will become larger than in the bulk of the system. But if the mole fraction of A is higher, then there will be more absorption of light and, again by a fast radiationless process, more energy is introduced into the small region than into the bulk of the system. This process clearly has a positive feedback loop and the nonlinear equations describing this process show an instability which is the onset of a spatial pattern.

A different class of spatial patterns is the one of periodic precipitations, of which Liesegang rings are one example. As far as I know all theories on Liesegang rings, except the most recent one by us, require a spatial concentration gradient for the occurrence of periodic precipitation. Yet if you form a colloid and watch the growth of the colloid in the absence of an external gradient, then you will find that growth does not occur homogeneously. Macroscopic regions of higher and lower colloid density will be formed. If you take an originally homogeneous colloid and impose a gradient of any coagulating species then well-defined periodic patterns occur. I believe that the role of gradient is that of sharpening the spatial pattern and polarizing the pattern; it is not essential for obtaining the pattern. Similar issues of the role of gradients arise in certain cases of asymmetric cell differentiation. In the eggs of fucus, for instance, the first differentiation is asymmetric: the axis of polarization in the first differentiation is random in a culture but can be oriented by a gradient of salinity or by a beam of light.

Finally, in regard to spatial patterns, let me come to the topic of chemical waves which are periodic variations in concentrations in space and time with the constraint of a wave pattern. So far a variety of different types of waves have been analyzed: the transient phenomena of kinematic waves which occur in an oscillating reaction under the imposition of, let us say, a temperature gradient; diffusion type (phase) waves; analogs of detonation type (relaxation oscillation) waves; plane, circular and spiral waves; waves emanating from or terminating in a center called a pace maker; surface waves on localized instabilities; wave-like drifts of spatial structures, etc. During the meeting we heard of the occurrence of waves in the Belusov–Zhabotinski reaction and in cell aggregation of slime molds.

The importance of the possibility of instabilities occurring in reaction mechanisms themselves was discussed by Eigen in detail.

F. O. SCHMITT: In starting the general discussion of Professor Ross's presentation,

I should like to say that in the 1920s and 30s much attention was indeed given to the so-called Liesegang-ring phenomena in all their forms. These are periodic structures which form by the chemical interaction of one substance with another reactant usually contained within a gel. In a book by E. S. Hedges and J. E. Myers (*The Problem of Physico-Chemical Periodicity*, New York: Longmans Green & Co., 95 pp., 1926), the conclusion was reached that whenever chemicals react in discontinuous media of large surface area, as in gels, capillary tubes, etc., the product of the reaction is likely to show aspects of periodicity in space or time (rings, bands, rhythmic beating, etc.). These instabilities were dynamic patterns which themselves arose from processes discussed by Professor Ross and which fall in the general framework of dissipative, instability patterns.

Professor Ross has raised some seven or eight points relevant to this symposium. These are now before you for discussion. Perhaps someone would like to discuss his first point which bears much food for thought about morphogenesis, i.e., the relation of the formation of ciliary structure to ionic strength.

I. R. MILLER: There is a special interest in periodic structures in space or in time formed in conjunction with surface phenomena. Aharon was very much interested in the precipitation membranes developed by Hirsh–Ayalon. The precipitation membranes are formed by diffusion of Ba^{++} and SO_4^- into a collodion or other inert membrane from the solutions on its two sides. When one examines the membrane closely one observes not one but several precipitation layers reminiscent of the Liesegang rings, symmetrically spaced within the membrane. The membrane thus obtained has interesting selectivity and rectification properties.

Another example, in this case of a structure changing periodically with time accompanied by periodically changing properties, can be observed when one applies an electrical field across a high-resistance surface layer. The high field causes a dielectric breakdown of the layer with subsequent decrease of resistance and, if the experiment is carried out at constant current, the electrical field across the interface drops also. Without the electrical field the surface layer is built up again with subsequent rise of the electrical field until the next breakdown is reached. Thus the periodical changes in surface structure are accompanied by changes in potentials or currents, depending on the experimental conditions.

I. PRIGOGINE: As Professor Ross has given a general survey of the subject of far-from-equilibrium phenomena and since several reviews of this subject have appeared recently (Portnow and Nicolis, *Chemical Reviews* **73**: 365–384 (1973); Prigogine and Nicolis, *Quart. Rev. Biophys.*, **4**, **2**, **3**: 107–148 (1971)) I shall limit myself to a specific remark.

Concerning evolving networks and the appearance of more and more organized states in such a system, I would like to say again that what is lacking is not so much the formal type of chemistry which may be involved as the understanding of the physical mechanism whereby a succession of instabilities can be generated. We are often confronted with the kind of stability problem appearing in the Martinez model for differentiation which was mentioned yesterday: each time that a new pattern of concentration appears in this system, conditions are realized simultaneously such that some cells in the system will undergo division. The question then arises whether or not the pattern of concentration is still stable with respect to growth. If not, a new pattern is produced and again a set of cells will divide. In a certain sense the same type of problem also occurs in the work of Eigen on the self-organization of macromolecules and in the related examples considered by Babloyantz. In each reaction or transcription step there is a non-vanishing probability that some mistake will be made. When such mistakes trigger the transition from one state to another, the new state again is tested against all mistakes compatible with the system properties. In order to permit a further increase of organization, there should then again exist a class of mistakes with respect to which the system is very sensitive, i.e., unstable.

Y. Y. ZEEVI: Discussing the example of the Zhabotinsky reaction in which pattern formation depends on diffusion type coupling, you conclude that coupling of reaction and diffusion is essential for the formation of patterns and propagation of waves. Such generalization is unjustified. To start with, we know from theoretical considerations that the existence of kinematic waves does not depend on coupling. For example, a linear array of identical *uncoupled* oscillators having certain phase relationships can give rise to an apparent wave-like pattern. Even more interesting is the case of an array of uncoupled oscillators whose frequencies vary monotonically. In fact, it was shown by Howard and Koppel that such a system has features of behavior very similar to those of the Belousov–Busse reaction. Howard and Koppel have also ruled out the importance of diffusion by demonstrating experimentally that physical partitioning of the reacting solution does not alter the patterns.

H. ATLAN: But you cannot get those waves without diffusion.

Y. Y. ZEEVI: Indeed, in the Busse reaction you do get propagation of bands in spite of the physical partitioning, when there is no diffusion.

K. KITAHARA: Zhabotinsky demonstrates the condition for appearance of oscillations. In the space of concentrations of bromate, malonic acid and cerium ions, the region in which oscillations exist is closed. How about your model?

G. WEISBUCH: A system of nonlinear differential equations giving oscillatory solutions can produce, according to the values of the parameters, either sinusoidal oscillations or relaxation oscillations. The reaction rates entering our equations are in fact dependent upon concentrations of the input reactants such as malonic acid and bromate ions. When these concentrations are changed, the shape of the oscillations changes from sinusoidal to relaxation oscillations.

From a mathematical point of view, the existence of instabilities depends on relations between the coefficients of the characteristic equation of the linear differential system. When one expresses these relations in terms of concentration of input reactants, one obtains the domains of concentration allowing the instabilities. These theoretical results are in partial agreement with the papers of Zhabotinsky (*Oscillatory Processes in Biological and Chemical Systems*—Ed. Nauka, Moscow, 1970).

D. COHEN: I want to comment about the logic of the evolving existence itself. It seems to me that these terms are not exactly the same and maybe one can formally deny the stability of self-organization in a strict sense; but I think in actual terms the evolving systems cannot. We think of a system as having fairly undefined states provided it has a rule of creation of new states. I leave to you whether it is a question of self-organization or not, but in practice this system can certainly get itself organized in a very deterministic manner.

H. ATLAN: In fact, there are such things as self-organizing systems but they are not isolated systems—there must be interaction with the surroundings. Therefore, in this strict sense, you can say that they are not self-organizing. However, we call them self-organizing systems by virtue of the fact that interactions with the surroundings are only interactions with random factors, which are not programmed. In this sense a self-organizing system is a system which, reacting with the environment, is driven to another state, which under some criteria could be called more organized than the previous one.

I. PRIGOGINE: The interactions with the exterior of a self-organizing dissipative system cannot simply be limited to random factors, i.e., in some sense to fluctuations. There must exist an overall driving force corresponding to the maintenance of a non-equilibrium situation; at equilibrium there is no possibility that a fluctuation triggers the appearance of a highly organized state. The simplest example of this, so often mentioned, is the Bénard transition; it is only beyond the Rayleigh number that small disturbances in the temperature will produce the spontaneous organization of the system. Essentially, you then see that you have not put in any information to organize the system. Therefore it seems very difficult to speak about such things as the flow of information, in spite of the fact that there is obviously here a simple example of self-organization.

In connection with the discussion of phenomena of the type found with the Zhabotinsky reaction, I would like to mention that on model systems the existence of time-independent spatial structures resulting from the coupling between diffusion and chemical reactions has now been established rigorously by Auchmutty and Nicolis (to appear in *Bull. Math. Biophys.*).

M. EIGEN: Then you have a true composition of states that you can populate. You have one distinct origin and if you repeat the experiment it might end up as a different state. It is simply something which you cannot calculate ahead. I would say that you could argue about it in a philosophical sense. If you could predict what will happen and could denote the outcome, then it is the system itself, but if you cannot, it is the result of the evolution of the system.

If the mechanism is such that it gives you a true alternative then that is also the sense of information theory which you change for probability distribution in an unpredictable way.

In the Bénard problem, of course, we have an extremely simple situation where only one transition is possible between two types of states, but one must realize that it is already the beginning of self-organization.

P. RICHTER: I would like to make some comments on what Professor Ross has said. One of his remarks refers to the dissipative structures. I have the feeling, hearing Professor Prigogine, that he thinks dissipation might be a necessary tool for the appearance of structure, but this is definitely not the case. I think one should mention this point and make clear that structures don't necessarily evolve by competition of dissipation and some other form of energy transport, but more generally just by competition of different types of energy which need not be dissipative. For example, you can have competition of volume and surface energies, and this might lead in the direction Oster indicated. It might also be something else, like magnetic and electric energy. In fact, you find a variety of structures already in equilibrium. For example, in the case of superconductors you have a special arrangement of magnetic flux lines which are in no way associated with dissipation; you know that supercurrents do not dissipate. This is one point I wanted to make.

Another point is related to the theory of nonlinear irreversible thermodynamics. If we are mainly interested in the various structures that may appear then the occurrence of different patterns, even in equilibrium, suggests that it may not always be necessary to search for a specific theory of irreversible thermodynamics. I have the feeling that many of the phenomena which are said to be far from equilibrium can be described by a theory very analogous to the usual equilibrium theory. For example, we have been studying the case of two-mode lasers. Of course, from the point of view of thermodynamics lasers are typical non-equilibrium systems. However, you can readily construct a theory which is very similar to the familiar Boltzmann theory in that it combines both aspects of usual statistical description: i) it provides the statistics, based on Boltzmann's principle; and ii) it contains at least part of the equations of motion in the form of a certain Hamiltonian. Of course, an adequate measure of the fluctuations will not always be given by the actual physical temperature. Instead, the internal stochastic processes produce some analogous quantity which takes the role of a temperature in the system. But once you have that you are left with a theory which I would call a thermodynamic theory of certain non-equilibrium systems; it might even include limit cycle phenomena as encountered in the case of a detuned laser.

One last remark. Nonlinearity in this connection is not an essential difficulty of the theory. You may easily incorporate nonlinear potentials into a Hamiltonian or its analogue which in phase transition theory, by the way, is generally called a Landau functional.

J. ROSS: My problem is how you can get cycles out of a system in which you have conserved quantities.

P. H. RICHTER: This is like in the case of a magnetic system where the velocity is not the canonical momentum. Instead you have a relation which involves the vector potential of the magnetic field. This gives rise to oscillations of the cyclotron type, with a definite frequency.

J. ROSS: But are these stable to small changes in frequency and amplitude?

P. H. RICHTER: Yes, these are determined by the external conditions only, at least in the laser case.

J. ROSS: There are changes in external conditions by modification of which you will change the frequency and amplitude.

P. H. RICHTER: Yes, but once the external conditions are given, I mean, for example, the flux through the system, then the internal fluctuations behave in the usual way.

I. PRIGOGINE: In response to the discussion remark by P. Richter, I would like to emphasize that I have never stated that dissipation is necessary for structure. On the contrary, I have emphasized the distinction—essential in my view—between *equilibrium* structures (crystals, superconductors) on the one hand and *dissipative structures* (Bénard cell, lasers . . .) on the other.

D. E. KOSHLAND: It seems to me, Manfred [Eigen], that what you are defining in your theory is an ideal replicating system which must have occurred at a time in evolution when enzymes were highly specific. Now we know that, for example, we can tolerate a fair amount of errors in making proteins, because some amino acids are not that essential. Would you say there could have been a more elementary form where you have the wobble hypothesis and it wasn't quite as necessary to have an error level as low as 1 in 10^5 or something? How early in evolution do we have to have a perfect copying as you were suggesting?

M. EIGEN: One of the principal results of this theory is that you cannot simply start with nucleic acids in the absence of any catalytic mechanism which greatly improves the accuracy of replication. I did not say that you can also turn the question around and from a random synthesis of protein arrive at any system which would replicate itself. The answer is that theoretically you could construct systems which have this property. They are networks, reaction networks, where a given protein makes peptides while other enzymes link peptides together to form longer units. You can construct reaction networks which have this property of replicating themselves. You know that Fox in Florida is doing such experiments. He takes random mixtures of amino acids, puts them together by some solid-state thermal polymerization and obtains proteins containing 100–200 amino acid residues, which he dissolves and tests for their properties. He gets some reproducible properties, including reproducible catalytic properties. Any estimate will reveal that this has to be so; all you need are two or three groups in a critical position. The calculation shows that the probability of three different groups assuming such a position is high. All this is obtained in a mixture of protenoids. Some functions are started, but this network has no chance of evolution. It should start from a given sequence with a certain catalytic advantage and substitute step by step until a highly sophisticated enzyme develops.

This does not work unless you build a cyclic network. However, it is nearly impossible for a cycle to close back on its origin, and the network has but little chance of evolution.

With the nucleic acids, due to the self-reproductive behavior, you have the possibility of evolution from the beginning, but it stops at a certain length and the information drifts away. Now if you could devise a system where both the protein and nucleic acids are utilized—for instance, proteins may recognize phenotypes on the RNA level and link a sufficient number of smaller pieces of RNA into a cyclic network—such a system could start an evolution if it involves translation. The first system certainly then will not be precise, but the precision has to be above a certain limit, given the amount of information to be copied, otherwise you invariably lose what you gain.

The specificity of the proteins has to be large enough to increase the resolution of the nucleic acids.

M. LEVITT: Professor Eigen, you mentioned that in your dice game on nucleic acid chains, you obtained cloverleaves. Do you refer to loops in the chain, or do you mean that the two ends of the loop-forming chain sections become linked together, forming a true cloverleaf?

M. EIGEN: If you use the chain applying the constraint of end-to-end matching, you can predict the number of loops. The simplest form would be a hairpin because it loses only the one region which does not pair, but the average pair matching is not high. There is only one way to make a hairpin. Of course, that structure will win which at the very outset has accidentally the largest number of pairs. By utilizing the large number of alternative foldings of a cloverleaf structure one definitely can start with a number of accidental pairs, which is above the average to be expected for a hairpin.

M. LEVITT: I wish to ask Professor Spiegelman whether the base pairs are part of a secondary structure.

S. SPIEGELMAN: I did not propose that as the secondary structure. That's the simplest thing that you get if you simply ask to pair neighboring interstrand complements. Obviously the structure is much more complicated: we have actually shown that the two ends come together and pair, we have demonstrated that the four C's at the end are part of a base pair because they are protected against conversion. It is quite clear that the molecule is more complicated.

M. LEVITT: It is interesting, Saul [Spiegelman], that you get this end group complementarily just so that the 5-prime ends of both strands look similar to the replicate; it is, however, strange that you also find the same end-to-end pairing in TRNA 6S RNAS, 5S RNAS, and it seems to be a general property of RNA secondary structure that you can pair them this way.

M. EIGEN: The simple pairing requirements are only there if the RNA is a phenotype—in other words, if you select for a functional RNA. As soon as you start to encode proteins with your RNA you obtain other constraints too.

S. SPIEGELMAN: The percentage of base pairing in the MDV mutant or variant is rather high, 75% of the bases are involved in base pairing, which is a good deal higher than you get in average viral RNA.

M. EIGEN: There is, by the way, one important point resulting from quantitative theory, which was not obvious to us; if right at the beginning one calculates the true replication rate from the differential equations, one finds that the rate at which the system finally replicates involves the geometric mean of the rate parameters for the plus and the minus strand. At first sight this is surprising because one would expect that plus makes minus and minus makes plus—hence the slowest one would be the rate controlling. However, if one is much slower than the other, it grows to a much higher concentration level; this balancing causes the overall rate to contain the square root of the product of rate constants for the plus-minus strand. This is a very clever trick of nature, since it takes care that mutations in both strands are selected with equal weight. Note that both strands have to include the recognition site.

S. SPIEGELMAN: There is a symmetry in these cube RNAs; not very much, but it is enough. There are things that I obviously did not do in this structure. For example, I did not use any GU pairing, which is feasible, and if you use GU pairing then you can make the plus-minus strands look rather different and thus provide a basis for selection between them and give one an advantage over the other.

A. M. KAYE: I would like to raise another question, one that we talked about a generation or so ago, stemming from the considerations that Oparin and others originally raised in defining the actual system, i.e., the size of the system, its boundaries and the provision of membranes to define an outside and inside. Have you considered in any of your games the size in terms of numbers of molecules that you would need in order to keep a system going? Have you attempted to put into any mathematical formulation the necessity for something like a membrane or boundary to keep what is evolving from simply dissipating in a larger structure? Have you put any of this into a type of mathematical formulation?

M. EIGEN: One can show, once a limitation is set by the accuracy of reproduction, that one cannot start with a very long piece of RNA. One rather has to link up several smaller pieces which still can be observed today in the gene structure. The next step is to build up a cyclic network, but such a cyclic network is very vulnerable if it does not become compartmentalized. In the cycle nucleic acids code for some enzymes which provide the coupling among all nucleic acids until the cycle is closed. Any mutation, let us say, which occurs in a given nucleic acid, could be of direct advantage to the cycle if it enhances the coupling, but it cannot be of advantage if it changes the enzyme which provides the next coupling. In order to favor the mutant, this mutation would have to propagate through the whole cycle in order to reproduce itself rather than the previous "wild type."

On the other hand, as soon as one encloses the cycle within a compartment, one can utilize any type of selective advantage. So one could visualize two things. One could say that whenever there is a possibility of compartmentation, the compartments will have an advantage over the dispersed system. It is important to start in the soup in order to test all possibilities, but as soon as the interaction starts, compartmentation is of advantage. The same holds for preventing damage from bad mutations which would kill the whole system. In such a case they would damage only a single compartment. We could even imagine that evolution started with compartmentalized systems, but the probability is low to find the right conditions for self-propagation inside a given compartment. Compartmentation is only a *consequence* of the evolution of any cyclic reaction network. Compartmentation and individualization—wherever they occur—have a selective advantage over the randomly dispersed state and therefore evolve. If dilution is a serious problem, which is certainly true for the bulk of the soup, one might imagine membrane or interface bound system to begin with.

As to your question about the size, we have—with a minimum size of translation apparatus—a classification of at least four different types of amino acids, let us say polar, non-polar-hydrophobic, positively and negatively charged. Therefore, we would need at least four translation enzymes, that means four enzymes which have specific recognition for given amino acids and link them to some carrier of an anticodon, i.e., a nucleic acid.

If this were the only requirement one could build a very nice cycle which does not contain more than four nucleic acids and four proteins. The four nucleic acids have eight functions. This means that four of them code for the proteins and are at the same time transfer factors, precursors of t-RNAs.

The protein, once it recognizes the transfer RNA, can be a translation protein, but at the same time it can be the coupling factor to link them together simply by enhancing selectively the survival chance of these four nucleic acids in competition with others. I doubt seriously that this minimum size system of four nucleic acids and four proteins could work; I doubt that a classification code could be as simple as specifying only four classes of amino acids. Furthermore, the first transfer factor might not have been the present t-RNAs. Which are the most abundant nucleic acids in the soup? The most abundant ones are mononucleotides, the next ones are dinucleotides, trinucleotides, tetranucleotides and so on. Now a tetranucleotide in the soup is present at appreciable concentration and with such an abundancy that there is no necessity to code for it; it is always around. In such a tetranucleotide three positions would be used for coding (anticodon) and the fourth position would couple with the amino acid. Then one can build up a whole cyclic network, but does not have to code for transfer factors, and one can also have enzymes with fairly small recognition sites. There is an interesting finding which was made recently. Dieter Soell and Don Crothers analyzed larger numbers of t-RNAs and found that the fourth nucleotide from the end always has a very definite relationship with the anticodon. In other words, they could identify such a correspondence and it might well have evolutionary significance.

We may, furthermore, say that in a primordial soup A probably has been the most abundant nucleotide, perhaps followed by inosine, its first oxidation product, and then decreasingly by G, U and C. Now if in a primordial soup A is most abundant, I would expect this to be reflected by the structure of the code. The mentioned correlation with the fourth nucleotide seems to indicate such an effect. But, of course, this all is speculation.

A. LEWIS: I would like to direct two questions to Saul [Spiegelman], that might be related. How much of the sequence of the MD-variant corresponds to known sequence of QB and how much does not? QB is separated by repeated transfers, serial transfers and by dilution from the MD-variant which was much shorter; how do you explain this?

S. SPIEGELMAN: In answer to the first question: above 25 nucleotides in the 5-prime end are almost the same in QB and the MD-variant. Now since only about 15% of the QB sequence has been worked out, we simply do not know what the rest is like. Now, I cannot answer your question why we never got anything smaller. We tried by all kinds of tricks, as you know, and Ed Rubin was the one who did the plus one experiments, but this did not work. We found the MD-variant as a contaminant of the QB enzyme and there are a number of them, some of which are smaller than NBV_1, but we did not use them because we could not separate the plus and the minus strands. They are made in exactly equal proportions and therefore you never get one without the other. That is why we did not use them, but I really do not know the answer to that question.

G. PIECZENIK: Would you tell us what these selective constraints are.

S. SPIEGELMAN: It is quite clear what must be going on at least with the RNA viruses. One of the things that puzzled us early in the game was finding that people who thought there were double-stranded replicative intermediates were wrong—though that was the simplest way to look at it. However, it turns out that the replicative intermediates which exist in C2 are open structures. This means, therefore, that the nucleic acids are expressing their structural phenotype while they are multiplying in the cell, and therefore selection is still going on in the same way that it did before cells were evolved, at least with this group of viruses.

Now the same is also going on in terms of the structure of messages, because messages which are RNA molecules also have structure, and they are expressing structural phenotype while they are being translated. There could be selection going on at that level which has obvious consequences requiring more than obvious analyses.

O. KEDEM: Can any fluctuation theory really describe a switch from one given state to another, when all the possible states form a preset pattern?

G. OSTER: Nonlinear thermodynamic systems can certainly exist in any one of a number of stationary states. In other words, there can be many equilibria in state space; some of them are stable and some of them are unstable. Fluctuations can certainly cause the system to switch from one equilibrium to another.

O. KEDEM: You mean fluctuations from one stationary state to another?

G. OSTER: Yes. Or even from one type of limit cycle to another limit cycle.

J. ROSS: Let's not call that a stationary state, because a stationary state is a non-equilibrium state in which the macroscopic parameters are invariant with time. This is merely a definition, but I think we had best adhere to it. The limit cycle is not a stationary state.

G. OSTER: Yes, of course. I meant that in state space one can have many stationary states and, in addition, various periodic orbits which, although they are not stationary states, are stable configurations. The system can switch from one of these stable configurations to another.

Y. Y. ZEEVI: With reference to the discussion on the multiple limit cycles of a biological system, it is unlikely that the system will depart far enough from one limit cycle—say a stable one—as a result of random fluctuations, to arrive at the next (unstable) limit cycle. Such "jumps" may, however, be mediated by enzymes which provide the system with a machinery for specific switching. The point is that rather than having order through fluctuations as Professor Prigogine suggests, one may have order through switching, which changes the topology of the system. This was in fact the point that George Oster was trying to make. Thus the evolution of a new molecule may change the topology of the network. Furthermore, the amount of information carried by a molecule is context dependent in the sense that there are topological constraints on the biochemical network level, which limit the subset of "acceptable" molecules, and in the sense that further optimization of sequencing is determined by the function of the network by means of natural selection. Thus there is more to a biological system than its stability, and more to a biological structure than its dissipative nature.

J. ROSS: I think the word fluctuations is being used here in a silly way, because if I have a limit cycle in a given situation or any reaction mechanism, be it stable or unstable, I will have certain fluctuations; they will either decay or, under certain conditions, they can grow.

G. PIECZENIK: With reference to Eigen's system, would you say it is theoretically possible from your conceptual framework to predict a nucleotide sequence *a priori*?

J. Ross: You are putting me now into a situation of defending that one particular theory.

G. Pieczenik: Is there anything in the physical chemical formalism that allows you to predict whether one nucleotide sequence will be the optimum or not?

J. Ross: I don't know the answer to that; the only thing I can tell you is I would be very surprised if there would be anything else.

H. Atlan: If I understand correctly what has been said, I think it's a difference in conceptual frameworks, which might be philosophical although it's important. The question is how to include newness in the evolution of biological systems, and not only biological systems on the scale of the evolution of species, but even at the cellular level. If one tends to think that there is no real newness in the evolution of biological systems, then one has to treat it as a machine, as you said, and as Jack Cowan said. However, there is a tendency to think that it is a machine, but a machine of a different kind in the sense that it can evolve toward new states which could not be predicted before, as Eigen attempted to show. More precisely, the evolution toward *some* state can be predicted, but which particular state cannot. Thus, if one tends to think that the evolution of biological systems, or at least of some biological system, can include newness, then the only way of including newness without calling upon some metaphysical force is to call upon fluctuations, i.e., to call upon randomness.

S. Spiegelman: I do not think anybody denies that the laws of physics and chemistry apply to living things. However, if a theory is proposed which is supposed to describe the evolution of something analogous at least to living material, the physical chemistry, if it doesn't contain the potentiality of genetics, is not going to give us what we want. And no matter how elegant the physical chemistry is, it can certainly not explain life.

G. M. Edelman: It's not fruitful to discuss such a diverse and vast set of entities in so abstract a fashion. We are really discussing the largest problems of biology using a bunch of abstract terms which are hopefully appropriately mapped onto some existing terms. This does not appear to be a very practical procedure. It would seem to me that sometimes abstractions are the most stupid way of saying something —it is preferable to pick an example such as Ora Kedem mentioned. Let me merely say that, although I do not think most biologists would admit that they fully understand everything, even in the genetic sense let alone the chemical sense, they have evolved a set of concepts which are operationally useful for the description of informational systems in biology. Those informational systems depend upon chemistry, and therefore the boundary conditions of chemistry must not be exceeded. But the chemistry *in itself,* as far as the biologist uses it, does *not* give him all the information.

It seems perfectly valueless to simply substitute the word fluctuation as a kind of metaphor for what is known to be an operational term in genetics called mutation. The information in molecular biology has to do with the sequences obeying a triplet code. If, in certain instances, you can describe a population and the phenotypic character that comes from the alteration of that code and if you then go down to the chemical level and can describe the alteration in the code to the letter, then in fact you have both necessary specifications, the structural and the informational. Now what we obviously do not have is a full description of the environment, which is uncoupled in some sense and through which the selections are made. The principle describing this (and certainly it is the largest one that I think biologists know) is called natural selection. It is not in itself, in the first instance, a chemical principle. The question might be raised whether one can reduce natural selection to principles of chemistry simply, but the sister question also arises as to whether that is operationally very useful.

I should like to make one other comment. Any phenomenological theory that you construct to describe a system is useful if it helps you to understand it. But it does not give you a way of specifying how to construct such a system—the question is whether the theory has very much value. In my opinion, genetics and evolutionary theory so far do not give us complete specifications on such a system.

Finally, I would think that in immunology, for example, what Ora [Kedem] said is quite to the point. We need no equivalent mapping language. We need chemistry but we do not need the language of dissipative functions and all of this in order to describe how information evolves in the immune system. What we need simply is—first of all, a specification of the kinds of information, that is to say the sequences and three-dimensional structures of the antibody molecules that have evolved within each species. Secondly, we need a description of the threshold conditions under which each cell can be triggered when a binding event takes place and, thirdly, we need (and here we are closest to your point) some description of the space in which the antigen molecules encounter a particular repertoire of antibodies.

J. Ross: Gerry [Edelman], I fully respect this, but when one is trying to build bridges between different types of information, one needs some simplifying theories even though the systems are very complex.

G. M. Edelman: Don't get me wrong. I think one should attempt such a theory. This is not the issue. The issue involves two things: the specification of the particular detailed structures in the system, and then to what degree you wish a correspondence with operationalism. That is all. Categorical denials of physics or of biology have nothing to do with it. To give you a very concrete example, very early in immunology, when we learned that the immune system could recognize so vast a variety of different antigenic structures, the feeling was that such a system could not operate except by instruction. That is to say, by the direct involvement of a template, of the geometric shape of that which is to be recognized in the building process. That theory turned out to be false. Then we said, well, we must have this very large prior reservoire, very much larger than that universe of objects which is to be recognized and all the specificity already resides in the antibody molecules. It turns out operationally that the system works by a series of selective events in which first the antigen binds to the antibody, and then a threshold phenomenon takes place followed by triggering of cells with high affinity antibodies. You need an elaborate cellular system as well as the antibodies to obtain so high a degree of chemical specificity.

We are now struggling with the parameters, to arrive at a parametric description of that specificity. I see no point at this stage of the inquiry either to make the language very much more abstract or even so precise that it exceeds the precision of our measures.

G. YAGIL: I would like to try to approach the same problem from a different angle. Information theorists have repeatedly pointed out that the information content of a physical system, as defined by them, is identical with the entropy, or rather negentropy, of the system. This means that information content is in principle a measurable quantity (e.g., by integration of heat capacity from absolute zero). To what extent is this true and helpful for the type of biological information discussed here? Recent direct measurements of the entropy of formation of several proteins by Hutchens and co-workers (*J. Biol. Chem.* **244**:26 (1969)) illustrate some important points relating to this question. The entropy of formation of the "informational" molecules, chymotrypsinogen and bovine serum albumins from amino acids (9.3 and 9.0 e.u. per peptide bond at 298°K), is found to be only slightly less than the entropy of formation of a simple dipeptide-like glycylglycine (10.3 e.u.). This means that there is no contribution of a term such as $-R \ln 20(-5.97$ e. u. per amino acid residue), commonly referred to as the informational contribution to ordering. One way of explaining this is that in a physical measurement one relates the entropy change at a certain temperature to that at $T = 0°K$ where, according to the third law of thermodynamics, $\Delta S = 0$, i.e., to a state of complete order. When considering a biological entity, such as a cell, a virus, or an ordered molecule, we are interested to know not how much *less* ordered it is than a completely ordered equivalent system, but rather how much *more* ordered it is than a sometimes ill-defined, equivalent non-living system. Such a reference system may be a mixture of random polypeptides and/or random polynucleotides, or a mixture of nutrients at suitable concentration or, as in the Morowitz treatment, merely a mixture of C, H, N, O, P and S atoms. There are two points to be noted. First, the further removed the reference state is from a biological state the more predominant will be the contribution to information content by physicochemical states of little specific biological interest (vibration-rotation states, etc.). Second, the definition of a non-biological reference state as close as possible to a biological state may require some arbitrary decisions, but will be a necessary step before biological information can be regarded as a biologically useful as well as a physically quantitative concept.

A. SILBERBERG: A question about the usefulness of network theory. I thought it would be appropriate since we have two representatives of this point of view here if they could summarize for us where they feel the network is important and useful.

G. OSTER: Let me first say that I am appalled by much of the loose talk that goes on about entropy, dissipation, etc. My feeling is that concepts such as entropy and entropy production are simply not discriminating enough to be of any real use in biology. In discussing dissipative structures I get the overwhelming impression that their relationship to biology is allegorical at best. I consider what we have done with network theory and thermodynamics as theoretical chemistry; I would not presume to call it biology.

Along these lines I would like to address a question to the audience: It is not clear to me what the role of mathematical theories in biology is or should be. Can anyone think of any fundamental advance in biology that was made solely as a result of a mathematical theory? I can't! The Hardy–Weinberg law is a trivial example; surely we can do better than that, at least on the molecular level.

Perhaps an example will clarify this point. A long time ago many people performed calculations that showed that mutation rates should be very high at the tops of mountains because of the high radiation load there. And yet when they went and examined the butterflies up in the Andes they found that mutation rates weren't any higher there than they were at sea level. The reason was, of course, that there was a little repair enzyme that went around and fixed up the DNA whenever it got broken. I maintain that something like that simply cannot be predicted on theoretical grounds. It's a special little device that evolved, and mathematical theory doesn't encompass this sort of framework. As a theorist, I find this very distressing. But it seems simply to be a fact of life that theories do not predict special gimmicks and that biology seems to be constructed out of special gimmicks!

J. ROSS: But that's true in chemistry too! There's not a single theoretical calculation of any chemical reaction rate. So what is the argument?

G. OSTER: When we talk about dissipation and patterned structures in space, I think we may be engaging in a sort of Gestalt thinking. We say it looks like biology but perhaps we're saying this to keep ourselves in business.

F. O. SCHMITT: Among NRP Associates, Aharon [Katzir-Katchalsky] took an active part in discussion of the biophysico-chemical basis of contemporary molecular neurobiology, including membrane bioelectric processes underlying the self-propagating action-potential wave, synaptic processes, and the role of specific cellular structures such as the fibrous constituents of neurons. In addition to these aspects of neoclassical neuroscience, the high function of the brain, including the permanent storage and recall of memory, learning, thinking, awareness, and consciousness—subjects with which NRP is deeply involved—also interested Aharon from the earliest beginnings of NRP. He sensed that there may be some underlying theory for such phenomena and that it would be of greatest importance for neuroscience, indeed for the whole of science, to discover such fundamental brain processes.

The situation in contemporary neuroscience resembles that during the classical period of genetics when much sophisticated information about the location and interaction of genes was available, but nothing was known about the chemical nature of genes or of the biochemistry of their replication and their role in cell synthesis. This was before the theoretical breakthrough of molecular genetics with its DNA–RNA–protein transcription and translation.

We seek a similarly clarifying theoretical breakthrough in the neurosciences which presently, in their neoclassical period, deal with partial systems that, though leading to penetrating new insights into the function of these systems, tell little about higher brain functions. Aharon was most helpful and stimulating in this search for new global principles, and for this reason he helped organize and co-chair a Work Session, held on May 14 to 16, 1972, entitled "Dynamic Patterns in Brain Cell Assemblies." The hope was that such advanced notions of biophysical chemistry, such as dissipative structures, dynamic patterns and instabilities, might bear on, if not explain, the steady, DC potential shifts in the brain observed when the animal correctly performs an act for which it was trained. Although the session fell short of this ambitious goal, the published monograph will stimulate many perceptive neuroscientists and will recruit interest in the problem on the part of many young biophysical chemists of high competency and originality.

Indeed, from results presented at a Work Session entitled "Local Circuit Neurons" (held on June 7–9, 1973), it is clear that standard neurophysiological concepts of synaptology and the role of neuronal circuits in brain function probably apply to but a fraction of the cells of the brain. A larger fraction, located in the cerebral cortex, retina, olfactory bulb and other cortical structures, may follow quite different biophysical and biochemical processes, in which relatively small potential gradients applied in both directions at neuronal junctions, particularly of the dendrodendritic type, may play a central, as yet unexplored role in information processing and possibly in producing the higher functions of the brain. Aharon would have been thrilled by these results in relation to his own ideas and to his Work Session and would doubtless have been stimulated to new and original ideas straight away.

The Work Session entitled "Dynamic Patterns of Brain Cell Assemblies," with its dramatic epilogue on May 17, 1972, was one of Aharon Katchalsky's last scientific efforts. Its impact on his audience was great and, in published form, constitutes a worthy part of a long list of original contributions from Aharon who was possessed of so many qualities of originality, rigorous science, charm and leadership.

Last night Jack Cowan gave a theoretical approach to the problem, and he spared us the mathematical details. In order to get this important subject before us this afternoon, I have asked him to epitomize briefly his network theory.

J. COWAN: Perhaps I didn't explain very well why I thought there was a problem at all. In physics and chemistry it seems to me that one goes from observations of molar systems to attempts to infer microscopic properties. In the nervous system one encounters a peculiar problem; it's different. You can stick microelectrodes into the nervous system and you can record cellular properties. You can also record large-scale electrical behavior in the nervous system and evoked potential electrocephalograms and so on. The single unit studies tell you very little about the detailed machinery of the nervous system. They provide much more important information than the global observations but there is an enormous gap between the behaving organism and these cellular properties. Therefore, one of the big problems is to go from the microscopic level, in which the logical atoms are not molecules but cells, to something more molar than that. I tried to show you the consequences of a theory based solely on the simpler properties of excitatory and inhibitory nerve cells spatially organized, connected together in aggregates into what I called sheets. I showed you that from the simpler properties of nerve cells, you obtain various constraints on the spatial responses of such sheets of tissue, and various constraints on the temporal responses, and I described how the existence of threshold properties of nerve cells made possible stable modes of behavior, which could account in molar terms for the existence of functional units. I talked a bit about the functional units, the components of the machine; what are they—are they single cells, are they subcellular, are they enormous populations of cells? We don't know anything really about the specificity of the nervous system in functional terms. We have no real hard information about this. I think it's not yet appropriate even to begin to attempt analysis in molecular terms when we don't know how the machine is constructed.

I then tried to give you some feeling for how one could apply such theories to sensory processes and I ended up by noting that for longer time scales than the 100 millisecond time scales that I talked about, even more complicated theories are required and that it is possible that molecular information will be extremely important for such time scales.

P. HILLMAN: First of all I think it should be recalled that there is an alternative way of narrowing the gap between single cell activity and animal behavior, and that is by choosing a relatively simple animal. There is now a large number of examples where, in simple vertebrates, we are able to relate the behavior of the animal as a whole to single cell triggers or receptors or motor units, and so on.

J. COWAN: You've told me something about invertebrate nervous systems, but what about real nervous systems?

F. O. SCHMITT: Early in the history of NRP we held a Work Session to determine the simplest living form which exhibits memory and learning. If, like the geneticists, we could find sufficiently simple organisms—comparable to the bacteria and viruses used by geneticists—the search for the physical and chemical bases of these psychological characteristics would be greatly simplified. T. H. Bullock, perhaps the world's leading neurobiologist of invertebrates, organized the Work Session which, held in 1965, was entitled "Simple Systems for the Study of Learning Mechanisms."

The conclusion of the Work Session was that the lowliest animal that shows what psychologists would call "learning" is a worm. The nervous systems of such animals are quite complicated. Therefore, no short cut to the study of the learning process is likely to come from a study of the lower organisms.

P. HILLMAN: I would dispute the unreality of invertebrate nervous systems, but let's not go into that. The other part of my comment is a very general plea I would offer to Jack Cowan and other theoreticians. I would be very happy if occasionally one of these theoreticians were not to say "I've got one theory which explains everything, but rather, I've got two theories—one of which doesn't explain everything."

J. COWAN: It took 300 years to produce a single theory that can account for the stability of atoms and molecules; it will take at least that length of time to produce a single theory that will begin to account for the behavior of the nervous system. It's trivial to produce a thousand theories to handle some bit of experimental data; it's very difficult to produce one, let alone two, that will handle a lot of data.

F. O. SCHMITT: I would like to emphasize that only a fraction—possibly only one-third—of the neurons of the brain reflect the standard textbook characteristics of synaptic function with excitation (or inhibition) passing post-synaptically from dendrite via the cell body and axon hillock, leading to the firing of an action wave in the axon, and this, in turn, to a reaction at subsequent synapses. This pattern characterizes "through" neurons which pass over preformed circuits and mediate excitation over large distances (through white and gray matter). Probably a much larger fraction of brain neurons never leave the gray matter and are called "local circuit" neurons because of their close interaction (largely dendrodendritic) by graded potentials, as shown by the work of Gordon M. Shepherd ("The Olfactory

Bulb as a Simple Cortical System: Experimental Analysis and Functional Implications," in *The Neurosciences: Second Study Program* (F. O. Schmitt, Ed.-in-Chief), pp. 539–552, The Rockfeller University Press, New York, 1970), and of Wilfrid Rall ("Dendritic Neuron Theory and Dendrodendritic Synapses in a Simple Cortical System," *ibid.*, pp. 552–565).

Axonless amacrine cells and short-axon neuronal cells play a large role in this close interaction which may reflect molecular computations involving transmitters still largely unknown. Prototypical of such local circuits is the retina (a piece of externalized cortex) with its five interacting cell-types which lead to no action potential until the last type, the ganglion cell, fires action potential into the optic nerve. Similarly prototypical are the olfactory lobe, the cerebellar cortex and, in all probability, the neocortex generally. As Shepherd puts it, the local circuit, cluster interaction of neurons may prove to be the *"E. Coli"* of neuroscience, meaning that it may lead to fruitful new data and concepts.

It is a question pending for a long time whether a special type of recognition protein, comparable to the gamma globulin of the lymphocyte, may be synthesized by such local circuit neurons; this possibility is raised the more strongly by your presentation.

N. K. JERNE: Let us say that the immune system—which after all cannot compete with the nervous system in importance—can make ten million different proteins. Might that not be beyond the nervous system?

F. O. SCHMITT: Perhaps not.

G. M. EDELMAN: Niels [Jerne], if I understand the idea correctly, another animal of the same litter does not produce the same kind of antibodies and it virtually never could have the same network. So what you are postulating is a degree of uniqueness which even exceeds that of individual brains. You are not going to find interconnections that are at all alike after a period of time. Is that correct?

N. K. JERNE: I think so, yes. Is it known that brains are alike?

G. M. EDELMAN: Well, that is another question! My second point related to your question of whether the system would collapse.

N. K. JERNE: Whether the system would spontaneously tend to divert? Or whether it would collapse?

G. M. EDELMAN: I am asking whether it would be a stable network. A couple of points that you made could be extended. In the first place, you mentioned that you have a number of the order of 10^{10} of each kind of molecule per ml of blood. That is certainly not questioned. If, however, on each cell you have 10^5 molecules of the same kind and these can cluster, then of course you can get enormous enhancement effects. If, for example, such recognizing molecules for idiotopes are on 'T' cells, are identical and can cluster, you have no entropy of mixing. You are not putting antibodies out into the body; you have them all concentrated on one cell which goes right to the other cell. So I would make the prediction that antibody molecules on cells and free antibodies would have entirely different effects on such a network.

N. K. JERNE: They would be more effective. But they would collide less often.

G. M. EDELMAN: Yes. And therefore I would make the closing proposal, in view of the first comment. A nude mouse—lacking 'T' cells to a large extent—should have a quite different structure for its network than an ordinary wild-type mouse. The question is, could you think of an experiment which might bring out such a difference? I would expect, for example, a nude mouse—lacking that 'T' cell repertoire—to really lack a very effective part of the network in the light of your ideas. So we could try to verify whether there are really these differences between the nude mouse and the wild-type mouse. What do you think?

N. K. JERNE: I cannot suggest any experiment. I heard that Abraham Weiler is doing cell transfer experiments to check the idea that the introduction of cells to another animal of the same mouse strain—that is not irradiated—usually doesn't succeed. The usual explanation has been that the animal is already full of cells so the incoming cells have no chance. But he is wondering whether they are perhaps not admitted because the network is already established. However, I cannot really suggest a good experiment at the moment.

G. M. EDELMAN: I would venture a guess that in fact the network is stable, or rather has buffering capacity. If the network was sitting, as it were, near a point where it was about to collapse, you would have disastrous behavior.

N. K. JERNE: If that is true there would be buffering, enormous buffering. But if you decrease one set you increase the set that is inhibited by it, and thereby the set tends to come back, in general.

M. EIGEN: There is an analogy if you replace your cells by enzymes. An enzyme may at the same time be the target for another enzyme, where the coupling is of a catalytic type. In other words, it is not that one substance makes another and disappears. Here one substance helps to make another. And if you then put them up to a network, you first can show that any self-organizing property, memory, whatever you want it to do, requires cyclisation. Secondly, you could say, if the network includes only stimulatory effects, that it could not evolve unless it is nonlinear. That is why we had to introduce for our case a hypercycle, which is inherently nonlinear. If, however, you include stimulatory and inhibitory cells you could build up simpler networks.

N. K. JERNE: However, we know that there are experiments which show that sometimes this interaction is inhibitory and sometimes it is stimulating.

J. ROSS: May I make an addition to that. If you think of every one of the circles as a localized instability, in a stable medium, then you find that, if you have a set of such instabilities, you can get wave phenomena going across, dependent on the density—that might just be another analogy between the two systems that you were talking about. This is a strictly density-dependent phenomenon.

J. COWAN: I think I understand this in general, but I want to see if I can clear up one point which I find a little obscure. There is another analogy which I think is deeply related to network models of the nervous system, and perhaps to the immune system. This is the analogy with error-corrected codes as used in information processing systems. In such systems one wants to minimize the equivocation or errors that might be made in response to noise either in the system, or in the stimulus. For example in storing programs in a computer memory bank, one wants to be able to retrieve such programs in as correct a form as possible despite the possibility of failure and malfunctions of components in the memory bank. To do so one has to build in special kinds of redundancies or degeneracies into the way programs are stored in the memory. It turns out that these redundancies have to be implemented by networks in which the mutual inhibition or suppression of elemental activity is primary. In neural net terms these would be called feedforward or feedback inhibition mechanisms.

My question in regard to the immune system is, what is the nature of the degeneracy, if any, that ought to exist in the immune network, to enable it to function in the reliable fashion it seems to exhibit?

F. O. SCHMITT: Last evening Dr. Werman discussed a subject of neurophysiological interest. In order to turn our attention this afternoon back to this subject, I suggest we discuss the problem of the role of receptors and their linkage to cellular effectors. The first step in the process is the binding of the ligand (e.g., hormone or transmitter) to the receptor, usually a protein with very high affinity for the ligand. The binding process, by intermolecular linkage, stimulates adjacent molecules of the enzyme adenylate cyclase to form 3′, 5′ cyclic alnosine monophosphate (cyclic AMP) which, in turn, forms the substrate for a cytoplasmic enzyme, protein kinase.

By processes still far from clear, the resultant phosphorylations activate many vital cellular processes, including, according to P. Greengard, the excitation of the neuronal membrane. The mechanism of the transduction from the ligand-bound receptor to the rest of the system is poorly understood. Dr. Stuart Edelstein has been purifying the cholinergic receptor protein and has ideas about how the system works.

S. EDELSTEIN: The basic point of view that I would like to convey reflects that of a biochemist who is interested in working with macromolecules but hopes to contribute something with regard to the nervous system. In particular, I want to talk for a few minutes about the acetylcholine receptor of the electric eel and describe two minumum criteria for studying the receptor. One is the necessity of having a quantitative theory of action to describe the receptor in the membrane, and the other is the need to have a way of relating the parameters of the quantitative theory for the membrane receptor to properties of the isolated macromolecular receptor.

Concerning the problem of obtaining a quantitative theory to describe the properties of the receptor, I would like to consider several pieces of data taken from the literature [1] as shown in the upper part of Fig. 1. Included are dose-response type curves presenting the course of depolarization as a function of concentration of ligand added. Data are shown for two different ligands, carbamylcholine, abbreviated CC, and decamethonium, abbreviated DM. The depolarization curves for both ligands are cooperative with Hill coefficients of approximately $n = 1.8$. However, a striking difference in the behavior with the two different ligands is the extent of depolarization: the extent of depolarization with decamethonium is considerably less than the maximum depolarization obtained with carbamylcholine. In addition, the absolute affinities are considerably different for the two ligands and, if the dissociation constant is estimated by the concentration at half maximal response, values of $3 \times 10^{-5} M$ and $10^{-6} M$ are obtained for carbamylcholine and decamethonium, respectively. Another striking feature of the system is that, in the presence of certain inhibitors such as the affinity label *p* (trimethylammonium) benzene diazonium difluoroborate (TDF), the depolarization curves are shifted to the right and the maximal depolarization is depressed, although again the extent of depolarization differs widely for these two ligands.

The various aspects of the depolarization curves from individual cells of the electric eel can be represented in a surprisingly simple and straightforward manner by the two-state or allosteric model [2]. The basis of the allosteric model is the existence of two conformational states (abbreviated T and R) with the two states in free equilibrium and the T state favored over the R state in the absence of ligand. The quantitative relationship between the population of the two states is given by the allosteric constant L, where $L = (T)/(R)$. Cooperative effects are generated if ligand is added and the R state has a considerably higher affinity for the ligand

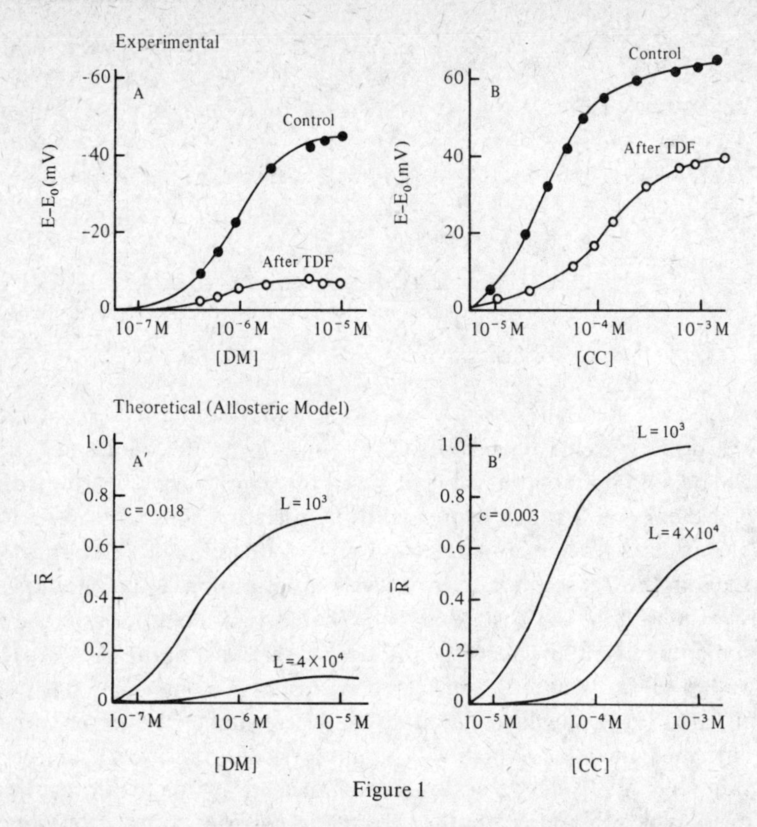

Figure 1

than the T state. In this case, although the population of macromolecules starts in the T state, with the addition of ligand the conformational equilibrium is shifted to the R state producing cooperativity.

With respect to the properties of the acetylcholine receptor, we assume that the important feature of the receptor is not the degree of saturation, but the extent to which molecules are either in the R or T state. We assume that the ion fluxes associated with depolarization are tightly coupled to the conformational state of the receptor. The function of state is given by the equation for $\bar{R}$, the fraction of molecules in the R state:

$$\bar{R} = \frac{(1 + \alpha)^n}{(1 + \alpha)^n + L(1 + c\alpha)^n}$$

The $\bar{R}$ function has a number of interesting properties which are summarized in Fig. 2. In the absence of ligand, the state function for very low values of L will be unity, indicating that all molecules are already present in the R state. As L is increased in the absence of ligand, a larger fraction of the population of molecules exists in the T state, and with values of L of about 100 effectively all of the molecules are in the T state.

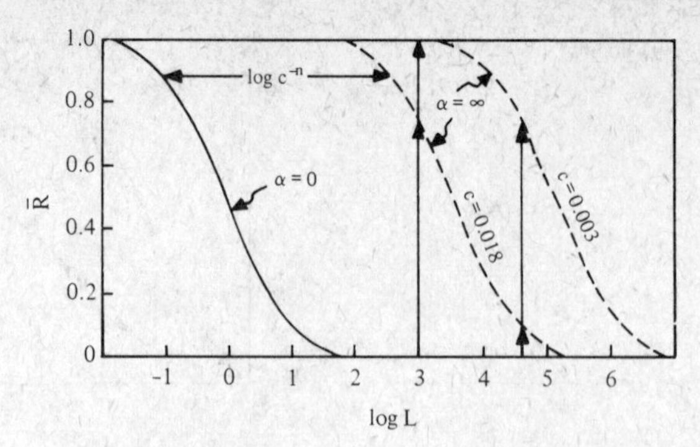

Figure 2

When the other extreme is considered, i.e., fully saturated molecules, the state function will depend both on the value of L and the parameter c, where c is the ratio of affinity for the R and T states, expressed by dissociation constants as $c = K_R/K_T$. When values of L are extremely large for a given value of c, the state function will remain locked in the T state even at high levels of saturation. This condition reflects the excessive stability of the T state and the inability of the advantage the R state gains from binding ligand to overcome the intrinsic stability of the T state. However, at lower values of L the addition of ligand will cause a conformational transition to give saturated molecules in the R state. Between these extremes of values of L, there are intermediate values which will permit a partial transition from the T state to the R state. In Fig. 2 such values of L are indicated by dotted vertical lines, and the positions at which they intercept the two curves (corresponding to different values of c) can be used to explain the depolarization data originally shown in Fig. 1. For example, if we assume that the two ligands, decamethonium and carbamylcholine, have slightly different relative affinities for the R and T states, with values of $c = 0.018$ for decamethonium and $c = 0.003$ for carbamylcholine, the extent of transition to the R state upon ligand binding will vary. For a value of $L = 10^3$ addition of ligand will result in only a 70% transition to the R state for decamethonium, whereas the same value of L will achieve 100% transition to the R state on the curve for carbamylcholine. This effect could therefore be used to explain the differing degrees of maximal depolarization with the two ligands. Also striking is the fact that a single change in L, raising its value to 4×10^4, would then bring the system to new positions on the two curves. In this case the extent of conformational change on the curve for decamethonium is only in the 10% range, whereas the curve for carbamylcholine is now in the 70% range, corresponding to levels of depolarization found in the presence of TDF.

When the various values suggested from Fig. 2 are used to generate model curves, the results shown in the lower part of Fig. 1 are obtained. The agreement between the experimentally obtained curves and the theoretically generated curves is excellent and supports the view that the varied effects in terms of differing degrees of depolarization with different ligand and addition of inhibitors can be readily explained by a simple formulation of the allosteric model based on the properties of the state function. Therefore, we can conclude that the first requirement for understanding the acetylcholine receptor has been met: the availability of a quantitative theory to describe the properties of the receptor when it is intact in the membrane of the cell. We can now turn to the second aspect of the problem: the relationship between the properties of the intact receptor and those of the isolated receptor. In this regard, we consider the possibility that the isolated receptor will spontaneously relax into the R state. The acquisition of the R state by many dissociated allosteric enzymes is well documented and similar changes occur with the allosteric protein hemoglobin [3]. The binding properties of the R state can be related to the value of L obtained for the system under conditions where cooperativity is maintained according to the relationship $L = (\alpha_{1/2})^2$ [3], where two sites are assumed to be involved in the interaction. For all of the analysis presented here, a two-site system has been assumed since Hill coefficients greater than 2 have not been observed. The term $\alpha_{1/2}$ refers to the concentration of ligand (X) at half maximal response, where the concentration is normalized to the dissociation constant of the R state, $(X)/K_R$. If the receptor is in the R state in the isolated condition, we could expect the affinity to be higher than that observed for the intact cell by the factor $\sqrt{L}$, or about 30. As shown in Table 1, the data obtained for the intact cell can be divided by 30 to give values expected for the apparent dissociation constants for binding of these ligands to the isolated receptor if it is relaxed in the R state, and values of $10^{-6}M$ for carbamylcholine and $3 \times 10^{-8}M$ for decamethonium are predicted. It is of interest to note that results very recently obtained by Meunier and Changeux [4] on the receptor isolated from the electric eel are in close agree-

Table 1

PROPERTIES OF ACETYLCHOLINE RECEPTOR

A) *In Electroplax Membrane:*

Carbamylcholine $K_D = 3 \times 10^{-5}M$

Decamethonium $K_D = 10^{-6}M$

Allosteric parameters ($n = 2$):

$L = 10^3$; $c = .003$ (CC); $c = 0.18$ (DM)

B) *Isolated Receptor:*

Allosteric parameters ($n = 2$):

$L = (\alpha_{1/2})^2$; $\alpha_{1/2} = 30$; $K_D = K_D$ (Electroplax)/30

	Predict	*Observe (Meunier + Changeux)*
Carbamylcholine	$K_D = 10^{-6}M$	$K_D = 2 \times 10^{-6}M$
Decamethonium	$K_D = 3 \times 10^{-8}M$	$K_D = 2 \times 10^{-8}M$

ment with these predicted values. Therefore, it appears that the second criterion we suggested, establishing a relationship between the receptor in the cell and the isolated receptor, can also be satisfied by the apparent agreement in the properties of the purified receptor with predictions based on the assumption that it exists in a relaxed (R) state.

It is of course of interest to attempt to ascertain the actual structure of the acetylcholine receptor and attempt to relate the functional properties to the structure in more detail. Work in this area is also progressing and evidence for an oligomeric structure of the receptor with a molecular weight of 15×10^5 has recently been obtained [5].

References

[1] CHANGEUX, J.-P., PODLESKI, T. and MEUNIER, J. C., 1969. *J. Gen. Physiol.* **54**, 2255.
[2] EDELSTEIN, S. J., 1972. *Biochem. Biophys. Res. Commun.* **48**, 1160.
[3] EDELSTEIN, S. J., 1971. *Nature* **230**, 224.
[4] MEUNIER, J.-C. and CHANGEUX, J.-P., 1973. *FEBS Letters* **32**, 143.
[5] CARROLL, R. C., ELDEFRAWI, M. E. and EDELSTEIN, S. J., 1973. *Biochem. Biophys. Res. Commun.* **55**, 864.

B. WERMAN: It is always amazing, to someone like myself who is essentially naive, I suspect, how sophisticated theories can explain artefacts, and I'm afraid that I must be unkind and say that in this case this is exactly what occurs. What you've measured is polarization. Now there are several problems relating to measuring polarization aside from conductance which I went into the other night. One of the major problems is that the conductance itself is a non-linear function and you get away from the resting potential, so the polarization, which is a non-linear function of this non-linear function, becomes even more complicated. Another is the observation that the conductance returns to normal while the potential hangs up. So, in other words, you obtain a depolarization in the presence of carbamylcholine while the conductance does not increase. It is not very likely that the polarization results from an interaction of a ligand with its receptor in the absence of a conductance change.

S. EDELSTEIN: I believe that for the data I have been working with there is, in fact, a linearity between conductance and voltage although I am aware that the complications you described have been found in some systems. Therefore I do not think that the reservations you are raising apply to the particular data that we are dealing with here.

M. SELA: Sometimes it's nice to look at old literature. I remember reading once the opening article in Volume 1 of the *Journal of Immunology,* 1916; it was a Presidential Address by Felix Weil who was then President of the American Association of Immunologists, and he summarized the status of the art of immunology by saying that every field passes through three stages. The first one is collecting data, the second is organizing the information into a coherent picture, and the third is getting to a unified theory of the field. He continued by saying—"we have just finished with the first and second stage and are on the verge of the unified theory." I think that today—more than 50 years later—we tend to say the same, and I think we honestly believe that the first and second stages now truly are finished, but I hesitate to say that we are on the verge of the unified theory. At least I would distinguish between the molecular part and the cellular part, because concerning the molecular part I think the information we have today is really solid, it's really translating immunology into good molecular chemical data. Perhaps the main outstanding point concerns the genetics, where we still do not know really, honestly, how the information is generated: whether it is all in the germ line or whether it is by somatic mutation.

When it comes to the cells, and their role in immunology, we are in the most exciting stage where there is a lot of material accumulating. There are enormous feuds. Everybody here tried to steer away from it, but it appeared in a discussion concerning the nature of the receptor on the so-called T cell, whether it is an immunoglobulin or not. I think that even there in a few years, if not earlier, we will know really where we are, and of course this will lead to some other questions. In this context it was very interesting to hear Dr. Jerne because, while he did not present a mathematical theory, I think that the importance of what he had to say was not so much in a mathematical formulation, as in stressing the awareness that you do not have cells running like a gas in an immunological space, but that there is regulation and suppression and so on.

R. Rein: I would like to take a few minutes to comment on some model calculations which we have performed. These calculations were concerned with structural models for receptor site complexes and with the energetics of formation of these complexes. The objective for my comments is to illustrate how quantum chemical calculations, when considered together with pertinent structural data, can lead to an understanding of the nature of forces and structural aspects of molecular interactions at receptor sites.

The system which we have studied consisted of a series of substituted trimethyl ammonium ions which bind to carboxylate-containing antibody sites. Using precipitation techniques, Pressman et al. have determined the free energy of binding for these series, and Burgen et al. studied the effect of binding on the methyl proton relaxation rates in these series. Now that we have this information, a number of questions can be asked. First, what is the functional group in the site which dominates the interaction with the ligand molecule? What is the nature of the force responsible for the binding? What is the geometry of the complex?

One can choose a number of plausible assumptions to answer these questions. The usefulness of the theoretical calculations lies in the fact that it becomes possible to test the consistency of a particular set of assumptions, by comparing calculated quantities, such as interaction energies and barriers to hindered rotations for a given model, with the values which are obtained from experimental data.

Assuming that the dominant force is electrostatic, an accurate calculation of both the interaction energy and the potential of hindered rotation of the methyl protons in the field of the binding site is possible. We performed these calculations in two steps. First, we calculated the wave functions and the lower multipole moments for the interacting ions; in the second step we used the multipole moments to calculate the ion pair binding energies and the electrostatic barrier for hindered rotation. The results of these calculations reproduce quantitatively the rotational barriers for the methyl groups in the field of the complex as obtained by Burgen et al. The experimental value of 2.1 kilo calories barrier height is in fair agreement with the values we obtained for a particular geometrical configuration 1.8 kilo calories. The relative free energies of bindings are small quantities of the order of several hundred small calories, and we obtained values which are of the same order of magnitude. We cannot fully reproduce the order; however, this is not surprising, because additional contributions not included in our model can easily offset the observed orders of small differences.

What does one learn from these calculations? First, these results seem to be fairly consistent with the assumption that the active site residue with which the ligand interacts is a carboxylate ion. Second, that the dominant interactions are electrostatic. Third, one can select a particular configuration that is to propose a structure for the complex, for which theory and experiment are in the best agreement.

In summary, with this illustration, I wanted to draw attention to the fact that chemical calculations can be a useful tool in the characterization of the molecular nature of receptors.

F. O. Schmitt: Yesterday Ephraim Katzir gave us a thought-provoking talk on the effect of the microenvironment on enzyme action. I would like to say a few words along similar lines, but on the microenvironment at the level of the cell. The biophysical chemistry of the cell membrane, a central topic of molecular biology, is highly complex and dynamic. The total ("greater") membrane may be thought of as being composed of two parts: an inner and an outer portion. The outer portion consists chiefly of glycoproteins, glycolipids, and various types of conjugated polysaccharides. These polysaccharide chains probably extend normal to the surface and are highly acidic, due primarily to the presence of one to three sialic acid residues in terminal positions on the chains. In the intercellular space, ca. 150–200 Å in width, the chains extending from the outer layer of the membrane of cells on one side of intercellular space probably interdigitate or interact with chains extending from cells on the other side of intercellular space. In addition, it is thought that, in some tissues, hyaluronic acid and other negatively charged polysaccharides exist. This strongly negative charge on the outer cell membrane is important in determining the microenvironment of the cell and the nature of reaction in intercellular space which obviously is not the simple saline-containing compartment visualized by electrophysiologists. Multivalent cations, particularly calcium ions, play a very important role especially in excitable tissue, but probably in most tissues. According to W. Ross Adey ("Neural Information Processing: Windows without and the Citadel within', in *Biocybernetics of the Central Nervous System* (L. D. Proctor, ed.), pp. 1–27, Little, Brown, Boston, 1969), the impedance of intercellular space responds dynamically to neurophysiological and psychological changes. When an animal correctly performs an act for which it has been trained, the impedance of intercellular space in the brain may fall by as much as 20%. Adey has indirect evidence that rapid displacement of calcium ions plays a role in this process.

Thus the microenvironment of the cell, especially the nerve cell, plays a very important role in determining the chemical reactions within and upon the membrane and, of course, in permeability processes.

Some very interesting data on chemotaxis were presented to us by Dr. Koshland. Does anyone wish to discuss this subject?

J. Ross: (To Professor E. Katzir): I am interested in your comments on heterogeneous catalysis, because we have been concerned with instabilities in heterogeneous catalysis. One can consider two cases assuming very different diffusion coefficients for the substrate (S) and the product (P) in the very system that you have considered. One can show by calculation that, if your membrane is bound, then next to that boundary there will be a stable undulatory spatial structure. Now, assuming equal diffusion coefficients for S and P, with a stable, let us say enzymic, but oscillatory reaction, then you can also show that you obtain in every single case a stable spatial structure around the heterogeneous catalyst.

The other comment that I wanted to make was in relation to your graph of enzyme activity versus pH. That is very interesting because it alone will give you multiple stationary states and permit cyclic behavior by an additional mechanism.

E. Katzir: Let me just add that in the literature there is only one paper on, let us say, the problems of stability as a result of these various reactions which have been discussed, a very vague paper. So perhaps it is worthwhile looking into the possibilities of stable states for various structures.

I. Prigogine: One example is the enzymatic membrane containing urease and glucose oxidase that Daniel Thomas has obtained. In this membrane, although the enzymes have simple Michaelian regulatory properties, several steady state solutions with different functional properties are possible.

O. KEDEM: For instance on the feedback image. If you are on the side of the pH activity curve, while the product increases activity, in principle you should be able to get stability oscillation, propagation and so on. And this has indeed been observed with the various systems which Ephraim [Katzir] has been describing now in Roy Caplan's lab. I understand that calculations along these lines are also being made by Dr. Goldbeter and Prof. Prigogine. You really have to hit just the right range to obtain oscillation between different states. If you are to the right or the left, it is stable.

M. EIGEN: A repressor binds to the operator with a binding constant of about 10^{12} or 10^{13} M^{-1}, and it binds also unspecifically to random DNA-sequences with a binding constant of around 10^6. Now the rate constant of recombination, measured by Suzanne Bourgeois and co-workers, was almost 10^{10} M^{-1} sec^{-1} which is a hundred times faster than expected for a diffusion controlled reaction. We were always wondering whether the effect could be because we have the whole DNA as a target for recombination or whether the increase of repressor concentration near the DNA molecules favors some higher rate of recombination.

J. D. COWAN: You should inhibit binding a repressor by simply adding the deletion DNA.

M. EIGEN: Effects of microenvironment could be tested in two ways. Either one shifts the concentration of repressor to much higher values and follows the rates—which may become quite high, or one reduces the length of DNA around the operator. Here one has to go as low as 100 to 1000 Å.

D. E. KOSHLAND: One of the things that has always puzzled me are Reads complexes, and some of these other complexes. Although people have always talked about diffusion as important, it never quite made sense because what happens is that the number of sub-units of one enzyme associate together and these are followed by another layer of sub-units, so the idea of an assembly line really doesn't work, and moreover the active sites are too far apart. I think the fatty acids complex is something different. The microenvironment effect really explains it. It may change not only the rate of production of the products but also the sequence in which they come off.

The second thing that occurred to me is that we might try to take some of those enzymes and put them on your immobilized layer and see if you get the same activity.

Another fact which fascinated me is that I have always looked for explanations why association-dissociation reactions should be so important. Their cooperativity really does not depend on association-dissociation, and yet there are very many enzymes which go from dimers to tetramers or monomers to dimers—right in the physiological range. Your suggestion at the end of your lecture that the molecular weight sieving effect might be an added control device superimposed on the ligand induced effects is a very attractive one.

A. SILBERBERG: Do you think that you could link two enzymes together by a polymeric chain and then vary the lengths of this chain, to study the microenvironment influence of one on the other?

E. KATZIR: Gershon Hazan and Elisha are trying to put simpler molecules and fluorescent donors and acceptors into oligo-peptides containing from 2 up to 14 or 15 amino acids. They have worked on this now for two years, and I am afraid that the attachment of enzymes to predetermined sites may still be a long way off.

But, one can infer that it is possible to change the distance and see how it will affect the reactivity. It is not so difficult to change the average density of an enzyme on a membrane, per area or per volume. Perhaps somebody has a suggestion how to orient an enzyme in a membrane? If we have an artificial membrane here, and if we could put an enzyme in a definite kind of section or direction, then it would be interesting to see how enzymes—or a battery of enxymes in different vectorial directions—act. However, we really do not have any method for doing this at the moment.

N. K. JERNE: Could one orient embedded enzyme molecules in membranes by simple mechanical stretching of the membranes, as was attempted 20–30 years ago by Werner Kuhn with other macromolecules?

E. KATZIR: I would assume that the difficulty is that most enzymes are more or less spherical and they are not very anistropic molecules.

I. R. MILLER: Orientation of the enzymes can be achieved by adsorption on a membrane (preferably a lipid membrane) surface. It does not have to be a membrane; it may be a hydrocarbon sphere in an emulsion, absorbing one enzyme, the second later, and so on and so forth. Since in many cases adsorption of proteins is quite irreversible, one could get a whole system or battery of oriented enzymes in a chosen sequence.

M. SELA: Neither of these suggestions makes use of specificity. I think that the orientation of enzymes may be achieved, thanks to this specificity, either by using a big substrate analogue, which is reacted reversibly, and then the rest of the enzyme is covered, e.g., by attaching a lipophilic chain, or, even more simply, by using selective antibody populations specific for different portions of the enzymic molecule. The antibodies against areas around the active site of the enzyme might be used to block reversibly the active site of the enzyme, and then hydrophilic groups might be introduced, followed by the removal of the antibody. A molecule would be obtained, possessing a more hydrophilic and a more hydrophobic moiety.

M. MEVARECH: Several years ago Aharon [Katzir-Katchalsky] had in mind the idea that a coupling between a flow of matter and an enzymatic activity can form a kind of feedback mechanism. The idea was that if a flow of matter induces an enzyme to change its state, let us say, from an inactive enzyme to an active one, then the activated enzyme acts to reduce that flow. The coupling between these two elements will produce a feedback device. Moreover, if the response of the enzyme to the flow has an hysteretic character the whole system will oscillate.

Looking for a biological system which might show a hysteresis phenomenon, we decided to study the response to changes in salt concentration of an enzyme isolated from extremely halophilic bacteria growing in the Dead Sea. These bacteria can grow only in very concentrated NaCl solutions and all the biochemical machinery is adapted to work under these extreme salt concentrations.

It was found that the enzyme malate-dehydrogenase responds to changes in NaCl concentration in an hysteretic manner; when the NaCl concentration is decreased the enzyme is inactivated; when the NaCl concentration is then increased, the enzyme is reactivated. But there is a range of NaCl concentrations in which the enzyme can exist in an inactive form as well as in an active form, depending on the direction of the NaCl changes.

E. KATZIR: It is a kind of hysteresis. It is a sharp transition. It is like reactivation of proteins by heat. You can reactivate. Many enzymes will lose activity. You could heat, and they will recover it but sometimes not at the same point; usually not at the same point. We are left with the question whether this phase is of any biological importance.

M. MEVARECH: I am not sure about the biological significance. The actual ionic concentration in solution in the bacteria is not yet clear. Although there are studies on the total Na^+ and K^+ concentration in the bacteria, the physical state of these ions is very unclear.

E. NEUMANN: The enzyme referred to by M. Mevarech represents an example of hysteresis in catalytic proteins. In general, hysteretic catalysts are capable of transforming stationarity in periodicity by chemidiffusional coupling (A. Katchalsky and R. Spangler, *Quart. Rev. Biophys.* **1**, 127 (1968)). For instance, an oscillatory change in the concentration of a reaction product can be obtained with a substance which participates both in a chemical reaction catalyzed by a hysteretic enzyme, and in a diffusion process which may be controlled by a membrane. Thus, the enzyme malate-dehydrogenase from Dead Sea halobacteria is a catalyst which, on a molecular level, is able to produce chemical oscillations based on the circular vectorial property of hysteresis.

One of the questions raised in the informative discussion on microenvironment and enzyme function was how one may orient enzyme molecules. I should like to draw your attention to orientations caused by electric fields. Commonly, proteins display permanent (or induced) dipole moments; in some cases these dipole moments are of the order of several hundred Debye units.

If you succeed in incorporating (dipolar) enzyme proteins in membranes without major reduction of the rotational mobility of the protein, you could achieve orientation by applying potential differences across these membranes. If the membranes are very thin, the potential differences are applied across small distances and can thus be equivalent to rather high fields. For large dipoles the degree of orientation may therefore be considerable. The potential difference may be achieved by external electrodes or by using permselective membranes and asymmetric ion distributions across the membrane. The degree of orientation could then be changed in a controlled way, by a variation of the electrode potentials or by a variation in the concentration of those ions which determine the membrane potential.

E. KATZIR: This idea of a lipid layer with electrodes may be an interesting one. There are many enzymes that will at least be soluble when mixed with liquid. Certainly in a normal aqueous buffer it is not a very attractive way to make such fields but I think your idea with membranes is very good.

Steinberg and Goldman tried to analyze the problem of enzyme kinetics in aqueous solution when the enzymes are either random or oriented. It seems that the kinetics will be identical. But if you have them in one layer or a sequence of enzymes, then there will be an enormous difference. I therefore think the main interest will be if you have sequential reactions and you can really orient the two or more enzymes, catalyzing one with respect to another.

J. R. FISHER: For some time I have been trying to understand the meaning of non-linear steady state enzyme kinetics obtained with homogeneous enzyme systems. As you know, this is one of the experimental systems in which sigmoid kinetics are often observed and are interpreted as indicating positive cooperativity. Also one often obtains what is called substrate activation, which is ascribed to negative cooperative interactions. I would simply like to point out that in such systems chemical cycles in the mechanism can account for all of these phenomena. It is unnecessary to conclude that there are cooperative effects in any system based solely on steady state kinetic evidence. All the evidence in this direction must come from other approaches. One can imagine quite a number of different processes which lead to chemical cycles, so in this field one needs to be very careful. If, for example, binding is an equilibrium process (as when oxygen binds hemoglobin), it is likely that cooperative mechanisms are involved. If, however, there is steady state flux, then one no longer needs to assume that nonlinearity is due to cooperativity but rather can consider a number of interpretations.

G. OSTER: Dr. Neumann spoke about hysteresis effects in macromolecules, and one of the ways one uses hysteresis in engineering is to store information. I was wondering if you could comment whether there is a reasonable possibility for storing biological information in such a fashion.

E. NEUMANN: Hysteresis was one of the phenomena which attracted the continuous attention of Aharon Katzir-Katchalsky. Thermodynamically, hysteresis comprises metastable states and nonequilibrium transitions; hysteresis thus reflects apparently *time-independent irreversibility*. Aharon Katzir-Katchalsky made major contributions to a conceptual clarification of this apparent contradiction and, furthermore, recognized the important role which metastabilities and hysteresis may play in central phenomena of life, such as memory recording or biological rhythms.

It has been previously mentioned that hysteresis is a means of converting a stationary flow of material into a periodic one. Thus in chemodiffusional flow coupling hysteresis can be the basis for chemical oscillations. Microscopic chemical oscillations are of fundamental importance for a mechanistic interpretation of biological clocks.

A further dynamic aspect of hysteresis comprises memory recording. Hysteresis is a well-known expression of physical memory. Modern computer technology utilizes hysteresis cycles of ferromagnetic and ferritic materials as memory devices for information storage. In biology, information storage, including the mechanism of memory recording, is still an unsolved problem. We may, however, conclude from biochemical and pharmacological studies that the multistaged process of biological memory formation involves the transformation of external stimuli into long-lived structures based on protein synthesis. For kinetic and energetic reasons it appears plausible that the initial step of recording nerve impulses, caused by external stimuli, into long-lived structures, could be based on electric-field induced conformational changes in macromolecules or membranes. In this context it has been demonstrated that electrical impulses similar to nerve impulses are indeed able to induce conformational changes in polyelectrolytic macromolecules and membranes. The results of these studies could be interpreted in terms of a counterion polarization mechanism characteristic of polyelectrolyte structures (E. Neumann and A. Katchalsky, *Proc. Natl. Acad. Sci. (USA)* **69**, 993 (1972)). Since large electric fields (up to 100 kV/cm) are present across the nerve membrane and transiently also across the synaptic junctions, any polyvalently charged system exposed to these fields could be affected in a way similar to the biopolymers studied recently.

We may, therefore, consider directed structural transitions induced by electrical impulses in metastable polyelectrolyte systems as a model reaction for the process of inprinting nerve impulses in the neuronal structures involved in a possible physical record of memory.

Further dynamic aspects of molecular hysteresis and the cybernetic significance of this phenomenon, for instance in a chemical control of electrical potential changes in neuronal membranes, are summarized in a review (E. Neumann, *Angew. Chem. intern. Edit.* **12**, 356 (1973)).

J. D. COWAN: Hysteresis is also a property that results from an autocatalytic reaction involving a large number of nerve cells. Given such a network of cells one can build flip flops which have spatial inhomogeneities which persist long after the stimulus has ceased. One can wipe out such states by using very large inhibitory pulses as stimuli. One can use such a network to store information, dynamically, for finite times. Obviously, in the model I described there wasn't any long-term dissipation, as there should have been since nerve cells do not maintain their activity for a very long time. There is undoubtedly some kind of mechanism like that in the frontal cortex. For example, conditioning experiments on monkeys show that if you do single cell recordings in monkeys, very long-lasting discharges in nerve cells are found after the conditioning experiments are done, but they do not persist for more than ten to twenty minutes or so.

G. OSTER: With regard to the chemotaxis mechanism presented by Dr. Koshland, I gain the impression that these bugs find their source of food by what is known as a "sample and hold" procedure. For example, if I want to find my way to the source of a river from its delta, I can take a sample of the water, go in a random direction for a fixed time, and then resample, If things are getting better (the water is getting fresher) I keep going in that direction; if they're not, I go in another randomly chosen direction. My question is: How does the "sample" device work on a molecular level if it samples the solution at discreet instants in time? How do the bacteria sample and hold? What is the sensor that takes a sample of local concentration and then does not change direction till it gets upstream to the next step, then resamples. I see that Prof. Koshland has proposed a very clever mechanism, but I do not see how it could work on a molecular level.

J. ROSS: I thought that the model consisted of coupled enzyme reactions with two very different relaxation times; the fast one was the sampling one, the slow one was the holding one.

G. OSTER: Perhaps this would work, but it is not clear to me that such a mechanism would not simply come into equilibrium with the local environment. Can you show that this kinetic scheme is a limit cycle oscillator? The problem of chemotactic behavior of bacteria can be viewed as sort of a generalized dissipative structure by treating it as a diffusion reaction problem. I am sure some of you have seen pictures of tetrahymena and how they swim in organized patterns if you put them in a petri dish. They get together and swim in regular hexagonal patterns; it looks rather like a Bénard cell phenomenon. The bacteria are in some way avoiding something toxic or seeking something that they like, but somehow their behavior organizes them into these very regular crystal-like swimming patterns.

J. ROSS: Hartman has done some experiments on this problem.

G. OSTER: Hartman's model, if I am not incorrect, is rather like Keller and Segal's model of a "dissipative structure" in ecology. However, the problem that Prof. Koshland's mechanism introduces is quite different: the diffusion coefficients are non-local. That is, this bug always remembers the way it was the last step back, and this "memory" appears to be the mechanism for creating the spatial order. It seems to me qualitatively different from the kind of spatial structure that Prof. Ross and Prof. Prigogine have spoken about, which does not depend on a memory effect.

J. D. COWAN: It seems to me that a simple polarization-depolarization model would actually work. The beast keeps swimming in a straight line as long as it receives the molecule that hyperpolarizes the membrane potential. As soon as it starts to receive noxious substances, they depolarize the membrane potential. They either polarize or depolarize—this would certainly discriminate between two doses of substances.

Concluding Remarks

F. O. SCHMITT (Chairman): I would like to close this roundtable by thanking all the participants in this session for their many contributions and for what seemed to me to be the fine spirit of give and take and intellectual "allostery"—if Dr. Fisher will pardon the word—that characterized this meeting. This ends the scientific part of the First Aharon Katzir-Katchalsky Conference. Future conferences will be held, as I understand it, annually, alternating between the Weizmann Institute and somewhere abroad. As in the past, when Aharon himself provided the yeast for the ferment of many conferences and congresses, the still warm and brightly glowing memory of Aharon has pervaded this conference, which was held in his honor.

Aharon's deep commitment to the best achievements of science and the human spirit in the service of humanity stimulated the participants in this conference to transcend their usual ways of thinking to achieve new interdisciplinary syntheses in the field in which Aharon contributed so much. He would have enjoyed every session of this conference. On behalf of all the participants and of myself I would like to convey our appreciation and thanks to our hosts of the Weizmann Institute, the Israel Academy of Science and Humanities, EMBO, and IUPAB for making this conference possible, to the organizing committee for creating the conference with the excellent choice of subjects and speakers and for, I must say, the very meticulous planning of every detail, and accommodation, and beautiful campus of the institute, meals and relaxation and, last but certainly not least, the wonderful sightseeing. No detail was overlooked. Dr. Miller and Mrs. Goldstein particularly deserve our warm thanks for all this, as do all the staff and personnel, too many to name individually.

G. OSTER: Professor Schmitt thought it would be appropriate if I said a few words in closing, as I was one of the organizers of the Berkeley symposium in Aharon's memory, and one of Aharon's last students. I find, however, that I can think of nothing to say about Aharon that others haven't said much more eloquently than I could. Each of us who knew and worked with Aharon felt we had a special relationship with him; this was a measure of the strength of his personality. Aharon was my teacher and I think the most important thing that he taught me was a love and enthusiasm for science; he had a way of looking at the world that was truly unique; I shall never forget the way he could excite me about a new idea he had.

Everyone has said of Aharon that he was a world citizen. I think this is certainly true. We in Berkeley felt that we owned a piece of Aharon, and that's why we also organized a symposium in his memory. The response to the memorial services in his honor was really astonishing; the chapel there was simply packed. But of course there was never any doubt where Aharon's heart lay. Although I know he looked forward very much to his annual visits to Berkeley, when it came time to leave once again, it was obvious where home was to him. And so I think that, while our symposium in Berkeley served his memory well, I feel that this symposium here in Israel really served his spirit much better than we ever could have.

In helping plan the Berkeley symposium I learned for the first time how very difficult it is to put together and organize one of these meetings, and I'd like to reiterate Prof. Schmitt's remarks in thanking all the people who put this symposium together. I especially wish we had had someone like Ruth Goldstein in Berkeley; her indomitable good humor in the face of a hundred scientific egos has astonished me. I hope that the symposia that will follow this one in subsequent years can measure up to this one. The spirit of Aharon has really permeated the meeting, just as it did in Berkeley.